AF358728

Alternative Splicing: From Abiotic Stress Tolerance to Evolutionary Genomics

Alternative Splicing: From Abiotic Stress Tolerance to Evolutionary Genomics

Editors

Bei Gao
Moxian Chen
Melvin J. Oliver

MDPI • Basel • Beijing • Wuhan • Barcelona • Belgrade • Manchester • Tokyo • Cluj • Tianjin

Editors

Bei Gao
Xinjiang Institute of Ecology
and Geography
Chinese Academy of Sciences
Urumqi
China

Moxian Chen
Center for R&D of Fine
Chemicals
Guizhou University
Guiyang
China

Melvin J. Oliver
Interdisciplinary Plant Group
University of Missouri
Columbia
United States

Editorial Office
MDPI
St. Alban-Anlage 66
4052 Basel, Switzerland

This is a reprint of articles from the Special Issue published online in the open access journal *International Journal of Molecular Sciences* (ISSN 1422-0067) (available at: www.mdpi.com/journal/ijms/special_issues/AS_ASTEG).

For citation purposes, cite each article independently as indicated on the article page online and as indicated below:

LastName, A.A.; LastName, B.B.; LastName, C.C. Article Title. *Journal Name* **Year**, *Volume Number*, Page Range.

ISBN 978-3-0365-7457-8 (Hbk)
ISBN 978-3-0365-7456-1 (PDF)

Contents

International Journal of
Molecular Sciences

MDPI

Editorial

Alternative Splicing: From Abiotic Stress Tolerance to Evolutionary Genomics

Bei Gao [1,2,†], Moxian Chen [3,†] and Melvin J. Oliver [4,*]

1   State Key Laboratory of Desert and Oasis Ecology, Key Laboratory of Ecological Safety and Sustainable Development in Arid Lands, Xinjiang Institute of Ecology and Geography, Chinese Academy of Sciences, Urumqi 830011, China
2   Xinjiang Key Laboratory of Conservation and Utilization of Plant Gene Resources, Xinjiang Institute of Ecology and Geography, Chinese Academy of Sciences, Urumqi 830011, China
3   State Key Laboratory Breeding Base of Green Pesticide and Agricultural Bioengineering, Key Laboratory of Green Pesticide and Agricultural Bioengineering, Ministry of Education, Research and Development Center for Fine Chemicals, Guizhou University, Guiyang 550025, China
4   Division of Plant Sciences and Interdisciplinary Group, University of Missouri, Columbia, MO 65211, USA
*   Correspondence: olivermj@missouri.edu
†   These authors contributed equally to this work.

Citation: Gao, B.; Chen, M.; Oliver, M.J. Alternative Splicing: From Abiotic Stress Tolerance to Evolutionary Genomics. *Int. J. Mol. Sci.* **2023**, *24*, 6708. https://doi.org/10.3390/ijms24076708

Received: 22 March 2023
Accepted: 31 March 2023
Published: 4 April 2023

The post-transcriptional regulation of gene expression, in particular alternative splicing (AS) events, substantially contributes to the complexity of eukaryotic transcriptomes and proteomes. The interest in AS events has increased as third-generation sequencing technology allowed researchers to obtain full-length transcripts efficiently and accurately, with or without a reference genome. Research interest surged with the advent of high-fidelity PacBio reads and reduced costs making alternative splicing studies affordable for many laboratories. Therefore, a large body of research has been generated and new areas are being explored. We focused on how variable (or not) the AS control of gene expression is across lineages or within phylogenetically related species with an emphasis on how AS has shaped the evolution of stress responses in plants: an important area which may have positive benefits for agriculture. Agricultural productivity has been threatened by the growing problems associated with climate change; therefore, molecular mechanisms that underpin the response of plants to various kinds of environmental stress have become a major focus for crop improvement.

Thus far, there are a considerable number of reports that have linked AS events to plant stress responses and tolerance [1,2], and this field is growing. It is this burgeoning field of research that provided the inspiration for this Special Issue, entitled "Alternative Splicing: From Abiotic Stress Tolerance to Evolutionary Genomics", which aims to recruit both original research and review papers that contribute to our understanding of the role of AS in the evolution of plant stress responses and tolerance mechanism.

The Special Issue contains seven original research articles and three reviews. The species investigated in each of the research articles are phylogenetically and agriculturally diverse; including the moss *Physcomitrium patens*, Arabidopsis, as well as crops such as rice, maize, rapeseed, Tartary buckwheat and *Lycoris longituba* (a Chinese medicinal plant). Most of the research articles concentrate on specific functional genes responsive to abiotic stresses. Notably, functional aspects of genes in plants with agricultural or economic values is a major trend, mirroring the urgent worldwide demand for crops with improved and desired agronomic traits. We are also deeply impressed by the enthusiasm and active participation of the contributing authors to this Special Issue, reflecting the widespread interest in this research topic in the plant sciences. Here, we summarize the contributing articles of this Special Issue.

Deprived nutrient availability can severely impact plant growth and crop yield and led Hua et al. to investigate the GARP transcription factor gene family in allotetraploid rapeseed (*Brassica napus* L.) [3]. They identified a total of 146 members of GARP transcription factor genes in the genome that were phylogenetically classified into five subfamilies. The genomic syntenic relationships within representative eudicot genomes revealed conserved genomic positions for the *BnaGARP* genes. Gene structure, evolutionary selection, *cis*-acting regulatory elements, miRNA-targeting predictions and chromosomal positions for GARP genes were analyzed and displayed. By gleaning insights from transcriptomic data of plants under nutrient stress, Hua et al. focused on *BnaA9.HHO1* and *BnaA1. HHO5* in further functional investigations, suggesting that they possess regulatory roles in response to nutrient stress.

With global warming, heat stress has become a major limiting factor for crop yield improvement. Yang et al. discovered two family members of RNase H1 in the moss *Physcomitrium patens* [4], a representative model for early-diverging land plants. They indicated the presence of homologues to the RNase H1 genes in chlorophytes and charophytes, suggesting an early origin for this gene family. The overexpression of *PpRNH1A* in *P. patens* resulted in OE lines that were more sensitive to heat stress along with a higher number of lipid droplets with less mobility in the transgenic cells. Transcriptomic analyses led the authors to suggest that *PpRNH1A* might be involved in regulating the expression of heat-related genes, such as *DNAJ* and *DNAJC*.

Alternatively, cold environments repress enzyme activities and impact a variety of biological processes. Hou et al. identified 14 histone deacetylase (HDACs) genes from the Tartary buckwheat (*Fagopyrum tataricum*) genome, which were phylogenetically clustered into three subfamilies with distinct tertiary structures [5]. Taking advantage of bioinformatic tools, the authors investigated gene structures, motif compositions, *cis*-acting elements, alternative splicing events, subcellular localizations and gene expression patterns. RNA-Seq read alignments and coverage depth analyses revealed that *FtHDA8-2* was likely subject to an alternative splice event under different low-temperature regimes, suggesting that AS played a role in the response to cold stress. A cold-tolerant variety, Dingku 1, was utilized to illustrate the phenotypic and transcript abundance changes of HDACs under different low temperatures. Overexpression lines, *OE-AtHDA6* and *OE-FtHDA6-1*, were constructed, enabling Hou et al. to demonstrate that *FtHDA6-1* positively regulated cold tolerance.

The membrane attack complex/perforin-like (MACPF) protein superfamily has been extensively studied in animal systems where molecular functions during pathogen infections have been revealed. However, these proteins have been rarely investigated in plants. Zhang et al. focused on the Arabidopsis MACP2 protein and investigated the responsiveness of this protein during both bacterial and fungal pathogen infection [6]. Their report revealed that MACP2 plays a role in promoting pathogen resistance by activating the biosynthesis of tryptophan-derived indole glucosinolates and the salicylic acid signaling pathway. By comparisons of MACP2-OE and -KO lines, indole glucosinolates were revealed to contribute to bacteria resistance in the MACP2-OE plants. Zhang et al., also demonstrated that alternatively spliced *MACP2* isoforms were differentially expressed under pathogen infection, suggesting different roles for AS isoforms, highlighting the prospect that functionally multifaceted expression products can be derived from a single gene.

Two research papers from this Special Issue focused on rice from Professor Ye's laboratory [7,8], where findings may directly assist in strategies for rice yield or quality improvements. In the first paper, Professor Ye's team studied the molecular effects of post-anthesis moderate soil drying (MD)n which is proven to be an effective method for promoting starch synthesis and inferior spikelet grain filling. In this paper, the authors probed 1840 AS events in 1392 genes that were differentially expressed between MD and well-water plants. RNA-Seq read alignment depth was used to illustrate the AS events of specific genes of interest. Specifically, the *OsAGPL2* transcript was alternatively spliced to generate transcripts with or without a binding site for the microRNA miR393b, suggesting a potential mechanism for the miRNA-mediated gene regulation of grain filling in inferior spikelets in response to MD. In the second paper, the authors analyzed quantitative trait loci (QTLs) for grain size in rice by developing a suite of F2 populations from a cross of 9311 and CG varieties. Bulked-segregant analysis-seq (BSA-seq) was employed to probe

QTLs for various agronomic traits. Within the analysis they detected more than 200 splicing-related loci using whole-genome sequencing, including a splicing-site mutation in the gene *Os03g0841800* (qGL3.3) that generated a truncated open reading frame. Their reported discoveries provide a valuable genetic resource for rice breeding.

Included in the Special Issue is a proteomic and metabolic research paper that focuses on three *Lycoris* species with different alkaloid contents [9], with *Lycoris longituba* exhibiting the highest alkaloid content. This study was included because *Lycoris* bulbs have long been used as traditional Chinese herbal medicine to treat various diseases such as sore throat, abscesses and suppurative wounds. In addition, the SWATH-MS (sequential window acquisition of all theoretical mass spectra)-based quantitative proteomic approach have been widely employed in animal systems but the merits of this technology remain yet to be widely recognized in plants. The comprehensive proteomic analyses presented by the authors were used to identify five candidate proteins encoding enzymes involved in alkaloid biosynthesis, with differential abundance in the three *Lycoris* species. However, particularly fascinating was the discovery of 11 RNA processing-related proteins, which indicated the role of AS in the variation between the three species in alkaloid production.

In addition to seven original research articles, three review papers are included in this Special Issue. Lam et al. summarized the current knowledge on molecular regulations of plant primary and specialized metabolism by alternative splicing [10], highlighting numerous genes that function in plant metabolism that were alternatively spliced at different developmental stages and under stress conditions. They proposed that AS may serve as a fine-tuning regulatory mechanism for plant metabolism. Li et al. concentrated on plant–microbe interactions and discussed the possible mechanisms that connect diverse plant ecotype-phenotype linkages, splicing isoforms, and commensal microbiomes [11]. Jia et al. reviewed the progress in the application of alternative splicing and genome-wide association analyses (GWAS). They proposed that the utilization of GWAS to investigate alternative splicing activity could provide novel insights that could lead to the elucidation of the relationship between AS and the regulation of agronomic traits [12].

In conclusion, as the global climate continues to change and deteriorate, various abiotic stressors and their adverse impacts on crop growth and productivity will present great risk for modern society. The risk of catastrophic crop failures drives the growing research interests in stress biology, especially for crops and plants with economic value. The original research and review articles assembled in the Special Issue provide valuable and novel knowledge of gene function and the potential role of alternative splicing in response to environmental stress. However, despite the well-tailored molecular experimentation for investigating gene function, we noticed the inadequate utilization of bioinformatic methods for comparative and evolutionary genomics to help identify suitable target genes and elucidate their evolutionary trajectories. This gap between "dry lab" and "wet lab" still exists and more collaborative projects need to be facilitated. The published articles in the current Special Issue cannot constitute a comprehensive collection of this interdisciplinary research topic but hopefully provide an impetus to generate more research in this field. With the support of the editorial office, we are delighted to launch a new Special Issue focusing on the same research theme, entitled "Alternative Splicing: From Abiotic Stress Tolerance to Evolutionary Genomics 2.0", representing our continuous effort on this topic, and we sincerely invite researchers in the field to contribute to this new collection.

Author Contributions: B.G. and M.C. writing—original draft preparation; M.J.O. writing—review and editing. All authors have read and agreed to the published version of the manuscript.

Funding: This work was supported by the National Natural Science Foundation for China (NSFC Grant No.: 32100256 to B.G.), Special funding schemes provided by the Human Resources and Social Security of China (N2022000005 and 2022000243 to B.G.) and the Third Xinjiang Scientific Expedition Program (2021xjkk0501).

Conflicts of Interest: The authors declare no conflict of interest.

References

1. Zhu, F.Y.; Chen, M.X.; Ye, N.H.; Shi, L.; Ma, K.L.; Yang, J.F.; Cao, Y.Y.; Zhang, Y.; Yoshida, T.; Fernie, A.R.; et al. Proteogenomic analysis reveals alternative splicing and translation as part of the abscisic acid response in Arabidopsis seedlings. *Plant J. Cell Mol. Biol.* **2017**, *91*, 518–533. [CrossRef] [PubMed]
2. Chen, M.X.; Zhu, F.Y.; Wang, F.Z.; Ye, N.H.; Gao, B.; Chen, X.; Zhao, S.S.; Fan, T.; Cao, Y.Y.; Liu, T.Y.; et al. Alternative splicing and translation play important roles in hypoxic germination in rice. *J. Exp. Bot.* **2019**, *70*, 817–833. [CrossRef] [PubMed]
3. Hua, Y.-P.; Wu, P.-J.; Zhang, T.-Y.; Song, H.-L.; Zhang, Y.-F.; Chen, J.-F.; Yue, C.-P.; Huang, J.-Y.; Sun, T.; Zhou, T. Genome-Scale Investigation of GARP Family Genes Reveals Their Pivotal Roles in Nutrient Stress Resistance in Allotetraploid Rapeseed. *Int. J. Mol. Sci.* **2022**, *23*, 14484. [CrossRef] [PubMed]
4. Yang, Z.; Duan, L.; Li, H.; Tang, T.; Chen, L.; Hu, K.; Yang, H.; Liu, L. Regulation of Heat Stress in Physcomitrium (Physcomitrella) patens Provides Novel Insight into the Functions of Plant RNase H1s. *Int. J. Mol. Sci.* **2022**, *23*, 9270. [CrossRef] [PubMed]
5. Hou, Y.; Lu, Q.; Su, J.; Jin, X.; Jia, C.; An, L.; Tian, Y.; Song, Y. Genome-Wide Analysis of the HDAC Gene Family and Its Functional Characterization at Low Temperatures in Tartary Buckwheat (*Fagopyrum tataricum*). *Int. J. Mol. Sci.* **2022**, *23*, 7622. [CrossRef] [PubMed]
6. Zhang, X.; Dai, Y.-S.; Wang, Y.-X.; Su, Z.-Z.; Yu, L.-J.; Zhang, Z.-F.; Xiao, S.; Chen, Q.-F. Overexpression of the Arabidopsis MACPF Protein AtMACP2 Promotes Pathogen Resistance by Activating SA Signaling. *Int. J. Mol. Sci.* **2022**, *23*, 8784. [CrossRef] [PubMed]
7. Hou, X.; Chen, M.; Chen, Y.; Hou, X.; Jia, Z.; Yang, X.; Zhang, J.; Liu, Y.; Ye, N. Identifying QTLs for Grain Size in a Colossal Grain Rice (Oryza sativa L.) Line, and Analysis of Additive Effects of QTLs. *Int. J. Mol. Sci.* **2022**, *23*, 3526. [CrossRef] [PubMed]
8. Teng, Z.; Zheng, Q.; Liu, B.; Meng, S.; Zhang, J.; Ye, N. Moderate Soil Drying-Induced Alternative Splicing Provides a Potential Novel Approach for the Regulation of Grain Filling in Rice Inferior Spikelets. *Int. J. Mol. Sci.* **2022**, *23*, 7770. [CrossRef] [PubMed]
9. Tang, M.; Li, C.; Zhang, C.; Cai, Y.; Zhang, Y.; Yang, L.; Chen, M.; Zhu, F.; Li, Q.; Li, K. SWATH-MS-Based Proteomics Reveals the Regulatory Metabolism of Amaryllidaceae Alkaloids in Three Lycoris Species. *Int. J. Mol. Sci.* **2023**, *24*, 4495. [CrossRef] [PubMed]
10. Lam, P.Y.; Wang, L.; Lo, C.; Zhu, F.-Y. Alternative Splicing and Its Roles in Plant Metabolism. *Int. J. Mol. Sci.* **2022**, *23*, 7355. [CrossRef] [PubMed]
11. Li, Y.-H.; Yang, Y.-Y.; Wang, Z.-G.; Chen, Z. Emerging Function of Ecotype-Specific Splicing in the Recruitment of Commensal Microbiome. *Int. J. Mol. Sci.* **2022**, *23*, 4860. [CrossRef] [PubMed]
12. Jia, Z.-C.; Yang, X.; Hou, X.-X.; Nie, Y.-X.; Wu, J. The Importance of a Genome-Wide Association Analysis in the Study of Alternative Splicing Mutations in Plants with a Special Focus on Maize. *Int. J. Mol. Sci.* **2022**, *23*, 4201. [CrossRef] [PubMed]

International Journal of
Molecular Sciences

Article

Genome-Scale Investigation of *GARP* Family Genes Reveals Their Pivotal Roles in Nutrient Stress Resistance in Allotetraploid Rapeseed

Ying-Peng Hua [1,†], Peng-Jia Wu [1,†], Tian-Yu Zhang [1], Hai-Li Song [1], Yi-Fan Zhang [1], Jun-Fan Chen [1], Cai-Peng Yue [1], Jin-Yong Huang [2], Tao Sun [1,*] and Ting Zhou [1,*]

[1] School of Agricultural Sciences, Zhengzhou University, Zhengzhou 450001, China
[2] School of Life Sciences, Zhengzhou University, Zhengzhou 450001, China
* Correspondence: suntao89815@zzu.edu.cn (T.S.); zhoutt@zzu.edu.cn (T.Z.); Tel.: +86-187-0271-0749 (T.Z.)
† These authors contributed equally to this work.

Abstract: The *GARP* genes are plant-specific transcription factors (TFs) and play key roles in regulating plant development and abiotic stress resistance. However, few systematic analyses of *GARPs* have been reported in allotetraploid rapeseed (*Brassica napus* L.) yet. In the present study, a total of 146 *BnaGARP* members were identified from the rapeseed genome based on the sequence signature. The *BnaGARP* TFs were divided into five subfamilies: *ARR, GLK, NIGT1/HRS1/HHO, KAN*, and *PHL* subfamilies, and the members within the same subfamilies shared similar exon-intron structures and conserved motif configuration. Analyses of the Ka/Ks ratios indicated that the *GARP* family principally underwent purifying selection. Several *cis*-acting regulatory elements, essential for plant growth and diverse biotic and abiotic stresses, were identified in the promoter regions of *BnaGARPs*. Further, 29 putative miRNAs were identified to be targeting *BnaGARPs*. Differential expression of *BnaGARPs* under low nitrate, ammonium toxicity, limited phosphate, deficient boron, salt stress, and cadmium toxicity conditions indicated their potential involvement in diverse nutrient stress responses. Notably, *BnaA9.HHO1* and *BnaA1.HHO5* were simultaneously transcriptionally responsive to these nutrient stresses in both hoots and roots, which indicated that *BnaA9.HHO1* and *BnaA1.HHO5* might play a core role in regulating rapeseed resistance to nutrient stresses. Therefore, this study would enrich our understanding of molecular characteristics of the rapeseed *GARPs* and will provide valuable candidate genes for further in-depth study of the GARP-mediated nutrient stress resistance in rapeseed.

Keywords: *Brassica napus*; transcription factors; nutrient stress; transcriptomic analysis; miRNA

Citation: Hua, Y.-P.; Wu, P.-J.; Zhang, T.-Y.; Song, H.-L.; Zhang, Y.-F.; Chen, J.-F.; Yue, C.-P.; Huang, J.-Y.; Sun, T.; Zhou, T. Genome-Scale Investigation of *GARP* Family Genes Reveals Their Pivotal Roles in Nutrient Stress Resistance in Allotetraploid Rapeseed. *Int. J. Mol. Sci.* **2022**, 23, 14484. https://doi.org/10.3390/ijms232214484

Academic Editor: Daniela Trono

Received: 16 October 2022
Accepted: 18 November 2022
Published: 21 November 2022

Publisher's Note: MDPI stays neutral with regard to jurisdictional claims in published maps and institutional affiliations.

1. Introduction

The transcriptional regulation of plant genes is a complex and accurate network system. In this process, transcription factors (TFs) play crucial roles in plant growth and development, species origin, and stress responses by precisely binding to the *cis*-acting regions of target genes [1]. After the identification of the Arabidopsis genome, the TFs were classified into 58 TF families [2]. Plant responses to nutrient stresses are regulated by complex signaling pathways and networks which are coordinated by TFs [3].

The *GARP* gene is a plant-specific TF and plays a key role in regulating plant development, disease resistance, hormone signaling, circadian clock oscillations, and abiotic stress resistance [4]. *GARP* is named from the Golden 2 (G2) protein in *maize*, the type B authentic response regulator (ARR-B) protein in *A. thaliana*, and the phosphate starvation response 1 (*PSR1*) protein in *Chlamydomonas* [5]. In the *GARP* family, the members can be classified if the derived protein contains the conserved signature motif called the B-motif (GARP motif) [6]. The B motif is a signature of type-B response regulators (ARRs) involved in His-to-Asp phosphorelay signal transduction systems in Arabidopsis, which contains an

HTH (helix-turn-helix) motif [5]. HTH motifs can regulate a variety of physiological processes, as well as participate in TF dimerization. The B motif is highly similar to MYB-DBD (Myb-DNA binding domain), and this also leads to frequent confusion with MYB-related TFs [5]. In contrast to MYB-related proteins characterized by the (SHAQK(Y/F) F) motif, GARP TFs contain a different consensus sequence (SHLQ(K/M) (Y/F)) [6].

The *GARP* TFs have been identified in Arabidopsis, rice, cotton, tea plant, and other species, and related studies have shown that they are involved in the regulation of plant growth and development, abiotic stress resistance, and other biological processes [7–10]. *AtGARPs* have been defined as important regulators of diverse nutrient stresses. The expression of *AtHHO3* (*NIGT1.1*) and *AtHHO2* (*NIGT1.2*) was induced in nitrogen (N) deficiency, while *AtHHO1(NIGT1.3)* is involved in primary root shortening under phosphate (Pi)-deficient conditions [11]. *AtKAN1* acts as a transcriptional repressor involved in auxin biosynthesis, auxin transport, and auxin response [12], and *AtKAN4* is shown to broadly control the flavonoid pathway in Arabidopsis seed [13]. In addition, *AtBOA* is a component of the Arabidopsis circadian clock [14]. In rice, *OsPHR1-4* has been linked to controlling Pi homeostasis-regulating sensing and signaling cascades in rice [15]. The expression of *OsARR-B5*, *OsARR-B22*, and *OsARR-B23* was upregulated under alkaline stress and was implicated in plant development modulation by controlling cellular processes, molecular activities, and biological functions [16]. *OsHHO2* inhibits Pi starvation response [17], whereas *OsHHO3* and *OsHHO4* play critical roles in the N deficiency response [18]. In cotton, *GhGLK1* is reported to be involved in the regulation of drought and cold stress responses [19]. Thus, GARP TFs play significant roles in the responses of different plant species to nutrient stress.

The GARP TFs are involved in the responses to nutrient stresses and include probable nutrient sensors of plants. CrPsr1 is the first reported GARP TF to be involved in the nutritional responses, and it is essential for the adaptation of *C. reinhardtii* to Pi starvation [20]. Then, NIGT1/HRS1/HHOs were found to be the most robustly and quickly nitrate (NO_3^-) regulated TFs [21]. N and Pi are essential macronutrients for the growth and development of plants. N participates in a variety of physiological and biochemical processes as a component of proteins, nucleic acids, and plant growth regulators [19]. Pi is an essential building block of Important compounds such as DNA, RNA, and proteins, and is involved in glycolysis, respiration, and photosynthesis [22]. In the process of N and Pi absorption and utilization, TFs play an important role in regulation [23–26]. A number of *GARPs*, particularly members of the *NIGT1/HRS1/HHO* subfamily, have been shown to play important roles in the regulation of plant responses to N and Pi stresses. The role of *NIGT1/HRS1/HHO* in response to N and Pi stresses can be summarized in two pathways: NRT-NLP-NIGT1/HRS1/HHO and NIGT1-SPX-PHR [27]. Specifically, the nitrate transporters *NRTs* can increase the content of N nutrients, enhancing the expression of the nitrate-responsive nodulin-like proteins (*NLPs*), and induce the expression of *GARPs* to suppress the N starvation response. The NIGT1 proteins repress the expression of *SPXs* by directly binding to the *SPX* promoters, and the SPX proteins function as the repressors of *PHR* TFs [28]. Under Pi-sufficient conditions, *PHR1* interacts with *SPXs* (SYG1/PHO81/XPR1), Pi sensor proteins, and inhibitors for *PHR1*, and the NIGT1-clade genes are not activated. Under Pi starvation conditions, *PHR1* is released from *SPXs* and promotes the expression of the NIGT1-clade genes [29]. In addition to *NIGT1/HRS1/HHO*, *PHR* of the *GARP* gene family has also been reported to affect N and Pi homeostasis [15].

Rapeseed (*Brassica napus* L.) is a major oilseed crop due to its economic value and oilseed production. However, its productivity has been repressed by many environmental adversities [30]. Under drought tolerance, the shoot and root growth of rapeseed seedlings is greatly inhibited, which ultimately will reduce crop production [31]. For rapeseed, salt stress severely affects all life stages from seed germination to yield production [32]. Cold stress has a negative impact on rapeseed germination and seedling establishment, causing wilting and plant death at the seedling stage [33]. Rapeseed production in the field is also often severely inhibited due to N deficiency [34] and is highly dependent on N fertilizer application, but its N use efficiency (NUE) is very low [35]. Rapeseed is also extremely sensitive to Pi deficiency [36]. A number of gene families, such as superoxide dismutase (*SOD*) [37], lipid phosphate phosphatases (*LPP*) [38], and B-box (*BBX*) [39] play critical roles in rapeseed growth, development, and response to stresses in rapeseed. To date, *GARPs* have also been identified to play important roles in plant growth and response to stress [16–18]. Given the importance of the *GARP* family in all aspects of plant developmental processes and stress responses, a comprehensive genome-wide investigation of *GARPs* is warranted in rapeseed.

However, few systematic analyses of *GARPs* in *B. napus* have been available so far. Thus, this study is aimed to (i) identify the genome-wide *GARPs* in *B. napus*, (ii) characterize the genomic characteristics and transcriptional responses of the *GARPs* to N stresses (including NO_3^- limitation and ammonium (NH_4^+) toxicity) and Pi limitation, and (iii) investigate the transcriptional responses of *GARPs* to other nutrient stresses, including boron deficiency, cadmium toxicity, and salt stress. This study would enrich our understanding of molecular characteristics of the rapeseed *GARPs* and will provide valuable candidate genes for further in-depth study of the *GARP*-mediated nutrient stress resistance in rapeseed.

2. Results

2.1. Genome-Wide Identification of the GARP Family Genes in B. napus

Since the GARP proteins are highly similar to MYB or MYB-like TFs in terms of both sequences and structures, the candidate *GARP* genes were compared and screened according to the methods reported by Safi et al. [6]. In this study, a total of 146 *BnaGARPs* were identified in the rapeseed genome ($A_nA_nC_nC_n$: A1-A10, C1-C9).

The physical and chemical characteristics, including the gene length and molecular weights (MW), of a total of 146 GARP proteins were analyzed and provided. The *BnaGARPs* have varying physicochemical characteristics (Table S1). The length of the GARP protein sequences ranged between 101 (*BnaA3.MYBC1a*) and 1022 (*BnaC8.GLK1*) amino acids in *B. napus*. The isoelectric point (pI) ranged between 4.74 (*BnaA2.PHL1*) and 11.06 (*BnaA5.MYBC1*). The molecular weight (MW) ranged from 11.62 to 111.69 kDa for *BnaA3.MYBC1a* and *BnaC8.GLK1*, respectively.

2.2. Phylogenetic Analysis and Ka/Ks Ratio Calculation

To elucidate the evolutionary relationships and functional divergence among Brassica GARP proteins, the sequences of 146 *B. napus* GARP proteins, and 56 *A. thaliana* GARP proteins were used to construct a phylogenetic tree (Figure 1). In general, the *GARP* homologs in *B. napus* significantly expanded compared to those in *A. thaliana*. Moreover, the number of *GARPs* in *B. napus* is much more than three times of those in *A. thaliana*.

Based on the topologies and bootstrap support values of the NJ phylogenetic tree, the candidate GARPs were divided into five subfamilies, which were identical to the previous study [7]. The distribution of *BnaGARPs* among different subfamilies was as follows: *ARR* (34 members), *GLK* (17 members), *NIGT1/HRS1/HHO* (27 members), *KAN* (17 members), and *PHL1* (52 members). The differences in the number of *BnaGARPs* within the five subfamilies indicated a distinct expansion trend among these subfamilies.

To explore the selective pressure on *BnaGARPs*, the non-synonymous/synonymous mutation ratio (Ka/Ks) was calculated; Ka/Ks > 1.0 indicates positive selection, Ka/Ks = 1.0 indicates neutral selection, and Ka/Ks < 1.0 indicates purifying selection [40]. The Ka/Ks ratio for all *BnaGARPs* was <1.0, ranging between 0.0697 (*BnaA5.ARR1*) and 0.5771 (*BnaA6.APRR4*), implying that the replicated *GARPs* could experience strong purification selection (Table S2).

2.3. Conserved Motif and Gene Structure Analyses

To further clarify the potential functions of *GARPs* in *B. napus*, MEME was used to identify 10 conserved motifs (Figure 2). Motif 1 and motif 2 in the *BnaARR-B* subfamilies, and the motif1 and motif 4 in the *BnaNIGT1/HRS1/HHO* subfamilies are the B-motif of the *GARP* signature motif and extensively distributed in *BnaGARPs* (Figure 2B). Furthermore, motif 7 and motif 10 only exist in *APRR2*, while motif 8 is specific to *BnaARR2 and BnaARR1*, and in the PHL subfamily, motif 8 only occurs in *PHL14* (Figure S1). However, the motif patterns of *BnaGARPs* within a subgroup are similar.

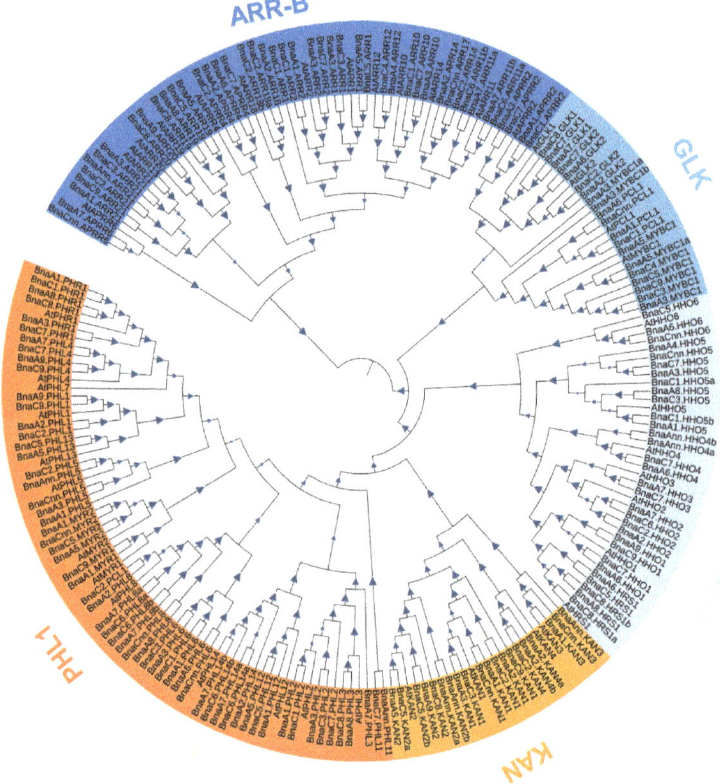

Figure 1. Phylogenetic tree of the *GARPs* retrieved from *B. napus* and *A. thaliana*. The phylogenetic tree was constructed according to the neighbor-joining method. The tree was generated using MEGA7.0 based on the *GARP* amino acid sequences retrieved from *B. napus* and *A. thaliana*. The genes from each group are indicated by different colors. The rectangle sizes at the nodes represent the bootstrap values.

To evaluate the sequence diversity of *BnaGARPs*, the exon–intron structures of each *BnaGARP* were detected. In detail, most of *BnaGARPs* had six exons and five introns, and several genes had five exons and four introns, while *BnaARR21s* contained 12 introns and *BnaMYBC1* contained one intron (Figure S2). Similarly, the majority of *BnaGARPs* in the same subgroups generally had similar gene structures (Figure 3).

We also found that the intron lengths are slightly different among different *BnaGARPs*. In comparison with *BnaPHL12s*, the introns within *BnaPHL5s* were relatively large. Although the exon-intron structures of most closely related genes exhibited high similarity and conservation, there still exist several differences.

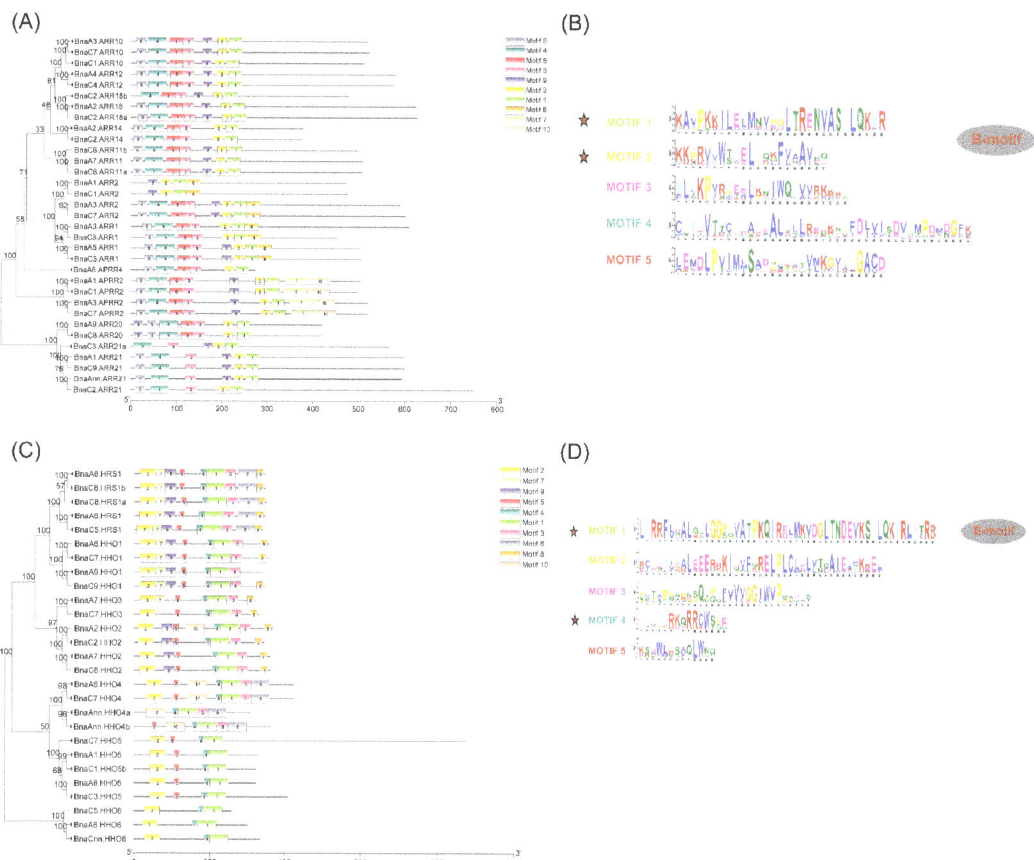

Figure 2. Identification and characterization of the conserved motifs in the GARP proteins in *B. napus*. (**A**) Molecular identification of *BnaARR-Bs*. (**B**) The sequence characterization of *BnaARR-Bs*. (**C**) Molecular identification of *BnaNIGT1/HRS1/HHOs*. (**D**) The sequence characterization of *BnaNIGT1/HRS1/HHOs*. In A and C, the boxes with different colors indicate different conserved motifs (motifs 1–10), and black lines represent the GARP protein regions without detected motifs. In C and D, the larger the fonts, the more conserved the motifs. Among them, the tagged motifs were identified as the B-motifs.

2.4. Gene Duplication and Synteny Analysis of GARP Gene Families

The genomic positions of the identified *GARPs* were physically mapped onto the rapeseed chromosomes using the MapGene2Chrom program. Ultimately, a total of 146 *GARPs* were mapped onto 20 chromosomes in *B. napus* (Figure 4). Evidently, there are only two *GARPs* on chromosomes chrA4, four on chrA10, while chrC4 has the most genes. The distribution of genes on the chromosomes is relatively scattered, and the genes on the same chromosome are far apart.

Gene duplication events can lead to the expansion of gene families and play crucial roles in the adaptation of plant species to the external environment by acquiring new gene functions. Given the importance of gene duplication in the evolution of plant gene families, the duplication patterns of 146 *GARP* family genes were analyzed in *B. napus*. Between the two sub-genomes of *B. napus*, 45 duplication events took place on the A subgenome, 36 events on the C subgenome, and 136 events across the A/C subgenomes (Table S3).

To better understand the evolution of *BnaGARPs*, the synteny of the *GARP* pairs between the genomes of *B. napus* and *A. thaliana, G. max, and M. truncatula* was constructed

(Figure 5 and Tables S4–S8). We found that 115 *BnaGARPs* exhibited syntenic relationships with *AtGARPs*. Some *AtGARPs* were associated with more than one orthologous copy in *B. napus*. For example, *KAN2/AT1G32240* showed a syntenic relationship with *BnaC5.KAN2b*, *BnaC5.KAN2a*, and *BnaC8.KAN2* (Table S4). As shown in Figure 5 and Figure S5, *BnaGARPs* shared 172 syntenic gene pairs with *G. max*, 66 with *M. truncatula*, and 3 with *T. aestivum* (Tables S5–S7). Additionally, syntenic gene pairs were identified between rapeseed and rice, which constituted the fewest number of background collinear blocks (Table S8). Interestingly, 46 genes were found in the comparative synteny maps between *B. napus* and other plant species (*A. thaliana*, *M. truncatula*, and *G. max*), and these collinear gene pairs were highly conserved within several syntenic blocks, such as *BnaA1.APRR2*, *BnaA1.HHO5*, *BnaA1.KAN3*, *BnaA1.PHR1*, and *BnaA10.MYR1* on the A1 chromosome and *BnaA3.APRR2*, *BnaA3.ARR10*, and *BnaA3.ARR2* on the A3 chromosome.

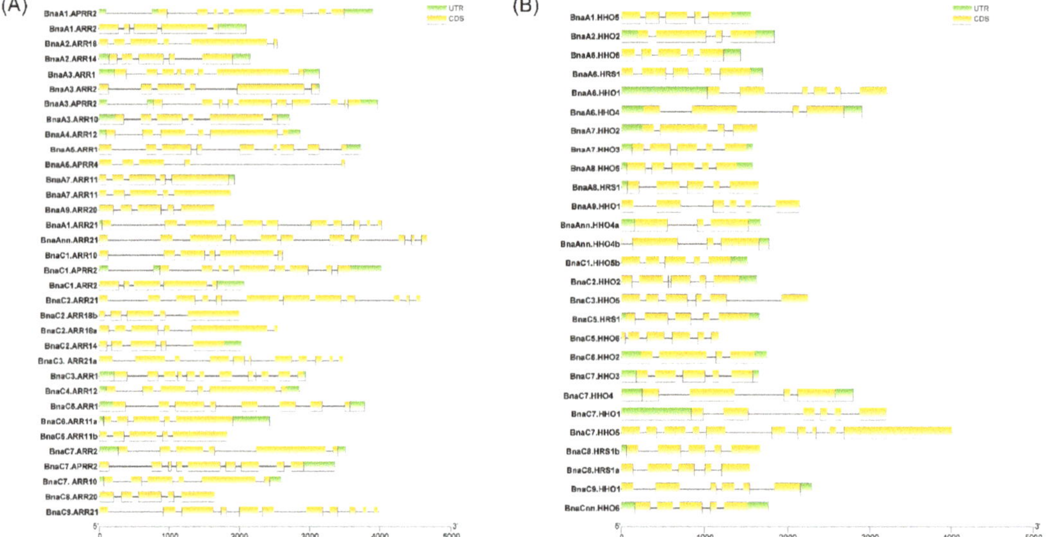

Figure 3. Exon-intron organizations of *BnaNIGT1/HRS1/HHOs* (**A**) and *BnaARR-Bs* (**B**). The green boxes represent untranslated regions, the yellow boxes represent exons, and the black lines represent the introns. The lengths of the exons and introns can be determined by the scale at the bottom.

2.5. Cis-Regulatory Element Prediction in the Promoter Regions of BnaGARPs

To investigate the potential regulatory mechanisms underlying *GARPs* in response to abiotic stresses and hormones, the *cis*-regulatory elements (CREs) in the 2000 bp upstream promoter sequences of each *GARP* gene were scanned by the plantCARE database. The results revealed that the promoter regions of each *BnaGARPs* have stress and hormone-related CREs.

In total, 20 types of CREs were detected, including 1848 light responsiveness CREs, 442 MeJA responsiveness CREs, 372 abscisic acid responsiveness CREs, and 365 anaerobic induction CREs (Table S9). The most and least CREs found in the promoter regions of the *GARP* genes were light responsiveness CREs (1848) and wound-responsive CREs (3), respectively. Meanwhile, a mass of putative CREs that were involved in hormone responses, such as GA, MeJA, and BHA, were found in a series of *BnaGARP* promoters. As well, many putative CREs associated with abiotic stress, such as the low-temperature responsive CREs, defense and stress responsive CREs, and drought inducibility CREs, were found in many *BnaGARP* promoter regions (Figure 6).

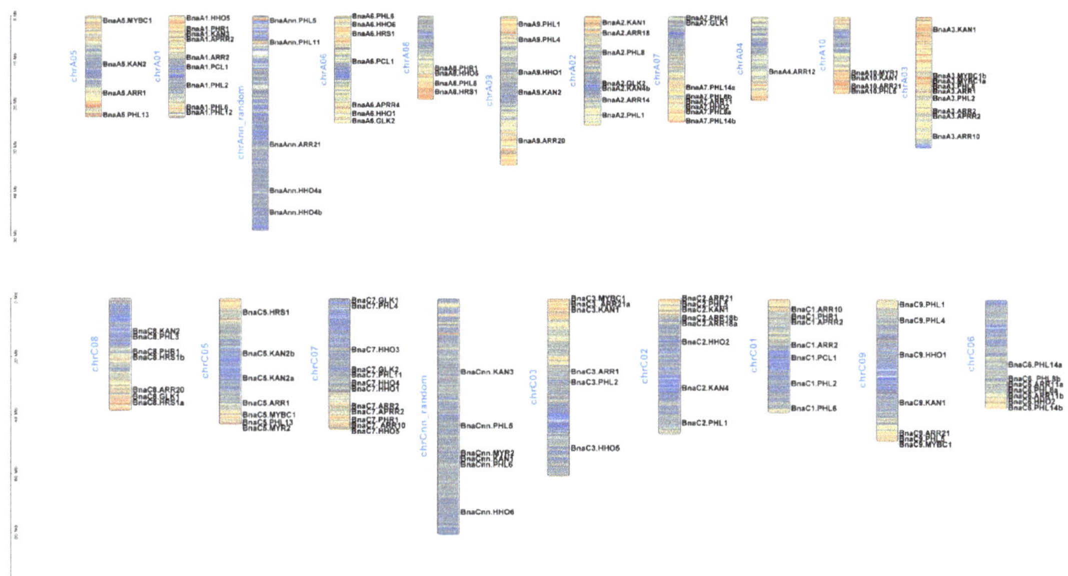

Figure 4. Chromosomal location of the 146 genes in *GARP* family genes in *B. napus*. The distribution of the 146 genes on the 20 chromosomes is presented. The Ann and Cnn chromosomes refer to the chromosome that is anchored to the A and C subgenomes, while they have been not the specific chromosome.

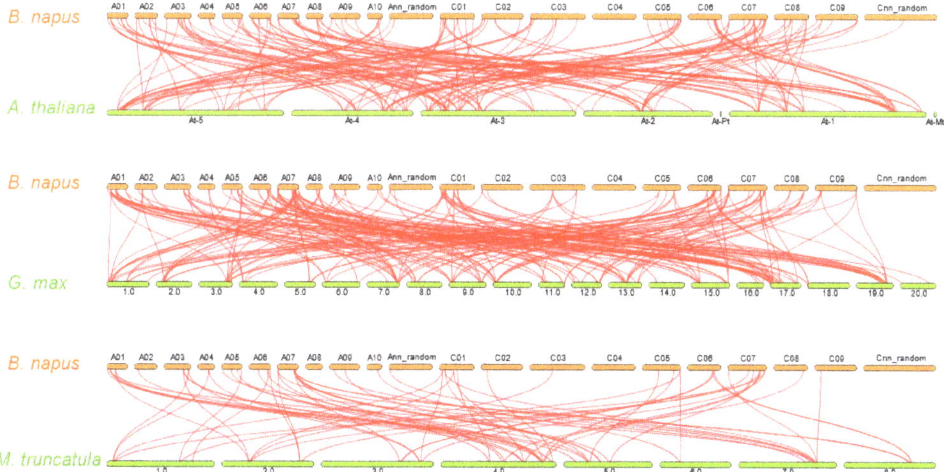

Figure 5. Synteny analysis of the *GARP* family genes between *B. napus* and *A. thaliana*, *G. max*, and *M. truncatula*. Gray lines indicate all collinear blocks within *B. napus* and *A. thaliana*, *G. max*, and *M. truncatula*. While the red lines depict the orthologous relationships.

2.6. Genome-Wide Analysis of miRNA Targeting BnaGARPs

In plants, miRNAs are important regulators of gene expression and play pivotal roles in abiotic stress responses [41]. To identify whether miRNAs are involved in the regulation of the *BnaGARP* expression, we identified 29 putative miRNAs targeting 34 *BnaGARPs*

(Figure 7). Some of the miRNA-targeted sites are presented in Figure S3, while the detailed information of all miRNAs targeted genes is presented in Table S10. The results showed that four members of the bna-miR164 family targeted three *BnaGARPs* (including *BnaC3.ARR1, BnaC6.HHO2,* and *BnaA7.HHO2*). Four members of the bna-miR172 family targeted two *BnaGARPs* (including *BnaC2.ARR18a* and *BnaC2.ARR18b*). Three members of the bna-miR390 family targeted two *BnaGARPs* (including *BnaA3.PHL2* and *BnaC3.PHL2*). Two members of the bna-miR397 family targeted three *BnaGARPs* (including *BnaC1.PHR1, BnaA1.PHR1,* and *BnaC7.PHR1*). Three members of the bna-miR156 family targeted *BnaC2.ARR18b*. One member of the bna-miR6029 family targeted four *BnaGARPs* (including *BnaA6.HRS1, BnaC5.HRS1, BnaA6.PHL6,* and *BnaCnn.PHL6*). One member of the bna-miR860 family targeted eight *BnaARRs* (Figure 7; Table S10). Predominantly, *BnaA3.ARR2, BnaC3.ARR1, BnaC6.HHO2* and *BnaC2.ARR18b* were predicted to be targeted by several miRNAs (Figure 7; Table S10).

2.7. Transcriptional Analysis of BnaGARPs under N and Pi Stresses

Nitrogen (N) is an essential macronutrient for plant growth and development, whereas rapeseed has a low NUE [42]. To improve the understanding of the role of *BnaGARPs* in NUE regulation in *B. napus*, the transcriptional responses of *BnaGARPs* were explored under low N conditions. Under limited NO_3^- conditions, 40 members of *BnaGARPs* were differentially expressed in rapeseed plants compared to sufficient NO_3^- (Figure 8). In the *BnaNIGT1/HRS1/HHOs* subfamily, most members were downregulated (87.88–98.12%) in the shoots under low NO_3^- supply. Notably, the expression levels of *BnaC7.HHO3* and *BnaA9.HHO1* decreased by 98.12% and 97.60% in the shoots, respectively. In the roots, the expression levels of *BnaC9.HHO1* and *BnaA9.HHO1* was reduced by 98.62% and 99.55% under low NO_3^- supply, respectively (Figure 8A). However, different *BnaGARPs* subfamilies showed distinct transcriptional responses under this circumstance. In detail, most (70%) of the differentially expressed genes (DEGs) of the *BnaGLKs* subfamily were upregulated in the shoots or roots under deficient NO_3^- conditions (Figure 8B). In particular, the expression level of *BnaA6.GLK2* decreased by 60.37% in the roots, whereas the expression level of *BnaA2.GLK2* was increased 1.15-fold in the shoots. In terms of the *ARR* subfamily, the expression level of *BnaA3.APRR2* and *BnaC3.ARR1* was repressed by 53.55% and 66.89% in the roots of rapeseed plants exposed to deficient NO_3^- conditions (Figure 8D). In the *BnaPHLs* subfamily, the expression level of *BnaAnn.PHL5* was increased 1.47-fold in the roots, whereas the expression level of *BnaC9.PHL1* was decreased by 79.90% in the shoots (Figure 8C).

To determine the core members that play a dominant role in the NO_3^- response, a co-expression network analysis of *BnaGARPs* was performed. The results showed that *BnaA9.HHO1* and *BnaC7.HHO3* might play a major role in the repression of N-starvation responses in the shoots (Figure 8E); whereas in the roots, *BnaA9.HHO1* and *BnaC9.HHO1* might play a core role in the adaptation of rapeseed plants to N limitation (Figure 8F).

Under the ammonium (NH_4^+) supply condition, a total of 23 *BnaGARP* DEGs were identified in the shoots and roots relative to the condition of NO_3^- sufficiency (Figure 9A). We found only the expression levels of *BnaA8.HHO5, BnaA1.HHO5,* and *BnaC3.HHO5* were increased by 1.51-fold, 1.24-fold, and 1.44-fold under the NH_4^+ supply condition than under NO_3^- sufficiency. Among all the down-regulated genes, particularly the expression level of *BnaA6.HHO1* was reduced by 99.28% in the roots under the NH_4^+ supply condition.

Gene co-expression network analysis showed that *BnaA9.HHO1, BnaA7.HHO3* and *BnaA6.HHO1* might play a core role in the responses of rapeseed plants to NH_4^+ as the sole N nutrient source (Figure 9B,C).

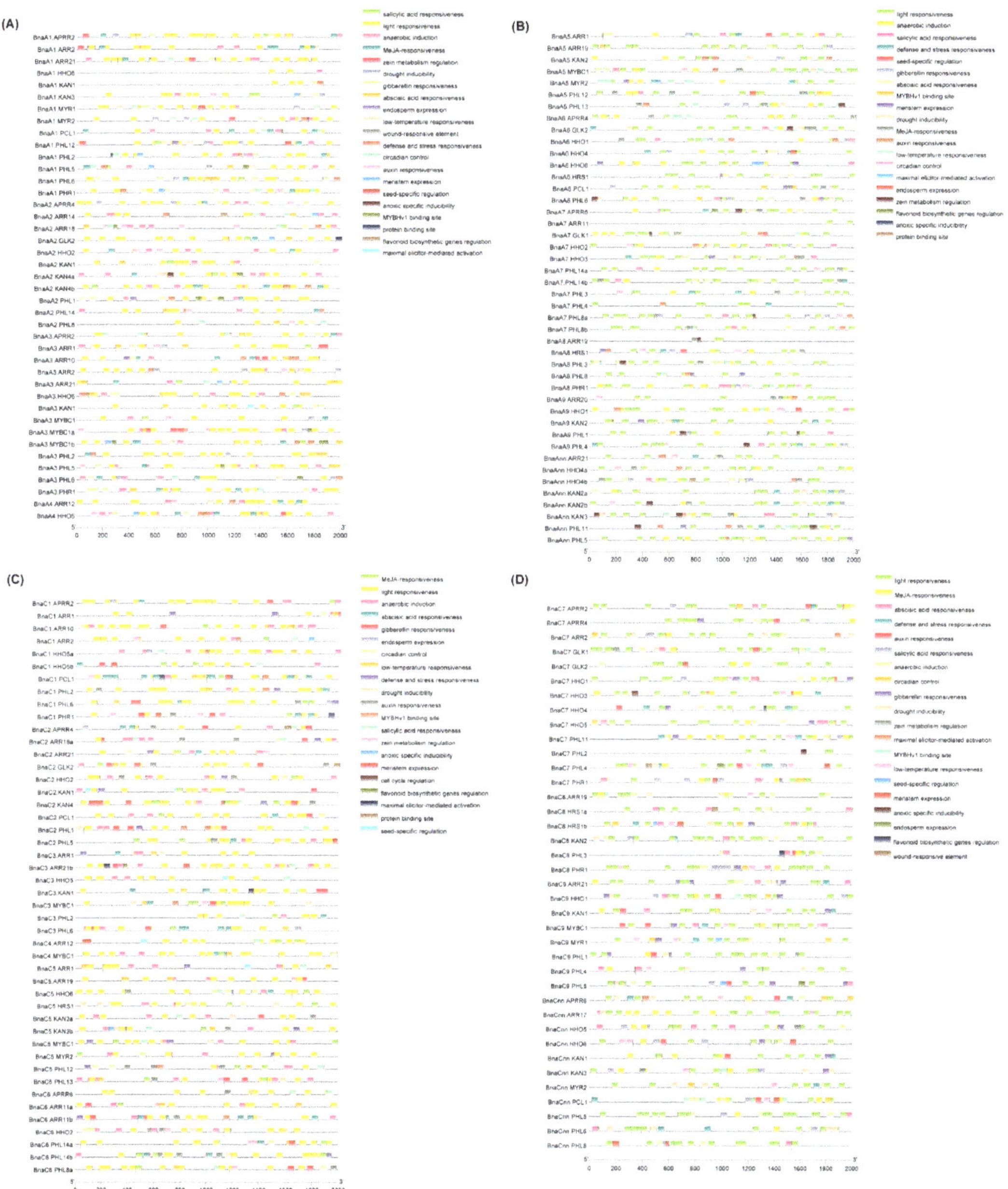

Figure 6. Predicted *cis*-regulatory elements (CREs) in the promoter regions of *BnaGARPs*. (**A**) Distribution of the CREs on chrA1 to chrA4. (**B**) Distribution of the CREs on chrA05 to chrAnn. (**C**) Distribution of the CREs on chrC1 to chrC6. (**D**) Distribution of the CREs on chrC7 to chrCnn. The CREs identified by PlantCARE are based on the sequence of 2000 bp upstream of the start codon of *BnaGARPs*. Different colored rectangles represent different CREs that are potentially involved in the regulation of stress resistance or phytohormone response.

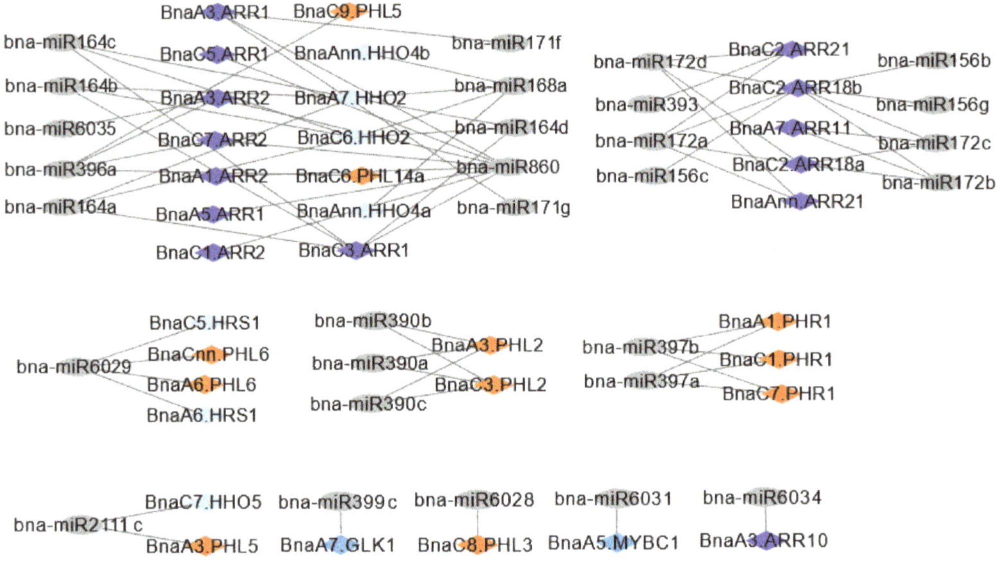

Figure 7. Network diagrams of predicted miRNAs targeting *BnaGARPs*. Different diamond colors represent *BnaGARPs*, and gray ellipse shapes represent potential regulatory miRNAs.

Based on expression pattern analysis and co-expression network analysis, we selected several key genes and analyzed their differential expression between the high-NUE (H12) and low-NUE (L73) rapeseed cultivars (Figure 10A). The results showed that these genes were upregulated (1.53 to 6.64-fold) in the L73 rapeseed cultivar under NO_3^- limitation condition (Figure 10B). In order to explore the role of these key genes involved in the regulation of differential NUE between the rapeseed genotypes, *BnaA9.HHO1* was selected to perform a functional analysis. The results showed that the *BnaA9. HHO1* fusion protein was mainly located in the nucleus and could colocalize with *OsGhd7* in the nucleus (Figure 10C).

Phosphate (Pi) performs a variety of biological functions, including structural elements in nucleic acids and phospholipids, signal transduction cascades, enzyme regulation, and so on [43]. Maeda et al., found that two independent transcriptional cascades for NO_3^- and Pi-starvation signaling are integrated via expression control of the *GARP*-clade genes [10]. Under Pi limitation conditions, a total of 45 *BnaGARP* DEGs were identified in the shoots or roots (Figure 11). In the shoots, most of the DEGs were upregulated except for *BnaC7.GLK2*, *BnaA6.HHO6*, and *BnaC1.PHL2*, which were downregulated. In the *BnaNIGT1/HRS1/HHO* subfamily, *BnaC7.HHO1*, *BnaC8.HRS1b* and *BnaA9.HHO1* were remarkably upregulated, increasing by 6.78, 8.26, and 5.03-fold, respectively (Figure 11A). The expression levels of *BnaA7.ARR11* and *BnaA6.PCL1* had higher expression levels that were increased by 2.86-fold and 2.50-fold in the shoots under Pi deficiency than Pi sufficiency (Figure 11B,C). In terms of the *BnaPHLs* subfamily, the expression level of *BnaA9.PHL1* was decreased by 59.61%, while the expression level of *BnaC6.PHL8b* was increased by 1.02-fold in the roots under low Pi (Figure 11D).

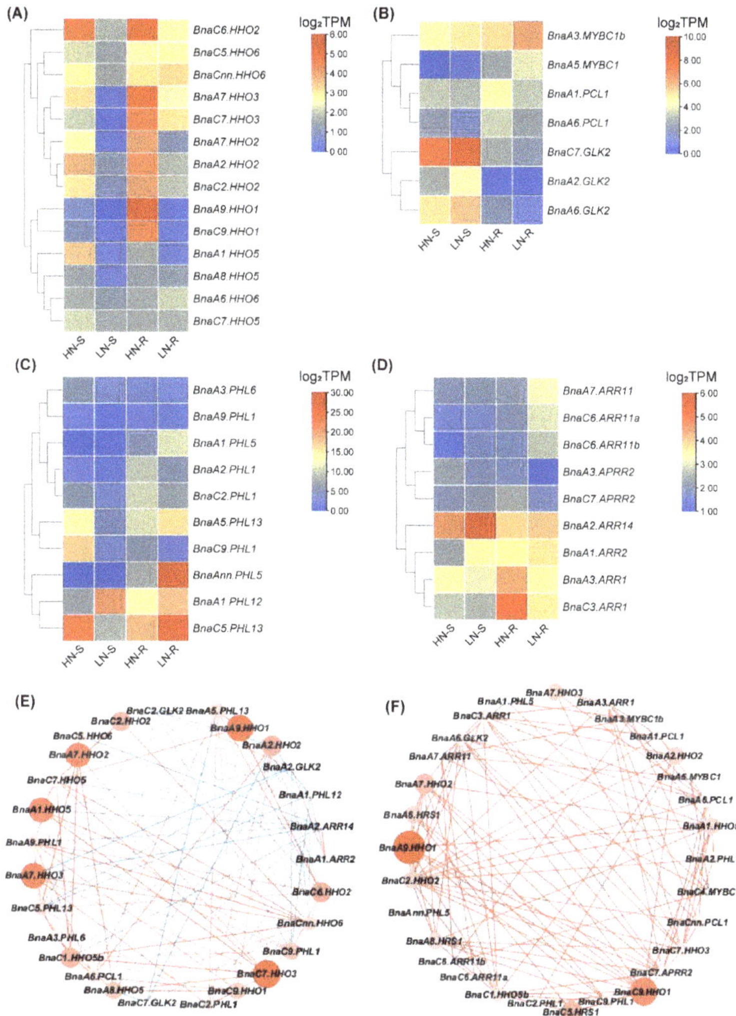

Figure 8. Expression profiles and co-expression network analysis of *BnaNIGT1/HRS1/HHOs* (**A**), *BnaARRs* (**B**), *BnaGLKs* (**C**), and *BnaPHLs* (**D**) in the shoots/S (**E**) and roots/R (**F**) under nitrate (NO_3^-) limitation conditions. HN, high N (6.0 mM NO_3^-); LN, low N (0.30 mM NO_3^-). In the heat maps, the expression levels are normalized by log_2 (TPM). TPM, transcripts per million (reads). The color scales represent relative expression levels from high (red color) to low (blue color). In the gene co-expression networks, the cycle nodes represent genes, and the size of the nodes represents the power of the interrelation among the nodes by log_2FC value. FC, fold change. The edges between two nodes represent interactions between genes.

2.8. Expression Profiles of BnaGARPs in Response to Diverse Nutrient Stresses

Further, the expression patterns of *BnaGARPs* under various nutrient stresses were studied, including deficient boron (B), salt stress, and cadmium (Cd) toxicity. The B requirement of plants varies from species to species, and *B. napus* is considered one of the highest B-requiring plants, which often suffers from yield and quality losses due to B deficiency, especially in Northern Europe, Canada, and China [44]. Under deficient B conditions, a total of 49 *BnaGARP* DEGs were identified in the shoots or roots. In the shoots, 39 DEGs were upregulated after B deficiency treatment (Figure 12). In particular, the expression level of *BnaAnn.PHL11* was increased 4.01-fold. In the *BnaARR-Bs* subfamily, the expression of three *BnaARRs* (including *BnaA2.ARR14*, *BnaA7.ARR11*, and *BnaC6.ARR11a*) was increased in the shoots after B deficiency treatment (Figure 12B). In the subfamily *BnaGLKs*, most of the genes had high expression levels (1.05 to 1.71-fold) under B deficiency than B sufficiency, whereas the expression of *BnaC1.PCL1* was reduced by 57.81% (Figure 12C). The expression pattern in subfamily *BnaNIGT1/HRS1/HHOs* and *BnaPHL1s* was similar to that in the subfamily of *BnaGLKs*. In the shoots, only *BnaA6.HHO6* and *BnaC1.PHL2* was downregulated (Figure 12D). Eight of 15 (53.33%) DEGs in the *BnaNIGT1/HRS1/HHO* subfamily and 15 of 18 (83.33%) DEGs in subfamily *BnaPHLs* were significantly induced by B deficiency.

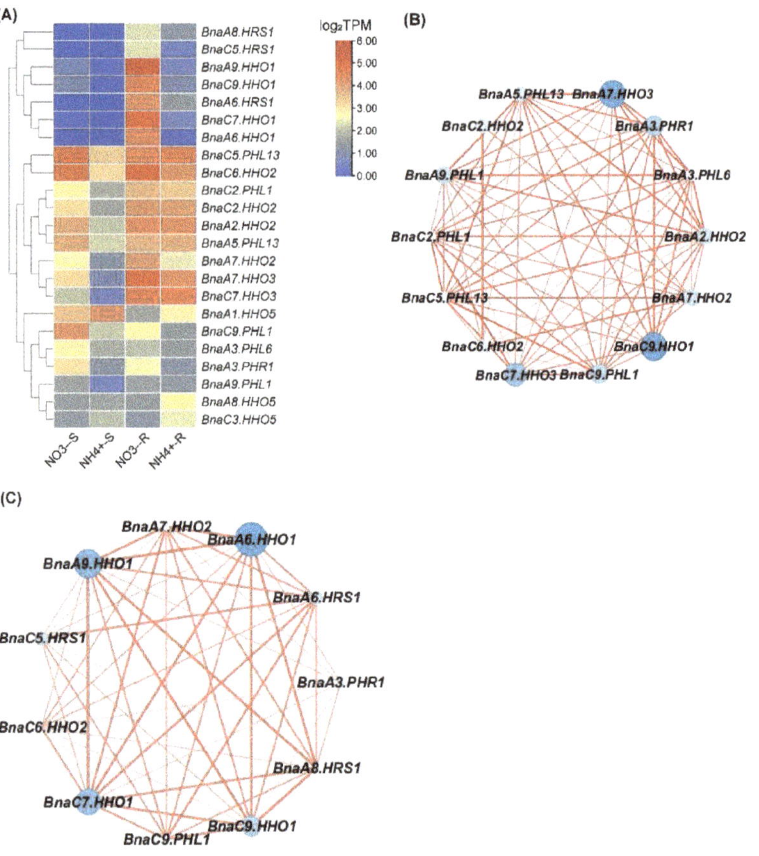

Figure 9. Expression profiles (**A**) and co-expression network analysis of *BnaGARPs* in the shoots/S (**B**) and roots/R (**C**) under different nitrogen (N) form conditions, including 6.0 mM nitrate (NO_3^-) and 6.0 mM ammonium (NH_4^+) conditions. The expression levels are normalized by $\log_2$(TPM). TPM, transcripts per million (reads). In the heat maps, the color scales represent relative expression levels from high (red color) to low (blue color). In the gene co-expression networks, the cycle nodes represent genes, and the size of the nodes represents the power of the interrelation among the nodes by $\log_2$FC value. FC, fold change. The edges between two nodes represent interactions between genes.

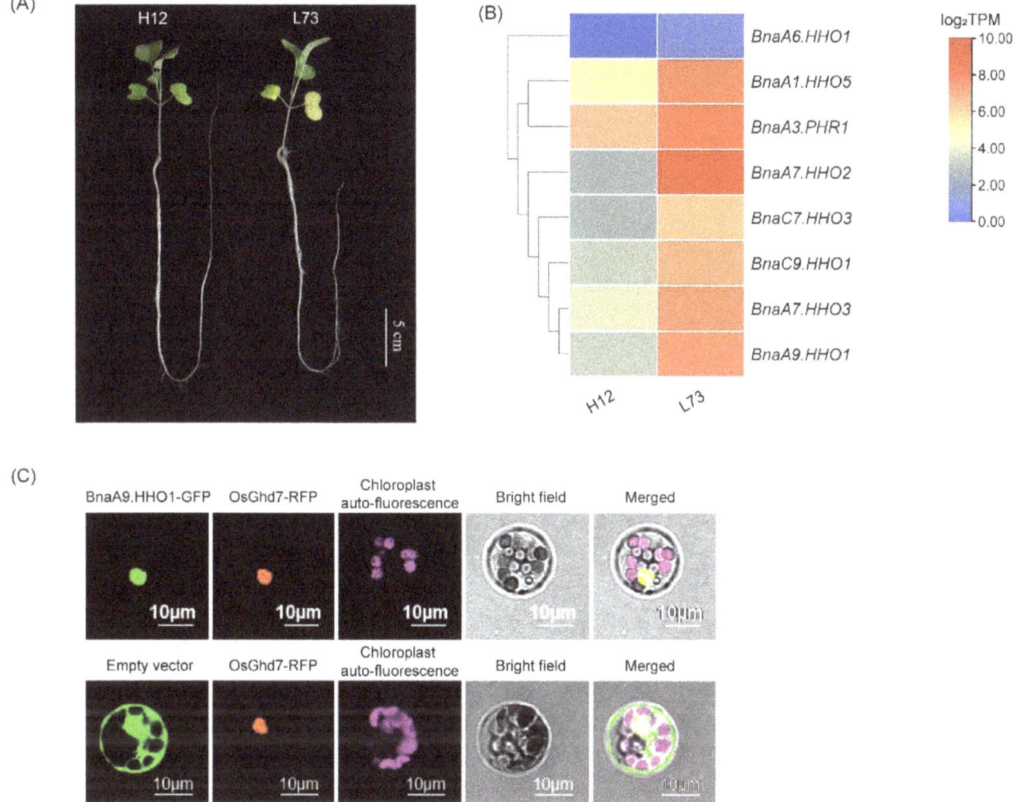

Figure 10. Characterization of the rapeseed genotypes H12 and L73 growing under nitrate (NO_3^-) limitation conditions and subcellular localization analysis of *BnaA9.HHO1*. (**A**) Growth performance of the high-NUE genotype H12 and the low-NUE genotype L73 growing under low (0.3 mM) NO_3^- condition. (**B**) Expression profiles of the *BnaGARP* DEGs between H12 and L73 under NO_3^- limitation conditions. (**C**) Subcellular localization analysis of *BnaA9.HHO1*. *OsGhd7* was used as a nuclear marker and fused with a red fluorescent protein sequence.

Cd is a non-essential heavy metal with high biotoxicity to many organisms, while oilseed rape has a high potential for the phytoremediation of Cd-polluted soils [45]. Under Cd toxicity, a total of 43 *BnaGARP* DEGs were identified in the shoots or roots (Figure 12F). Most genes were downregulated in the roots in response to Cd toxicity, particularly the expression of *BnaC5.KAN2b* and *BnaCnn.PHL6* was reduced by 88.12% and 89.38%. In the shoots, Cd toxicity resulted in an obvious decrease in the expression of *BnaA1.PHL5*. Under Cd toxicity condition, *BnaA8.HHO5* was significantly increased by 2.92-fold in the roots.

Salt stress is one of the most important abiotic factors affecting global agricultural productivity, inhibiting plant growth, development and productivity by disrupting many physiological and biochemical processes [46]. In the salt stress-treated group, the expression of two *BnaARRs* (including *BnaA1.ARR2 and BnaA2.ARR14*) was induced by 1.53 and 1.14-fold in the shoots, whereas the expression of *BnaC3.ARR1* was significantly decreased (Figure 13). In the roots, *BnaA2.ARR14*, *BnaA3.APRR2*, and *BnaC7.APRR2* was upregulated by salt stress, while the expression of *BnaC3.ARR1* was decreased by 61.96% (Figure 13B). Under salt stress, *BnaA3.MYBC1*, *BnaA6.GLK2* and *BnaC9.MYBC1* in the *BnaGLK* subfamily showed a low expression level in both roots and shoots; however, *BnaA3.MYBC1b*,

BnaA5.MYBC1 and *BnaC4.MYBC1* in this subfamily shared higher expression levels under salt stress (Figure 13C). In terms of *BnaNIGT1/HRS1/HHO* subfamilies, most of them had lower expression levels (40.90% to 87.24%) under salt stress (Figure 13A). After salt treatment, the expression of 16 *BnaPHLs* was distinctly upregulated in the shoots, while the expression of eight *BnaPHLs* was obviously downregulated in the roots (Figure 13D).

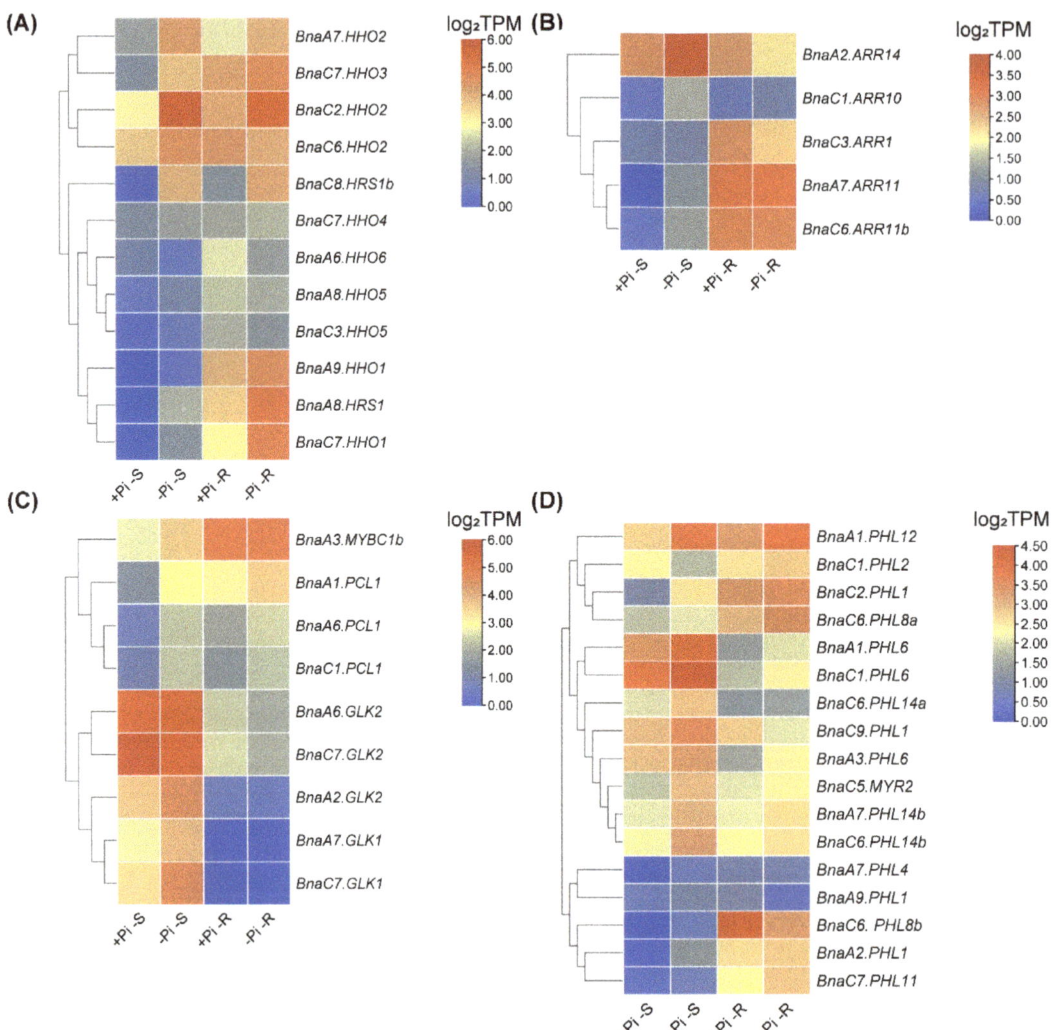

Figure 11. Expression profiles of *BnaNIGT1/HRS1/HHOs* (**A**), *BnaARRs* (**B**), *BnaGLKs* (**C**), and *BnaPHLs* (**D**) in the shoots/S and roots/R under different phosphate (Pi) levels. conditions. +Pi, high Pi (250 µM); -Pi, low Pi (5 µM), The expression levels are normalized by log₂ (TPM). TPM, transcripts per million (reads). The color scales represent relative expression levels from high (red color) to low (blue color).

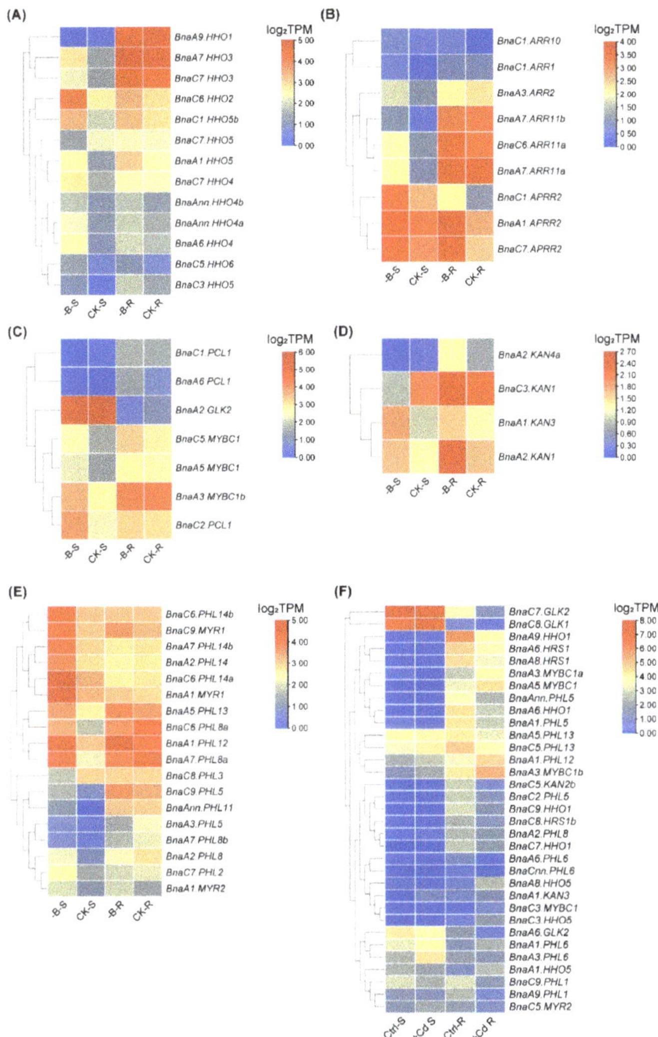

Figure 12. Expression profiles of *BnaNIGT1/HRS1/HHOs* (**A**), *BnaARRs* (**B**), *BnaGLKs* (**C**), *BnaKANs* (**D**), and *BnaPHLs* (**E**) in the shoots/S and roots/R under different boron (**B**) and cadmium (Cd) toxicity (**F**) conditions. -B, low B (0.25 µM); CK, high B (25 µM); Ctrl, Cd-free; +Cd, high Cd (10 µM). In the heat maps, the expression levels are normalized by $\log_2$ (TPM). TPM, transcripts per million (reads). The color scales represent relative expression levels from high (red color) to low (blue color).

To characterize the common genes responsive to nutrient stresses, a Venn diagram was constructed with the DEGs identified, respectively, under the diverse nutrient stresses above-mentioned. As shown in Figure 14, *BnaA9.HHO1* and *BnaA1.HHO5* was simultaneously regulated by low NO_3^-, NH_4^+ toxicity, limited Pi, deficient B, salt stress, and Cd toxicity in the shoots and roots (Figure 14). This result indicated that *BnaA9.HHO1* and *BnaA1.HHO5* might play a multifaceted role in regulating rapeseed resistance to nutrient stresses.

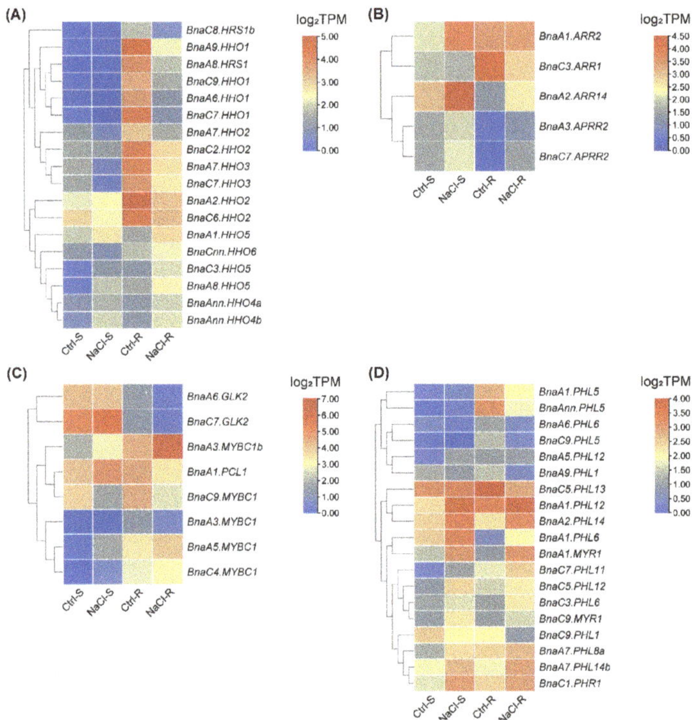

Figure 13. Expression profiles of *BnaNIGT1/HRS1/HHOs* (**A**), *BnaARRs* (**B**), *BnaGLKs* (**C**), and *BnaPHLs* (**D**) in the shoots/S and roots/R under salt stress conditions. Ctrl, control, NaCl-free; NaCl, +NaCl, 200 mM. In the heat maps, the expression levels are normalized by $\log_2$ (TPM). TPM, transcripts per million (reads). The color scales represent relative expression levels from high (red color) to low (blue color).

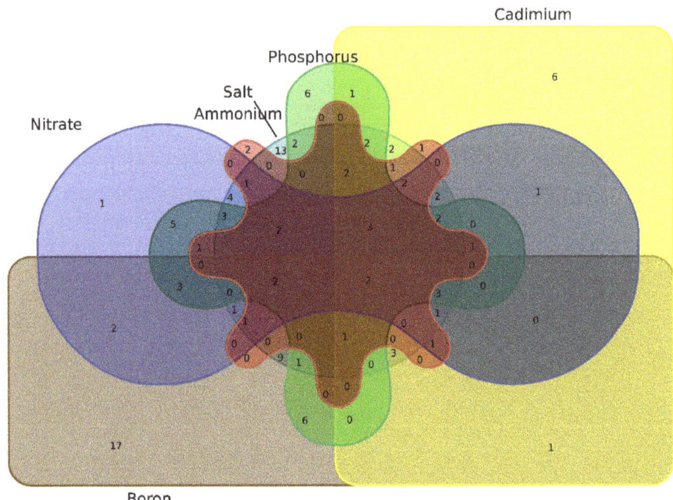

Figure 14. Venn diagram showing the common and specific differentially expressed genes of *Bna-GARPs* under diverse nutrient stresses.

3. Discussion

Previous studies have shown that the *GARP* family members play critical roles in phytohormone transport and signaling, plant organ development, and nutrient responses [47–49]. However, there have been few systematic studies on *GARPs* in *B. napus*. In the present study, the genome-scale *GARP* family genes were identified in *B. napus* and their phylogenetic relationships, conserved motif and domain, gene structures, duplication and synteny relationships, CREs, and chromosomal locations were performed. In addition, we delineated the differential expression profile of *BnaGARPs* under low NO_3^-, NH_4^+ toxicity, limited Pi, deficient B, salt stress, and Cd toxicity. The global identification of *BnaGARPs* provides the foundation for further in-depth functional studies of these genes.

3.1. An Integrated Bioinformatics Analysis Provided Comprehensive Insights into the Molecular Features of BnaGARPs

In this study, a total of 146 *BnaGARPs* were identified (Table S1). A previous study has revealed 56 *GARPs* in *A. thaliana*, 69 *GARPs* in *Camellia sinensis*, and 35 *GARPs* in *S. polyrhiza* [8,10,50], suggesting that the *GARP* TF family is ubiquitous in plants (Table S11). The *GARP* gene family in rapeseed is larger than those in other plant species, which might be due to complex whole genome duplication and subsequent evolution of the rapeseed genome. [51]. Phylogenetic analysis showed that the *B. napus* genome retains the orthologs of *AtGARPs* and the gene phylogeny roughly followed the species phylogeny (Figure 1). Furthermore, the phylogenetic tree also showed that all subfamilies have expanded during the evolution process. A lineage-specific expansion of *BnaGARP* via the partial alteration of the genome is used to adapt to internal and external environments during evolution [52,53]. Generally, the Ka/Ks ratios for all the homologous *GARP* pairs were less than 1.0, indicating that *BnaGARPs* might have undergone purifying selection pressure (Table S2). Arabidopsis and Brassica diverged about 20 million years ago, and evolutionary selection pressure analysis suggested that the divergence of *GARPs* also occurred during this period.

Due to the similarity between the B-motif and the MYB-like domain, the GARP TFs were frequently mistaken for MYB or MYB-like TFs. However, the MYB TFs contain the (SHAQK(Y/F) F) motif, while the *GARP* TFs contain a different consensus sequence (SHLQ (K/M) (Y/F)) [5]. All the *BnaGARPs* were predicted to contain some conserved motifs, which are components of the B-motif and are important for DNA binding (Figure 2). In this study, the conserved motifs in each subfamily of *BnaGARPs* are essentially similar, indicating that their amino acid residues are very conserved in terms of evolution, and have essential roles in gene function or structure. In addition, we found that *BnaNIGT1/HRS1/HHO* subfamily contains two different motifs EAR-like at their N or C terminal. The EAR-like motifs play an important role in inhibiting gene expression as transcription repressors or recruit corepressors [54]. In this study, different gene structures were found among *BnaGARPs*, and *BnaNIGT1/HRS1/HHO* subfamily had fewer exons than the *BnaPHL* subfamily, implying structural diversification among the *BnaGARP* subfamilies (Figure 3). The differences in the intron lengths suggested a possible role in the functional diversification of *BnaGARPs*. Chromosomal localization results showed that 146 genes are unevenly distributed on 20 chromosomes, presumably due to multiple polyploidization events in the genome of oilseed rape during its evolution [55]. Previous research revealed that tandem duplication events or segmental duplication events act as a mechanism for family expansion, and it also could promote the emergence of new functional genes that plants can better cope with abiotic stress during evolution [56,57].

To further elucidate the synteny relationships of *BnaGARPs* with *GARPs* in other model plants, we identified 172, 152, 66, 3, and 1 orthologous gene pairs between *BnaGARPs* with other *GARPs* in *G. max*, *A. thaliana*, *M. truncatula*, *O. sativa*, and *T. aestivum*, respectively (Figure 5). Synteny analysis results suggested that some *BnaGARPs* possibly came into being during gene duplication, and the segmental duplication events could play key roles in the expansion of *GARP* genes in *B. napus* [58]. In addition, *B. napus* and *A. thaliana* shared 152 syntenic gene pairs within the *GARP* family, indicating that *B. napus* and *A. thaliana* are closely evolutionarily related. Additionally, the allotetraploid *Brassica napus* L. ($A_nA_nC_nC_n$, 2n = 4x = 38) was formed by natural distant hybridization of diploid *Brassica rapa* L. (A_rA_r, 2n = 2x = 20) and diploid *Brassica oleracea* L. (C_oC_o, 2n = 2x = 18) [59]. In the present study, 45 duplication events took place on the A_n sub-genome, 36 events on the C_n sub-genome, and 136 events across A_n/C_n sub-genomes. Therefore, we proposed that the *BnaGARP* expansion is a synergistic effect of polyploidization and hybridization working together [60].

The CREs in the promoter regions play an important role in regulating and functioning genes [61]. In this study, the CRE analysis confirmed the potential roles of *BnaGARPs* in the regulation of stress resistance (Figure 6). Many stresses and phytohormone-related CREs were identified in the promoter regions of most *GARPs*, including the ARE, G-box, MBS, and LTR elements. The most common CREs were light responsiveness CREs. Studies have confirmed that *AtHHO4* can interact with *JMJ30*, which is the *H3K36Me2* demethylase and is involved in light-responsive circadian clock [62].

MicroRNAs (miRNAs) are crucial non-coding regulators of gene expression in plants [63] and play essential roles in plant–environment interactions [64]. Over the past few years, a number of miRNAs have been recognized through genome-wide examination in rapeseed to participate in diverse nutrient stresses [38,65]. In this study, we identified 29 miRNAs targeting 34 *BnaGARPs* (Figure 7; Tables S4–S8). miRNA164 has been reported to be involved in lateral root development in maize (*Zea mays* L.) [66]. miRNA156 has been reported to be significantly upregulated under dehydration stress responsiveness in different species [67]. Similarly, miR172 has also been found to regulate drought escape and drought tolerance by affecting sugar signaling in *A. thaliana* [68]. miR396 is a conserved miRNA and is involved in plant growth, development, and abiotic stress response in various plant species through regulating its targets, *Growth Regulating Factor* (*GRF*) TFs [69]. Some miRNAs have also been reported in rapeseed, playing a significant role in rapeseed genetic improvement [70,71]. These findings suggest that these bna-miRNAs might play pivotal roles against a variety of stresses by modifying the transcriptional or translational levels of *BnaGARPs*.

3.2. Differential Expression Profiling of BnaGARPs Implied Their Potential Involvement in the Responses of Rapeseed to Diverse Nutrient Stresses

TFs regulate gene expression by recognizing and combining *CREs* on the promoter regions of target genes [72]. TFs play key roles in plant developmental processes, phytohormone signaling pathways, and disease resistance responses. Given that expression patterns can lead to the estimation of gene functions [73]. For example, through analyzing the expression profile of *TaWRKY* family members under drought, cold, and high-temperature conditions, a considerable number of *TaWRKY* genes are shown to respond to drought stresses [3]. When exposed to $ZnSO_4$ and $FeCl_3$ solutions, the *TaZIP* genes showed differential expression patterns [74].

Previous studies have confirmed that *GARPs* play an important role in nutrient sensing [6]. The first *GARP* TF shown to be involved in nutritional responses was the Chlamydomonas phosphorus-stress response 1 (Psr1) [75]. Under Pi starvation, *OsPHR2* binds to a CRE (P1BS) in the promoter of various PSI genes and upregulates their transcription, thus optimizing rice Pi acquisition and utilization [76]. Another *GARP* subfamily that attracted recently lots of attention was *NIGT1/HRS1/HHO* subfamily. *NIGT1/HRS1/HHOs* have recently been confirmed to be involved in the perception and transduction of N and Pi nutritional signals in plant transcriptional regulatory networks [27].

In this study, we found that most *BnaGARPs* were significantly downregulated in the shoots or roots under NO_3^- limitation conditions, among which the downregulated levels of *BnaNIGT1/HRS1/HHOs* were the highest (Figure 8). This finding highlighted the crucial role of *BnaNIGT1/HRS1/HHOs* in the regulation of NO_3^- starvation. It has been demonstrated that *NIGT1(HHO1-HHO3, HRS1)* expression was induced by NO_3^- signaling, and it also inhibited N starvation response (NSR) genes (*NRT2.1* and *NRT2.4*) under N sufficient conditions [10]. The *GARP* TFs modulate the expression of target genes by positive or negative feedforward mechanisms under abiotic stress [10]. For example, *AtNIGT1/HRS1* binds to the promoter of *NRT2.4* and represses an array of N starvation-responsive genes under conditions of high N availability [77]. *HRS1* and *HHO1* control ROS accumulation in response to NSR and directly repress NSR sentinel genes (*NRT2.5*) [78]. *NLPs* (including *NLP5* and *NLP7*) expression were downregulated by NLP-induced *NIGT1s* [10]. *SPX1*, *SPX2*, and *SPX4* are putative Pi-dependent inhibitors of Arabidopsis PHOSPHATE STARVATION RESPONSE1 (*PHR1*) [79]. To improve the understanding of *BnaGARP*-mediated transcriptional networks under abiotic stress responses, the transcriptional responses of 25 target genes were explored under these circumstances (Figure S4). Under NO_3^- limitation conditions, *BnaAn.NRT2.4*, *BnaC9.NRT2.4*, and *BnaA8.NRT2.5* were upregulated, while *BnaA7.NLP5* and *BnaC6.NLP5* also shared higher expression levels. It indicates that *BnaNRT2.4* and *BnaNLP5* might play key roles in the *BnaGARP*-mediated transcriptional networks.

NH_4^+ is also a major N source for plants, and it is also an indispensable intermediate in the biosynthesis of essential cellular components [80]. In general, compared with NO_3^-, NH_4^+ as the sole N nutrient source had a weakened effect on the transcriptional responses of *BnaGARPs*. Under Pi limitation conditions, most of the *BnaGARP* DEGs were upregulated in the shoots or roots, among which the upregulated levels of *BnaC8.HRS1b* was the highest (Figure 11). Previous studies have reported Pi deprivation increased the *HRS1* expression level and expanded its expression domain [81]. Transcripts of *SPX1* and *SPX2* accumulate in the roots and shoots of Pi-limited plants in a PHR1-dependent manner [82]. In this study, *BnaSPX1* and *BnaSPX2* were upregulated in the roots and shoots under Pi-limited conditions. Moreover, we found no differences in the expression of *BnaKANs* under both NO_3^- limitation and NH_4^+ toxicity conditions. A previous study has revealed that *AtKANs* regulate auxin biosynthesis, transport, and signaling [12]. Therefore, *BnaKANs* might be not involved in N absorption and utilization.

The expression patterns of *BnaGARPs* were also studied under various nutrient stresses. Under deficient B conditions, most *BnaGARP* DEGs were upregulated. Therefore, it could be concluded that *BnaGARPs* are also involved in response to B deficiency and might play important roles in B absorption in *B. napus* (Figure 12). Most members of the *BnaGARPs* have been shown to play a role in salt stress [18,83,84]. For instance, *HRS1* has transcriptional repressive activity and appears to suppress the expression of factors that negatively regulate salt tolerance, *ZmGLK3*, *SlGLK7*, and *SlGLK15* were upregulated under salt stress. In our results, we found that the expression level of *BnaHRS1* was significantly downregulated after salt stress (Figure 12). In the roots, *BnaA6.GLK2* and *BnaC7.GLK2* were upregulated. These results suggested that homologous genes should have similar expression patterns under abiotic stress. Moreover, most *BnaGARP* DEGs were downregulated under Cd toxicity (Figure 12F). It is worth noting that although some differential genes of *BnaGARPs* have been found, there are still some genes that have not been identified, which may be the problem of variety, expression site, and genome assembly.

In short, *BnaGARPs* were responsive to diverse nutrient stresses, which implied the essential roles of *BnaGARPs* in the resistance or adaptation of rapeseed to stresses.

3.3. BnaNIGT1/HRS1/HHOs Might Be Major Regulators of N-Starvation Responses

It has been reported that *NIGT1/HRS1/HHOs* were key regulators involved in plant response to limited NO_3^- conditions. In Arabidopsis, the *NIGT1/HRS1/HHO* subfamily directly represses the expression of the *NRT2* genes (including *NRT2.1*, *NRT2.4*, and *NRT2.5*), *NLP* TFs directly activate genes encoding *NIGT1/HRS1/HHO* family TFs [10]. In rice, the overexpression of NIGT1 might have a negative effect on NUE and thus reduce the chlorophyll content [85]. In this study, *BnaNIGT1/HRS1/HHOs* were significantly downregulated under N-starvation responses (Figure 7). Among all the *BnaNIGT1/HRS1/HHO* DEGs, the transcription levels of *BnaA9.HHO1* and *BnaC9.HHO1* was most obviously downregulated. Furthermore, GFP-assisted subcellular localization analysis showed that *BnaA9.HHO1* was localized in the nucleus (Figure 9). Based on the co-expression network analysis and

Venn diagram, we proposed that it was *BnaHHO1s* that might be the core genes in the N starvation response. However, functional validation is needed to reveal the in-depth functional roles of *BnaNIGT1/HRS1/HHOs*.

4. Materials and Methods

4.1. Identification of GARP Family Genes in Plants

In this study, the genomic, coding sequences, and protein sequences from *A. thaliana* and *B. napus* (Brana_ Dar_V5 genome) were downloaded from the Arabidopsis Information Resource (TAIR10, https://www.arabidopsis.org/, accessed on 1 October 2022) [86] and the Brassica Database (BRAD V3.0, http://brassicadb.cn/#/, accessed on 1 October 2022) [87]. To identify the *GARP* genes in these species, 56 *GARP* protein sequences from Arabidopsis were used as queries in a reciprocal Basic Local Alignment Search Tool (BLAST) analysis using the threshold and minimum alignment coverage parameters described previously [6,88]. All the *GARP* protein sequences were confirmed by comparison with *GARP* member sequences through searches of the Pfam (V35.0, http://pfam.xfam.org/, accessed on 1 October 2022) [89] and NCBI-CDD (https://www.ncbi.nlm.nih.gov/Structure/bwrpsb/bwrpsb.cgi, accessed on 1 October 2022) [90] database. The protein length, molecular weight (MW), and isoelectric point (pI) of each *GARP* protein were predicted using the ExPASy server (https://web.expasy.org/protparam/, accessed on 1 October 2022) [91].

The genes in Brassica species were named as follows: abbreviation of species name + chromosome + the name of gene homologs in *A. thaliana*. For example, *BnaC1.APRR2* represents a gene homologous to *APRR2* in *A. thaliana* on the C1 chromosome of *B. napus*.

4.2. Phylogenetic Analysis of the GARP Family in B. napus

Multiple sequence alignments of the *GARP* coding sequences between *B. napus* and *A. thaliana* were conducted using ClustalW2 (http://www.genome.jp/tools-bin/clustalw, accessed on 1 October 2022) [92] with default parameters. The phylogenetic trees were generated using the Molecular Evolutionary Genetics Analysis (MEGA) 7.0 program (https://megasoftware.net/home, accessed on 1 October 2022) [93] with the NJ method, the p-distance + G substitution model, 1000 bootstrap replications, and conserved sequences with a coverage of 70%. The phylogenetic trees were visualized using iTOL (V5, https://itol.embl.de/, accessed on 1 October 2022) [94]. The coding sequence alignments were imported into KaKs_calculator (https://ngdc.cncb.ac.cn/biocode/tools/BT000001, accessed on 1 October 2022) [95] to calculate the synonymous mutation rate (Ks) and non-synonymous mutation rate (Ka) using the NG method The time (T) of duplication in millions of years (Mya)was estimated with the formula T = Ks/2λ (λ = 1.5 × 10^{-8}) [96].

4.3. Motif Identification and Gene Structure Analysis

Conserved motifs in the proteins were identified using the Expectation Maximization for Motif Elucidation program (MEME v4.12.0, https://meme-suite.org/meme/, accessed on 1 October 2022) [97] with the following parameter settings: the maximum number of motifs was 10. The conserved domains of *GARPs* were confirmed by NCBI-CDD search. TBtools was used to visualize the motifs and conserved domains of candidate genes. The gene structure was visualized by Gene Structure Display Server (2.0, http://gsds.gao-lab.org/, accessed on 1 October 2022) [98].

4.4. Chromosomal Locations and Synteny Analyses

Information about the physical locations of the *GARP* genes in the genomes of *B. napus* was collected from the BRAD database, and their positions were drafted to chromosomes by using MapGene2Chrom (http://mg2c.iask.in/mg2c_v2.1/, accessed on 1 October 2022) [99].

To uncover the evolutionary linear relationships within species and with ancestral species, the MCScanX plugin in TBtools V1.098 [100] was used to perform a collinearity analysis of *B. napus*. The circos plots of *BnaGARs* were generated by the Circos plugin in TBtools [101].

4.5. CRE Analysis

The CREs in the promoter regions of genes are considered to be related to the regulation of genes. In order to further investigate the potential regulatory network of *GARPs*, the 2000 bp upstream genomic DNA sequences of these genes' start codon were submitted to PlantCARE (http://bioinformatics.psb.ugent.be/webtools/plantcare/html/, accessed on 1 October 2022) [102] to obtain CREs.

4.6. Prediction of Putative miRNA Targeting BnaGARPs

The cDNA sequences of *BnaGARPs* were used to identify possible target miRNAs in the psRNATarget database (V. 2017, Available online: https://www.zhaolab.org/psRNATarget/, accessed on 1 October 2022) [103] with default parameters, except maximum expectation (E) = 5.0. The targeted sites with high degrees of complementarity were selected. Cytoscape software (V3.8.2, https://cytoscape.org/download.html, accessed on 1 October 2022) [104] was used to create the interaction network between the prophesied miRNAs and the equivalent target BnaGARPs.

4.7. Plant Materials and Treatments

The *B. napus* seedlings (Darmor-bzh) germinated in this experiment. "Darmor-bzh" are a French winter oilseed rape variety, whose reference genome sequence was first published in 2014 [59].

First, plump *B. napus* seeds were selected, disinfected with 1% NaClO for 10 min, cleaned with ultra-pure water, soaked overnight at 4 °C, and then sown on the seedling tray. The 7-d old uniform *B. napus* seedlings after seed germination were transplanted into black plastic containers with 10 L Hoagland nutrient solution. The basic nutrition solution contained 1.0 mM KH_2PO_4, 5.0 mM KNO_3, 5.0 mM $Ca(NO_3)_2 \cdot 4H_2O$, 2.0 mM $MgSO_4 \cdot 7H_2O$, 0.050 mM EDTA-Fe, 9.0 µM $MnCl_2 \cdot 4H_2O$, 0.80 µM $ZnSO_4 \cdot 7H_2O$, 0.30 µM $CuSO_4 \cdot 5H_2O$, 0.10 µM $Na_2MoO_4 \cdot 2H_2O$, and 46 µM H_3BO_3. The rapeseed seedlings were cultivated in an illuminated chamber following the growth regimes: light intensity of 300–320 µmol m^{-2} s^{-1}, temperature of 25 °C daytime/22 °C night, light period of 16 h photoperiod/8 h dark, and relative humidity of 70% [105].

To further analyze the expression patterns of *BnaGARPs* under different nutrient stresses, five treatments were set. For the NO_3^- depletion treatment, the 7-d old uniform *B. napus* seedlings were hydroponically cultivated under high (6.0 mM) NO_3^- for 10 d, and then were grown under low (0.30 mM) NO_3^- for 3 d until sampling. For the NH_4^+ toxicity treatment, the 7-d-old uniform *B. napus* seedlings after seed germination were hydroponically cultivated under high NO_3^- for 10 d and then were grown under N-free conditions for 3 d. Finally, the plants were grown under excess (9.0 mM) NH_4^+ for 6 h until sampling. For the inorganic Pi starvation treatment, the 7-d-old uniform *B. napus* seedlings after seed germination were first hydroponically grown under 250 µM Pi (KH_2PO_4) for 10 d, and then were grown under 5 µM Pi for 3 d until sampling. For the salt stress treatment, the 7-d-old uniform *B. napus* seedlings after seed germination were hydroponically cultivated in a NaCl-free solution for 10 d and then were transferred to 200 mM NaCl for 1 d until sampling. In the B deficiency treatment, the 7-d-old uniform *B. napus* seedlings after seed germination were first hydroponically grown under 10 µM H_3BO_3 for 10 d, and then were transferred to 0.25 µM H_3BO_3 for 3 d until sampling. For the Cd toxicity treatment, the 7-d-old uniform *B. napus* seedlings after seed germination were hydroponically cultivated in a Cd-free solution for 10 d and then were transferred to 10 µM $CdCl_2$ for 12 h until sampling. In addition, a high-NUE (H73) and a low-NUE (L12) rapeseed cultivar were also used for the experiment under nitrate limitation conditions [106].

The shoots and roots of fresh rapeseed seedlings above-mentioned were sampled separately and were immediately stored at 80 °C. Each sample contained three independent biological replicates for the transcriptional analyses of *BnaGARPs* under diverse nutrient stresses.

4.8. Transcriptional Analysis of BnaGARPs under Diverse Nutrient Stresses

A total of 12 RNA samples from each treatment were subjected to an Illumina HiSeq X Ten platform (Illumina Inc., San Diego, CA, USA). The illumine RNA-seq data were analyzed to reveal the transcriptional responses of *BnaGARPs* under diverse nutrient stresses. To identify the DEGs between different samples/groups, the expression level of each gene was calculated according to the TPM method. RSEM (http://deweylab.biostat.wisc.edu/rsem/, accessed on 1 October 2022) [107] was used to quantify gene abundances. Essentially, differential expression analysis was performed using DESeq2 [108], and the DEGs with $|\log_2 (FC)| \geq 1$ and P-adjust ≤ 0.05.

4.9. Subcellular Localization Assay

Subcellular localization of target genes was determined using polyethylene glycol-mediated protoplast transformation in Arabidopsis [109]. *OsGhd7* was used as a nuclear marker and fused with a red fluorescent protein sequence [110]. Fluorescence was observed using a Nikon C2-ER confocal laser-scanning microscope with emission filters set at 510 nm (GFP) and 580 nm (RFP), and excitation was achieved at 488 nm (GFP) and 561 nm (RFP).

5. Conclusions

In this study, a systematic genome-wide analysis and molecular characterization of the 146 *GARP* members in allotetraploid rapeseed was completed. In addition, RNA-seq data showed that *BnaGARPs* respond to various nutritional stresses Among all DEGs, *BnaA9.HHO1* and *BnaA1.HHO5* might play a core role in regulating rapeseed resistance to nutrient stresses. However, additional investigations are required to confirm the functional roles of these core genes. The present results would increase the understanding of the evolution of the *GARP* family genes and provide valuable candidate genes for further study of the transcriptional regulation mechanism in response to nutrient stresses in rapeseed.

Supplementary Materials: The following supporting information can be downloaded at: https://www.mdpi.com/article/10.3390/ijms232214484/s1, Table S1: Basic Physicochemical properties of GARP transcription factors in *B. napus*; Table S2: Non-synonymous and synonymous nucleotide substitution rates between AtGARPs and the corresponding orthologs in *B. napus*; Table S3: Segmentally duplicated *B. napus* GARP genes; Table S4: One-to-one orthologous relationships of the GARP genes between *B. napus* and *A. thaliana*; Table S5: One-to-one orthologous relationships of the GARP genes between *B. napus* and *G.max*; Table S6: One-to-one orthologous relationships of the GARP genes between *B. napus* and *M. truncatula*; Table S7: One-to-one orthologous relationships of the GARP genes between *B. napus* and *T. aestivum*; Table S8: One-to-one orthologous relationships of the GARP genes between *B. napus* and *O. sative*; Table S9: Type and number of cis-acting elements; Table S10: Prediction of miRNAs target sites; Table S11: Copy number analysis of GARP family genes in *A. thaliana*, *B. napus*, *Oryza sativa*, *Camellia sinensis*, *Spirodela polyrhiza*, and *Gossypium hirsutum*; Table S12: GARP members and protein sequences in *B. napus*; Figure S1: Identification and characterization of the conserved motifs in the GARP proteins in *B. napus*; Figure S2: Exon-intron organizations of *BnaGARPs*; Figure S3: miRNA targeting sites of *BnaGARPs*; Figure S4: Expression profiles of *BnaGARP* target genes under diverse nutrient stresses; Figure S5: Synteny analysis of GARP genes between *B. napus* and *O. sativa* and *T. aestivum*.

Author Contributions: Y.-P.H. and T.Z. designed the experiments; Y.-P.H. and P.-J.W. conceived the project, analyzed the data, and wrote the article with contributions of all the authors; T.-Y.Z., J.-Y.H., C.-P.Y., H.-L.S., Y.-F.Z., T.-Y.Z. and J.-F.C., provided technical assistance to Y.-P.H.; T.S. and T.Z. supervised and complemented the writing. All authors have read and agreed to the published version of the manuscript.

Funding: This study was financially supported by the National Natural Science Foundation of China (U21A20236), National Key R&D Program of China (2021YFD1700900), and National Natural Science Foundation of China (31801923).

Institutional Review Board Statement: Not applicable.

Informed Consent Statement: Not applicable.

Data Availability Statement: All the data and plant materials in relation to this work can be obtained through contacting with the corresponding author Dr. Ting Zhou (zhoutt@zzu.edu.cn).

Acknowledgments: We are very grateful to the editor and reviewers for critically evaluating the manuscript and providing constructive comments for its improvement.

Conflicts of Interest: The authors declare that they have no competing interest.

Abbreviations

At: *Arabidopsis thaliana*; Bna: *Brassica napus*; Bol: *Brassica oleracea*; Bra: *Brassica rapa*; BRAD: Brassica Database; CDS: Coding sequence; CRE: *cis*-acting regulatory element; DEGs: differentially expressed genes; MEME: Multiple expectation maximization for motif elicitation; MW: Molecular weight; N: Nitrogen; NCBI: National Center for Biotechnology Information; NH_4^+: ammonium; NO_3^-: nitrate; Pi: Inorganic phosphate; TF: transcription factor; NRT: NO_3^- transporter; NLP: Nodule Inception-like Protein.

References

1. Yang, T.; Hao, L.; Yao, S.; Zhao, Y.; Lu, W.; Xiao, K. TabHLH1, a bHLH-type transcription factor gene in wheat, improves plant tolerance to Pi and N deprivation via regulation of nutrient transporter gene transcription and ROS homeostasis. *Plant Physiol. Biochem.* **2016**, *104*, 99–113. [CrossRef] [PubMed]
2. Jin, J.; Tian, F.; Yang, D.-C.; Meng, Y.-Q.; Kong, L.; Luo, J.; Gao, G. PlantTFDB 4.0: Toward a central hub for transcription factors and regulatory interactions in plants. *Nucleic Acids Res.* **2017**, *45*, D1040–D1045. [CrossRef] [PubMed]
3. Ye, H.; Qiao, L.; Guo, H.; Guo, L.; Ren, F.; Bai, J.; Wang, Y. Genome-Wide Identification of Wheat WRKY Gene Family Reveals That TaWRKY75-A Is Referred to Drought and Salt Resistances. *Front. Plant Sci.* **2021**, *12*, 663118. [CrossRef] [PubMed]
4. Riechmann, J.L.; Heard, J.; Martin, G.; Reuber, L.; Jiang, C.; Keddie, J.; Adam, L.; Pineda, O.; Ratcliffe, O.J.; Samaha, R.R.; et al. *Arabidopsis* transcription factors: Genome-wide comparative analysis among eukaryotes. *Science* **2000**, *290*, 2105–2110. [CrossRef] [PubMed]
5. Hosoda, K.; Imamura, A.; Katoh, E.; Hatta, T.; Tachiki, M.; Yamada, H.; Mizuno, T.; Yamazaki, T. Molecular structure of the GARP family of plant Myb-related DNA binding motifs of the *Arabidopsis* response regulators. *Plant Cell* **2002**, *14*, 2015–2029. [CrossRef]
6. Safi, A.; Medici, A.; Szponarski, W.; Ruffel, S.; Lacombe, B.; Krouk, G. The world according to GARP transcription factors. *Curr. Opin. Plant Biol.* **2017**, *39*, 159–167. [CrossRef]
7. Liu, J.; Mehari, T.G.; Xu, Y.; Umer, M.J.; Hou, Y.; Wang, Y.; Peng, R.; Wang, K.; Cai, X.; Zhou, Z.; et al. GhGLK1 a Key Candidate Gene From GARP Family Enhances Cold and Drought Stress Tolerance in Cotton. *Front. Plant Sci.* **2021**, *12*, 759312. [CrossRef]
8. Yue, C.; Chen, Q.Q.; Hu, J.; Li, C.; Luo, L.; Zeng, L. Genome-Wide Identification and Characterization of GARP Transcription Factor Gene Family Members Reveal Their Diverse Functions in Tea Plant (*Camellia sinensis*). *Front. Plant Sci.* **2022**, *13*, 947072. [CrossRef]
9. Ueda, Y.; Ohtsuki, N.; Kadota, K.; Tezuka, A.; Nagano, A.J.; Kadowaki, T.; Kim, Y.; Miyao, M.; Yanagisawa, S. Gene regulatory network and its constituent transcription factors that control nitrogen-deficiency responses in rice. *New Phytol.* **2020**, *227*, 1434–1452. [CrossRef]
10. Maeda, Y.; Konishi, M.; Kiba, T.; Sakuraba, Y.; Sawaki, N.; Kurai, T.; Ueda, Y.; Sakakibara, H.; Yanagisawa, S. A NIGT1-centred transcriptional cascade regulates nitrate signaling and incorporates phosphorus starvation signals in *Arabidopsis*. *Nat. Commun.* **2018**, *9*, 1376. [CrossRef]
11. Medici, A.; Marshall-Colon, A.; Ronzier, E.; Szponarski, W.; Wang, R.; Gojon, A.; Crawford, N.M.; Ruffel, S.; Coruzzi, G.M.; Krouk, G. AtNIGT1/HRS1 integrates nitrate and phosphate signals at the *Arabidopsis* root tip. *Nat. Commun.* **2015**, *6*, 6274. [CrossRef]
12. Huang, T.B.; Harrar, Y.; Lin, C.F.; Reinhart, B.; Newell, N.R.; Talavera-Rauh, F.; Hokin, S.A.; Barton, M.K.; Kerstetter, R.A. Arabidopsis KANADI1 acts as a transcriptional repressor by interacting with a specific cis-element and regulates auxin biosynthesis, transport, and signaling in opposition to HD-ZIPIII factors. *Plant Cell* **2014**, *26*, 246–262. [CrossRef]
13. Gao, P.; Li, X.; Cui, D.J.; Wu, L.; Parkin, I.; Gruber, M.Y. A new dominant *Arabidopsis* transparent testa mutant, sk21-D, and modulation of seed flavonoid biosynthesis by KAN4. *Plant Biotechnol. J.* **2010**, *8*, 979–993. [CrossRef]
14. Dai, S.; Wei, X.; Pei, L.; Thompson, R.L.; Liu, Y.; Heard, J.E.; Ruff, T.G.; Beachy, R.N. Brother of lux arrhythmo is a component of the *Arabidopsis* circadian clock. *Plant Cell* **2011**, *23*, 961–972. [CrossRef]
15. Sun, Y.; Luo, W.; Jain, A.; Liu, L.; Ai, H.; Liu, X.; Feng, B.; Zhang, L.; Zhang, Z.; Guohua, X.; et al. OsPHR3 affects the traits governing nitrogen homeostasis in rice. *BMC Plant Biol.* **2018**, *18*, 241. [CrossRef]
16. Rehman, O.U.; Uzair, M.; Chao, H.; Fiaz, S.; Khan, M.R.; Chen, M. Role of the type-B authentic response regulator gene family in fragrant rice under alkaline salt stress. *Physiol. Plant.* **2022**, *174*, e13696. [CrossRef]
17. Obertello, M.; Shrivastava, S.; Katari, M.; Coruzzi, G.M. Cross-Species Network Analysis Uncovers Conserved Nitrogen-Regulated Network Modules in Rice. *Plant Physiol.* **2015**, *168*, 1830–1843. [CrossRef]

18. Zhao, Z.L.; Shuang, J.R.; Li, Z.G.; Xiao, H.; Liu, Y.; Wang, T.; Wei, Y.; Hu, S.; Wan, S.; Peng, R. Gossypium hirsutumIdentification of the Golden-2-like transcription factors gene family in *Gossypium hirsutum*. *PeerJ* **2021**, *9*, e12484. [CrossRef]
19. Frans, J.M. Physiological functions of mineral macronutrients. *Curr. Opin. Plant Biol.* **2009**, *12*, 250–258.
20. Wykoff, D.D.; Grossman, A.R.; Weeks, D.P.; Usuda, H.; Shimogawara, K. Psr1, a nuclear localized protein that regulates phosphorus metabolism in Chlamydomonas. *Proc. Natl. Acad. Sci. USA* **1999**, *96*, 15336–15341. [CrossRef]
21. Krouk, G.; Mirowski, P.; LeCun, Y.; Shasha, D.E.; Coruzzi, G.M. Predictive network modeling of the high-resolution dynamic plant transcriptome in response to nitrate. *Genome Biol.* **2010**, *11*, R123. [CrossRef] [PubMed]
22. Wang, F.; Deng, M.; Xu, J.; Zhu, X.; Mao, C. Molecular mechanisms of phosphate transport and signaling in higher plants. *Semin. Cell Dev. Biol.* **2018**, *74*, 114–122. [CrossRef] [PubMed]
23. O'Brien, J.A.; Vega, A.; Bouguyon, E.; Krouk, G.; Gojon, A.; Coruzzi, G.; Gutiérrez, R.A. Nitrate Transport, Sensing, and Responses in Plants. *Mol. Plant* **2016**, *9*, 837–856. [CrossRef] [PubMed]
24. Hua, Y.-P.; Zhou, T.; Song, H.-X.; Guan, C.-Y.; Zhang, Z.-H. Integrated genomic and transcriptomic insights into the two-component high-affinity nitrate transporters in allotetraploid rapeseed. *Plant Soil* **2018**, *427*, 245–268. [CrossRef]
25. Verma, P.; Sanyal, S.K.; Pandey, G.K. Ca-CBL-CIPK: A modulator system for efficient nutrient acquisition. *Plant Cell Rep.* **2021**, *40*, 2111–2122. [CrossRef] [PubMed]
26. Jagadhesan, B.; Sathee, L.; Meena, H.S.; Jha, S.K.; Chinnusamy, V.; Kumar, A.; Kumar, S. Genome wide analysis of NLP transcription factors reveals their role in nitrogen stress tolerance of rice. *Sci. Rep.* **2020**, *10*, 9368. [CrossRef]
27. Li, Q.; Zhou, L.; Li, Y.; Zhang, D.; Gao, Y. Plant NIGT1/HRS1/HHO Transcription Factors: Key Regulators with Multiple Roles in Plant Growth, Development, and Stress Responses. *Int. J. Mol. Sci.* **2021**, *22*, 8685. [CrossRef]
28. Ueda, Y.; Kiba, T.; Yanagisawa, S. Nitrate-inducible NIGT1 proteins modulate phosphate uptake and starvation signalling via transcriptional regulation of SPX genes. *Plant J.* **2020**, *102*, 448–466. [CrossRef]
29. Puga, M.I.; Mateos, I.; Charukesi, R.; Wang, Z.; Franco-Zorrilla, J.M.; de Lorenzo, L.; Irigoyen, M.L.; Masiero, S.; Bustos, R.; Rodríguez, J.; et al. SPX1 is a phosphate-dependent inhibitor of Phosphate Starvation Response 1 in *Arabidopsis*. *Proc. Natl. Acad. Sci. USA* **2014**, *111*, 14947–14952. [CrossRef]
30. Raza, A.; Razzaq, A.; Mehmood, S.S.; Hussain, M.A.; Wei, S.; He, H.; Zaman, Q.U.; Xuekun, Z.; Hasanuzzaman, M. Omics: The way forward to enhance abiotic stress tolerance in *Brassica napus* L. *GM Crop. Food* **2021**, *12*, 251–281. [CrossRef]
31. Ayyaz, A.; Miao, Y.; Hannan, F.; Islam, F.; Zhang, K.; Xu, J.; Farooq, M.A.; Zhou, W. Drought tolerance in *Brassica napus* is accompanied with enhanced antioxidative protection, photosynthetic and hormonal regulation at seedling stage. *Physiol. Plant.* **2021**, *172*, 1133–1148. [CrossRef]
32. Wassan, G.M.; Khanzada, H.; Zhou, Q.; Mason, A.S.; Keerio, A.A.; Khanzada, S.; Solangi, A.M.; Faheem, M.; Fu, D.; He, H. Identification of genetic variation for salt tolerance in *Brassica napus* using genome-wide association mapping. *Mol. Genet. Genom.* **2021**, *296*, 391–408. [CrossRef]
33. Lei, Y.; He, H.; Raza, A.; Liu, Z.; Xiaoyu, D.; Guijuan, W.; Yan, L.; Yong, C.; Xiling, Z. *Brassica napus* Exogenous melatonin confers cold tolerance in rapeseed (*Brassica napus* L.) seedlings by improving antioxidants and genes expression. *Plant Signal. Behav.* **2022**, *17*, 2129289. [CrossRef]
34. Shen, X.; Yang, L.; Han, P.; Gu, C.; Li, Y.; Liao, X.; Qin, L. Metabolic Profiles Reveal Changes in the Leaves and Roots of Rapeseed (*Brassica napus* L.) Seedlings under Nitrogen Deficiency. *Int. J. Mol. Sci.* **2022**, *23*, 5784. [CrossRef]
35. Li, Q.; Ding, G.D.; Ningmei Yang, N.; White, P.J.; Ye, X.; Cai, H.; Lu, J.; Shi, L.; Xu, F. Comparative genome and transcriptome analysis unravels key factors of nitrogen use effi-ciency in *Brassica napus* L. *Plant Cell Environ.* **2020**, *43*, 712–731. [CrossRef]
36. Li, Y.; Yang, X.; Liu, H.; Wang, W.; Wang, C.; Ding, G.; Xu, F.; Wang, S.; Cai, H.; Hammond, J.P.; et al. Local and systemic responses conferring acclimation of *Brassica napus* roots to low phosphorus conditions. *J. Exp. Bot.* **2022**, *73*, 4753–4777. [CrossRef]
37. Su, W.; Raza, A.; Gao, A.; Jia, Z.; Zhang, Y.; Hussain, M.A.; Mehmood, S.S.; Cheng, Y.; Lv, Y.; Zou, X. Genome-Wide Analysis and Expression Profile of Superoxide Dismutase (SOD) Gene Family in Rapeseed (*Brassica napus* L.) under Different Hormones and Abiotic Stress Conditions. *Antioxidants* **2021**, *10*, 1182. [CrossRef]
38. Su, W.; Raza, A.; Zeng, L.; Gao, A.; Lv, Y.; Ding, X.; Cheng, Y.; Zou, X. Genome-wide analysis and expression patterns of lipid phospholipid phospholipase gene family in *Brassica napus* L. *BMC Genom.* **2021**, *22*, 548. [CrossRef]
39. Singh, S.; Chhapekar, S.S.; Ma, Y.; Rameneni, J.J.; Oh, S.H.; Kim, J.; Lim, Y.P.; Choi, S.R. Genome-Wide Identification, Evolution, and Comparative Analysis of B-Box Genes in *Brassica rapa*, *B. oleracea*, and *B. napus* and Their Expression Profiling in B. rapa in Response to Multiple Hormones and Abiotic Stresses. *Int. J. Mol. Sci.* **2021**, *22*, 10367. [CrossRef]
40. Zhang, Z.; Li, J.; Zhao, X.-Q.; Wang, J.; Wong, G.K.-S.; Yu, J. KaKs_Calculator: Calculating Ka and Ks through Model Selection and Model Averaging. *Genom. Proteom. Bioinform.* **2006**, *4*, 259–263. [CrossRef]
41. Begum, Y. Regulatory role of microRNAs (miRNAs) in the recent development of abiotic stress tolerance of plants. *Gene* **2022**, *821*, 146283. [CrossRef] [PubMed]
42. McAllister, C.H.; Beatty, P.; Good, A.G. Engineering nitrogen use efficient crop plants: The current status. *Plant Biotechnol. J.* **2012**, *10*, 1011–1025. [CrossRef] [PubMed]
43. Ren, F.; Zhao, C.-Z.; Liu, C.-S.; Huang, K.-L.; Guo, Q.-Q.; Chang, L.-L.; Xiong, H.; Li, X.-B. A *Brassica napus* PHT1 phosphate transporter, BnPht1;4, promotes phosphate uptake and affects roots architecture of transgenic *Arabidopsis*. *Plant Mol. Biol.* **2014**, *86*, 595–607. [CrossRef] [PubMed]

44. Dinh, A.Q.; Naeem, A.; Sagervanshi, A.; Wimmer, M.A.; Mühling, K.H. Boron uptake and distribution by oilseed rape (*Brassica napus* L.) as affected by different nitrogen forms under low and high boron supply. *Plant Physiol. Biochem.* **2021**, *161*, 156–165. [CrossRef] [PubMed]
45. Zhou, T.; Yue, C.P.; Zhang, T.Y.; Liu, Y.; Huang, J.; Hua, Y. Integrated ionomic and transcriptomic dissection reveals the core transporter genes responsive to varying cadmium abundances in allotetraploid rapeseed. *BMC Plant Biol.* **2021**, *21*, 372. [CrossRef]
46. Raza, A.; Tabassum, J.; Fakhar, A.Z.; Sharif, R.; Chen, H.; Zhang, C.; Ju, L.; Fotopoulos, V.; Siddique, K.H.M.; Singh, R.K.; et al. Smart reprograming of plants against salinity stress using modern biotechnological tools. *Crit. Rev. Biotechnol.* **2022**, 1–28. [CrossRef]
47. Wu, C.; Feng, J.; Wang, R.; Liu, H.; Yang, H.; Rodriguez, P.L.; Qin, H.; Liu, X.; Wang, D. HRS1 Acts as a Negative Regulator of Abscisic Acid Signaling to Promote Timely Germination of *Arabidopsis* Seeds. *PLoS ONE* **2012**, *7*, e35764. [CrossRef]
48. Krouk, G.; Tranchina, D.; Lejay, L.; Cruikshank, A.A.; Shasha, D.; Coruzzi, G.M.; Gutiérrez, R.A. A systems approach uncovers restrictions for signal interactions regulating genome-wide responses to nutritional cues in *Arabidopsis*. *PLoS Comput. Biol.* **2009**, *5*, e1000326. [CrossRef]
49. Canales, J.; Moyano, T.C.; Villarroel, E.; Gutiérrez, R.A. Systems analysis of transcriptome data provides new hypotheses about Arabidopsis root response to nitrate treatments. *Front. Plant Sci.* **2014**, *5*, 22. [CrossRef]
50. Zhao, X.; Yang, J.; Li, X.; Li, G.; Sun, Z.; Chen, Y.; Chen, Y.; Xia, M.; Li, Y.; Yao, L.; et al. Identification and expression analysis of GARP superfamily genes in response to nitrogen and phosphorus stress in *Spirodela polyrhiza*. *BMC Plant Biol.* **2022**, *22*, 308. [CrossRef]
51. Perumal, S.; Waminal, N.; Lee, J.; Lee, J.; Choi, B.-S.; Kim, H.H.; Grandbastien, M.-A.; Yang, T.-J. Elucidating the major hidden genomic components of the A, C, and AC genomes and their influence on *Brassica* evolution. *Sci. Rep.* **2017**, *7*, 17986. [CrossRef]
52. Kesawat, M.S.; Kherawat, B.S.; Singh, A.; Dey, P.; Routray, S.; Mohapatra, C.; Saha, D.; Ram, C.; Siddique, K.H.M.; Kumar, A.; et al. Genome-Wide Analysis and Characterization of the Proline-Rich Extensin-like Receptor Kinases (PERKs) Gene Family Reveals Their Role in Different Developmental Stages and Stress Conditions in Wheat (*Triticum aestivum* L.). *Plants* **2022**, *11*, 496. [CrossRef]
53. Kumar, M.; Kherawat, B.S.; Dey, P.; Saha, D.; Singh, A.; Bhatia, S.K.; Ghodake, G.S.; Kadam, A.A.; Kim, H.-U.; Manorama; et al. Genome-Wide Identification and Characterization of PIN-FORMED (PIN) Gene Family Reveals Role in Developmental and Various Stress Conditions in *Triticum aestivum* L. *Int. J. Mol. Sci.* **2021**, *22*, 7396. [CrossRef]
54. Wang, W.; Wang, X.T.; Wang, Y.T.; Zhou, G.; Wang, C.; Hussain, S.; Adnan; Lin, R.; Wang, T.; Wang, S. SlEAD1, an EAR motif-containing ABA down-regulated novel transcription re-pressor regulates ABA response in tomato. *GM Crop. Food* **2020**, *11*, 275–289. [CrossRef]
55. Jiao, Y.; Wickett, N.J.; Saravanaraj, A.; Chanderbali, A.S.; Landherr, L.; Ralph, P.E.; Tomsho, L.P.; Hu, Y.; Liang, H.; Sotis, P.S.; et al. Data from: Ancestral polyploidy in seed plants and angiosperms. *Nature* **2011**, *473*, 97–100. [CrossRef]
56. Cannon, S.B.; Mitra, A.; Baumgarten, A.; Young, N.D.; May, G. The roles of segmental and tandem gene duplication in the evolution of large gene families in *Arabidopsis thaliana*. *BMC Plant Biol.* **2004**, *4*, 10. [CrossRef]
57. Soylev, A.; Le, T.M.; Amini, H.; Alkan, C.; Hormozdiari, F. Discovery of tandem and interspersed segmental duplications using high-throughput sequencing. *Bioinformatics* **2019**, *35*, 3923–3930. [CrossRef]
58. Das Laha, S.; Dutta, S.; Schäffner, A.R.; Das, M. Gene duplication and stress genomics in *Brassicas*: Current understanding and future prospects. *J. Plant Physiol.* **2020**, *255*, 153293. [CrossRef]
59. Chalhoub, B.; Denoeud, F.; Liu, S.; Parkin, I.A.P.; Tang, H.; Wang, X.; Chiquet, J.; Belcram, H.; Tong, C.; Samans, B.; et al. Early allopolyploid evolution in the post-Neolithic *Brassica napus* oilseed genome. *Science* **2014**, *345*, 950–953. [CrossRef]
60. Xu, J.; Zhao, X.; Bao, J.; Shan, Y.; Zhang, M.; Shen, Y.; Abubakar, Y.S.; Lu, G.; Wang, Z.; Wang, A. Genome-Wide Identification and Expression Analysis of SNARE Genes in *Brassica napus*. *Plants* **2022**, *11*, 711. [CrossRef]
61. Hernandez-Garcia, C.M.; Finer, J.J. Identification and validation of promoters and cis-acting regulatory elements. *Plant Sci.* **2014**, *217*, 109–119. [CrossRef] [PubMed]
62. Yan, Y.; Shen, L.; Chen, Y.; Bao, S.; Thong, Z.; Yu, H. A MYB-Domain Protein EFM Mediates Flowering Responses to Environmental Cues in *Arabidopsis*. *Dev. Cell* **2014**, *30*, 437–448. [CrossRef] [PubMed]
63. Li, M.; Yu, B. Recent advances in the regulation of plant miRNA biogenesis. *RNA Biol.* **2021**, *18*, 2087–2096. [CrossRef] [PubMed]
64. Song, X.; Li, Y.; Cao, X.; Qi, Y. MicroRNAs and Their Regulatory Roles in Plant–Environment Interactions. *Annu. Rev. Plant Biol.* **2019**, *70*, 489–525. [CrossRef] [PubMed]
65. Raza, A.; Su, W.; Gao, A.; Mehmood, S.; Hussain, M.; Nie, W.; Lv, Y.; Zou, X.; Zhang, X. Catalase (CAT) Gene Family in Rapeseed (*Brassica napus* L.): Genome-Wide Analysis, Identification, and Expression Pattern in Response to Multiple Hormones and Abiotic Stress Conditions. *Int. J. Mol. Sci.* **2021**, *22*, 4281. [CrossRef]
66. Li, J.; Guo, G.; Guo, W.; Guo, G.; Tong, D.; Ni, Z.; Sun, Q.; Yao, Y. miRNA164-directed cleavage of ZmNAC1 confers lateral root development in maize (*Zea mays* L.). *BMC Plant Biol.* **2012**, *12*, 220. [CrossRef]
67. González-Villagra, J.; Kurepin, L.V.; Reyes-Díaz, M.M. Evaluating the involvement and interaction of abscisic acid and miRNA156 in the induction of anthocyanin biosynthesis in drought-stressed plants. *Planta* **2017**, *246*, 299–312. [CrossRef]
68. Han, Y.; Zhang, X.; Wang, Y.; Ming, F. The Suppression of WRKY44 by GIGANTEA-miR172 Pathway Is Involved in Drought Response of *Arabidopsis thaliana*. *PLoS ONE* **2013**, *8*, e73541. [CrossRef]
69. Yuan, S.; Zhao, J.; Li, Z.; Hu, Q.; Yuan, N.; Zhou, M.; Xia, X.; Noorai, R.; Saski, C.; Li, S.; et al. MicroRNA396-mediated alteration in plant development and salinity stress response in creeping bentgrass. *Hortic. Res.* **2019**, *6*, 1–13. [CrossRef]

70. Hua, Y.-P.; Zhou, T.; Huang, J.-Y.; Yue, C.-P.; Song, H.-X.; Guan, C.-Y.; Zhang, Z.-H. Genome-Wide Differential DNA Methylation and miRNA Expression Profiling Reveals Epigenetic Regulatory Mechanisms Underlying Nitrogen-Limitation-Triggered Adaptation and Use Efficiency Enhancement in Allotetraploid Rapeseed. *Int. J. Mol. Sci.* **2020**, *21*, 8453. [CrossRef]
71. Liu, Y.; Hua, Y.-P.; Chen, H.; Zhou, T.; Yue, C.-P.; Huang, J.-Y. Genome-scale identification of plant defensin (PDF) family genes and molecular characterization of their responses to diverse nutrient stresses in allotetraploid rapeseed. *PeerJ* **2021**, *9*, e12007. [CrossRef]
72. Franco-Zorrilla, J.M.; López-Vidriero, I.; Carrasco, J.L.; Godoy, M.; Vera, P.; Solano, R. DNA-binding specificities of plant transcription factors and their potential to define target genes. *Proc. Natl. Acad. Sci. USA* **2014**, *111*, 2367–2372. [CrossRef]
73. Xiao, G.; He, P.; Zhao, P.; Liu, H.; Zhang, L.; Pang, C.; Yu, J. Genome-wide identification of the GhARF gene family reveals that GhARF2 and GhARF18 are involved in cotton fibre cell initiation. *J. Exp. Bot.* **2018**, *69*, 4323–4337. [CrossRef]
74. Li, S.; Liu, Z.; Guo, L.; Li, H.; Nie, X.; Chai, S.; Zheng, W. Genome-Wide Identification of Wheat ZIP Gene Family and Functional Characterization of the TaZIP13-B in Plants. *Front. Plant Sci.* **2021**, *12*, 748146. [CrossRef]
75. Shimogawara, K.; Wykoff, D.D.; Usuda, H.; Grossman, A.R. *Chlamydomonas reinhardtii* Mutants Abnormal in Their Responses to Phosphorus Deprivation. *Plant Physiol.* **1999**, *120*, 685–694. [CrossRef]
76. Guan, Z.; Zhang, Q.; Zhang, Z.; Zuo, J.; Chen, J.; Liu, R.; Savarin, J.; Broger, L.; Cheng, P.; Wang, Q.; et al. Mechanistic insights into the regulation of plant phosphate homeostasis by the rice SPX2–PHR2 complex. *Nat. Commun.* **2022**, *13*, 1581. [CrossRef]
77. Kiba, T.; Inaba, J.; Kudo, T.; Ueda, N.; Konishi, M.; Mitsuda, N.; Takiguchi, Y.; Kondou, Y.; Yoshizumi, T.; Ohme-Takagi, M.; et al. Repression of Nitrogen Starvation Responses by Members of the *Arabidopsis* GARP-Type Transcription Factor NIGT1/HRS1 Subfamily. *Plant Cell* **2018**, *30*, 925–945. [CrossRef]
78. Safi, A.; Medici, A.; Szponarski, W.; Martin, F.; Clément-Vidal, A.; Marshall-Colon, A.; Ruffel, S.; Gaymard, F.; Rouached, H.; Leclercq, J.; et al. GARP transcription factors repress *Arabidopsis* nitrogen starvation response via ROS-dependent and -independent pathways. *J. Exp. Bot.* **2021**, *72*, 3881–3901. [CrossRef]
79. Osorio, M.B.; Ng, S.; Berkowitz, O.; De Clercq, I.; Mao, C.; Shou, H.; Whelan, J.; Jost, R. SPX4 Acts on PHR1-Dependent and -Independent Regulation of Shoot Phosphorus Status in *Arabidopsis*. *Plant Physiol.* **2019**, *181*, 332–352. [CrossRef]
80. Zhou, T.; Hua, Y.; Yue, C.; Huang, J.; Zhang, Z. Physiologic, metabolomic, and genomic investigations reveal distinct glutamine and mannose metabolism responses to ammonium toxicity in allotetraploid rapeseed genotypes. *Plant Sci.* **2021**, *310*, 110963. [CrossRef]
81. Liu, H.; Yang, H.; Wu, C.; Feng, J.; Liu, X.; Qin, H.; Wang, D. OverexpressingHRS1Confers Hypersensitivity to Low Phosphate-Elicited Inhibition of Primary Root Growth in *Arabidopsis thaliana*. *J. Integr. Plant Biol.* **2009**, *51*, 382–392. [CrossRef]
82. Bustos, R.; Castrillo, G.; Linhares, F.; Puga, M.I.; Rubio, V.; Pérez-Pérez, J.; Solano, R.; Leyva, A.; Paz-Ares, J. A central regulatory system largely controls transcriptional activation and re-pression responses to phosphate starvation in *Arabidopsis*. *PLoS Genet.* **2010**, *6*, e1001102. [CrossRef] [PubMed]
83. Mito, T.; Seki, M.; Shinozaki, K.; Ohme-Takagi, M.; Matsui, K. Generation of chimeric repressors that confer salt tolerance in *Arabidopsis* and rice. *Plant Biotechnol. J.* **2011**, *9*, 736–746. [CrossRef] [PubMed]
84. Liu, F.; Xu, Y.; Han, G.; Zhou, L.; Ali, A.; Zhu, S.; Li, X. Molecular Evolution and Genetic Variation of G2-Like Transcription Factor Genes in Maize. *PLoS ONE* **2016**, *11*, e0161763. [CrossRef] [PubMed]
85. Sawaki, N.; Tsujimoto, R.; Shigyo, M.; Konishi, M.; Toki, S.; Fujiwara, T.; Yanagisawa, S. A Nitrate-Inducible GARP Family Gene Encodes an Auto-Repressible Transcriptional Repressor in Rice. *Plant Cell Physiol.* **2013**, *54*, 506–517. [CrossRef]
86. Rhee, S.Y.; Beavis, W.; Berardini, T.Z.; Chen, G.; Dixon, D.; Doyle, A.; Garcia-Hernandez, M.; Huala, E.; Lander, G.; Montoya, M. The *Arabidopsis* Information Resource (TAIR): A model organism database providing a centralized, curated gateway to *Arabidopsis* biology, research materials and community. *Nucleic Acids Res.* **2003**, *31*, 224–228. [CrossRef]
87. Chen, H.; Wang, T.; He, X.; Cai, X.; Lin, R.; Liang, J.; Wu, J.; King, G.; Wang, X. BRAD V3.0: An upgraded Brassicaceae database. *Nucleic Acids Res.* **2022**, *50*, D1432–D1441. [CrossRef]
88. Altschul, S.F.; Madden, T.L.; Schäffer, A.A.; Zhang, J.; Zhang, Z.; Miller, W.; Lipman, D.J. Gapped BLAST and PSI-BLAST: A new generation of protein database search programs. *Nucleic Acids Res.* **1997**, *25*, 3389–3402. [CrossRef]
89. El-Gebali, S.; Mistry, J.; Bateman, A.; Eddy, S.R.; Luciani, A.; Potter, S.C.; Qureshi, M.; Richardson, L.J.; Salazar, G.A.; Smart, A.; et al. The Pfam protein families database in 2019. *Nucleic Acids Res.* **2019**, *47*, D427–D432. [CrossRef]
90. Lu, S.; Wang, J.; Chitsaz, F.; Derbyshire, M.K.; Geer, R.C.; Gonzales, N.R.; Gwadz, M.; I Hurwitz, D.; Marchler, G.H.; Song, J.S.; et al. CDD/SPARCLE: The conserved domain database in 2020. *Nucleic Acids Res.* **2020**, *48*, D265–D268. [CrossRef]
91. Artimo, P.; Jonnalagedda, M.; Arnold, K.; Baratin, D.; Csardi, G.; de Castro, E.; Duvaud, S.; Flegel, V.; Fortier, A.; Gasteiger, E.; et al. ExPASy: SIB bioinformatics resource portal. *Nucleic Acids Res.* **2012**, *40*, W597–W603. [CrossRef]
92. Larkin, M.A.; Blackshields, G.; Brown, N.P.; Chenna, R.; McGettigan, P.A.; McWilliam, H.; Valentin, F.; Wallace, I.M.; Wilm, A.; Lopez, R.; et al. Clustal W and Clustal X version 2.0. *Bioinformatics* **2007**, *23*, 2947–2948. [CrossRef]
93. Kumar, S.; Stecher, G.; Li, M.; Knyaz, C.; Tamura, K. MEGA X: Molecular Evolutionary Genetics Analysis across Computing Platforms. *Mol. Biol. Evol.* **2018**, *35*, 1547–1549. [CrossRef]
94. Letunic, I.; Bork, P. Interactive Tree Of Life (iTOL) v5: An online tool for phylogenetic tree display and annotation. *Nucleic Acids Res.* **2021**, *49*, W293–W296. [CrossRef]
95. Wang, D.; Zhang, Y.; Zhang, Z.; Zhu, J.; Yu, J. KaKs_Calculator 2.0: A Toolkit Incorporating Gamma-Series Methods and Sliding Window Strategies. *Genom. Proteom. Bioinform.* **2010**, *8*, 77–80. [CrossRef]

96. Koch, M.A.; Haubold, B.; Mitchell-Olds, T. Comparative evolutionary analysis of chalcone synthase and alcohol dehydro-genase loci in *Arabidopsis*, Arabis, and related genera (*Brassicaceae*). *Mol. Biol. Evol.* **2000**, *17*, 1483–1498. [CrossRef]
97. Bailey, T.L.; Boden, M.; Buske, F.A.; Frith, M.; Grant, C.E.; Clementi, L.; Ren, J.; Li, W.W.; Noble, W.S. MEME SUITE: Tools for motif discovery and searching. *Nucleic Acids Res.* **2009**, *37*, w202–w208. [CrossRef]
98. Hu, B.; Jin, J.; Guo, A.-Y.; Zhang, H.; Luo, J.; Gao, G. GSDS 2.0: An upgraded gene feature visualization server. *Bioinformatics* **2015**, *31*, 1296–1297. [CrossRef]
99. Jiangtao, C.; Yingzhen, K.; Qian, W.; Yuhe, S.; Daping, G.; Jing, L.; Guanshan, L. MapGene2Chrom, a tool to draw gene physical map based on Perl and SVG languages. *Hereditas* **2015**, *37*, 91–97. [CrossRef]
100. Chen, C.J.; Chen, H.; Zhang, Y.; Thomas, H.R.; Frank, M.H.; He, Y.H.; Xia, R. TBtools: An Integrative Toolkit Developed for Interactive Analyses of Big Biological Data. *Mol. Plant* **2020**, *13*, 1194–1202. [CrossRef]
101. Buchfink, B.; Xie, C.; Huson, D.H. Fast and sensitive protein alignment using DIAMOND. *Nat. Methods* **2015**, *12*, 59–60. [CrossRef] [PubMed]
102. Sánchez, J.-P.; Duque, P.; Chua, N.-H. ABA activates ADPR cyclase and cADPR induces a subset of ABA-responsive genes in *Arabidopsis*. *Plant J.* **2004**, *38*, 381–395. [CrossRef] [PubMed]
103. Dai, X.; Zhuang, Z.; Zhao, P.X. psRNATarget: A plant small RNA target analysis server (2017 release). *Nucleic Acids Res.* **2018**, *46*, W49–W54. [CrossRef] [PubMed]
104. Shannon, P.; Markiel, A.; Ozier, O.; Baliga, N.S.; Wang, J.T.; Ramage, D.; Amin, N.; Schwikowski, B.; Ideker, T. Cytoscape: A software environment for integrated models of Biomolecular Interaction Networks. *Genome Res.* **2003**, *13*, 2498–2504. [CrossRef]
105. Zhou, T.; Yue, C.P.; Huang, J.Y.; Cui, J.; Liu, Y.; Wang, W.; Tian, C.; Hua, Y. Genome-wide identification of the amino acid permease genes and molecular characterization of their transcriptional responses to various nutrient stresses in allotetraploid rapeseed. *BMC Plant Biol.* **2020**, *20*, 151. [CrossRef]
106. Han, Y.-L.; Song, H.-X.; Liao, Q.; Yu, Y.; Jian, S.-F.; Lepo, J.E.; Liu, Q.; Rong, X.-M.; Tian, C.; Zeng, J.; et al. Nitrogen Use Efficiency Is Mediated by Vacuolar Nitrate Sequestration Capacity in Roots of *Brassica napus*. *Plant Physiol.* **2016**, *170*, 1684–1698. [CrossRef]
107. Li, B.; Dewey, C.N. RSEM: Accurate transcript quantification from RNA-Seq data with or without a reference genome. *BMC Bioinform.* **2011**, *12*, 323. [CrossRef]
108. Love, M.I.; Huber, W.; Anders, S. Moderated estimation of fold change and dispersion for RNA-seq data with DESeq2. *Genome Biol.* **2014**, *15*, 550. [CrossRef]
109. Voelker, C.; Schmidt, D.; Mueller-Roeber, B.; Parkin, I.A.P.; Tang, H.; Wang, X.; Chiquet, J.; Belcram, H.; Tong, C.; Samans, B. Members of the *Arabidopsis* AtTPK/KCO family form homomeric vacuolar channels in planta. *Plant J.* **2006**, *48*, 296–306. [CrossRef]
110. Li, L.; He, Y.; Zhang, Z.; Shi, Y.; Zhang, X.; Xu, X.; Wu, J.-L.; Tang, S. OsNAC109 regulates senescence, growth and development by altering the expression of senescence- and phytohormone-associated genes in rice. *Plant Mol. Biol.* **2021**, *105*, 637–654. [CrossRef]

International Journal of
Molecular Sciences

MDPI

Article

Regulation of Heat Stress in *Physcomitrium* (*Physcomitrella*) *patens* Provides Novel Insight into the Functions of Plant RNase H1s

Zhuo Yang [1,†], Liu Duan [1,†], Hongyu Li [1], Ting Tang [2], Liuzhu Chen [1], Keming Hu [3,4], Hong Yang [1,*] and Li Liu [1,*]

[1] State Key Laboratory of Biocatalysis and Enzyme Engineering, Hubei Collaborative Innovation Center for Green Transformation of Bio-Resources, Hubei Key Laboratory of Industrial Biotechnology, School of Life Sciences, Hubei University, Wuhan 430062, China
[2] School of Life and Health Sciences, Hunan University of Science and Technology, Xiangtan 411201, China
[3] Jiangsu Key Laboratory of Crop Genomics and Molecular Breeding/Key Laboratory of Plant Functional Genomics of the Ministry of Education, College of Agriculture, Yangzhou University, Yangzhou 225009, China
[4] Co-Innovation Center for Modern Production Technology of Grain Crops of Jiangsu Province/Key Laboratory of Crop Genetics and Physiology of Jiangsu Province, Yangzhou University, Yangzhou 225009, China
[*] Correspondence: yang@hubu.edu.cn (H.Y.); liuli2020@hubu.edu.cn (L.L.)
[†] These authors contributed equally to this work.

Abstract: RNase H1s are associated with growth and development in both plants and animals, while the roles of RNase H1s in bryophytes have been rarely reported. Our previous data found that *PpRNH1A*, a member of the *RNase H1* family, could regulate the development of *Physcomitrium* (*Physcomitrella*) *patens* by regulating the auxin. In this study, we further investigated the biological functions of *PpRNH1A* and found *PpRNH1A* may participate in response to heat stress by affecting the numbers and the mobilization of lipid droplets and regulating the expression of heat-related genes. The expression level of *PpRNH1A* was induced by heat stress (HS), and we found that the *PpRNH1A* overexpression plants (*A*-OE) were more sensitive to HS. At the same time, *A*-OE plants have a higher number of lipid droplets but with less mobility in cells. Consistent with the HS sensitivity phenotype in *A*-OE plants, transcriptomic analysis results indicated that *PpRNH1A* is involved in the regulation of expression of heat-related genes such as *DNAJ* and *DNAJC*. Taken together, these results provide novel insight into the functions of RNase H1s.

Keywords: PpRNH1A; heat stress (HS); lipid droplets; *Physcomitrium* (*Physcomitrella*) *patens*

Citation: Yang, Z.; Duan, L.; Li, H.; Tang, T.; Chen, L.; Hu, K.; Yang, H.; Liu, L. Regulation of Heat Stress in *Physcomitrium* (*Physcomitrella*) *patens* Provides Novel Insight into the Functions of Plant RNase H1s. *Int. J. Mol. Sci.* **2022**, *23*, 9270. https://doi.org/10.3390/ijms23169270

Academic Editors: Melvin J. Oliver, Bei Gao and Moxian Chen

Received: 11 June 2022
Accepted: 13 August 2022
Published: 17 August 2022

Publisher's Note: MDPI stays neutral with regard to jurisdictional claims in published maps and institutional affiliations.

1. Introduction

With global warming, heat stress (HS) has gradually become a major limiting factor affecting plant growth, plant geographic distribution, crop yield, and quality [1–3]. Research analysis showed that the negative impact of high temperature on crop yield is becoming more severe [4]. Plants do not move and have evolved a series of self-protective mechanisms against external stresses to adapt to the environment [5]. Heat shock proteins (HSPs) and molecular chaperones are commonly found to be involved in plants' responses to HS [6,7]. DnaJ (HSP40) proteins are important molecular chaperones involved in signal transduction, cellular proteostasis, and tolerance to various stresses in plants [8,9]. The expression levels of these genes are up-regulated to improve heat tolerance in plants [4,6]. For example, *GmDNJ1* (a major *HSP40*) is highly induced at high temperature, and *Gmdnj1*-knockout mutants have more severe browning and lower chlorophyll content, and higher reactive oxygen species (ROS) content under HS, which suggest that *GmDNJ1* plays an important role in response to heat stress in soybean [10].

Cytosolic lipid droplets play important roles in plant growth and responses to stress. Lipid droplets (or so-called lipid bodies in some cases) are subcellular structures that are capable of storing neutral lipids and various hydrophobic compounds in eukaryotes, which are present in almost all green plant lineages and is widely explored in various kind of tissues such as seeds, pollen, and leaves in different plant species [11–14]. Feeney et al. found that a large ectopic accumulation of lipid droplets in vegetative organs produced abnormal embryogenic structures in plants [15]. Triacylglycerol (TAG) is one of the main components of lipid droplets. Abnormally accumulating TAG in plant leaves usually affects plant growth and causes cell death [15], which was also observed when subjected to various stresses [16]. Abiotic stresses such as heat stress, high light, drought, and cold induce lipid droplets accumulation in plants [17], including not only land plants but also chlorophyte algae such as *Haematococcus pluvialis*, *Chlamydomonas reinhardtii*, and cyanobacterium *Synechocystis* sp. PCC6803, among others [18–20]. Furthermore, the accumulation and mobilization of lipid droplets were reported to be positively related to plant stress resistance [17,21,22]. In the salt-sensitive variety, salt stress resulted in lipid droplets accumulation and higher lipid droplets retention, whereas the tolerant variety exhibited faster lipid droplets mobilization [22]. In *P. patens*, lipid droplets were found to be present in dehydrated spores and photosynthetic gametophytes [14]. Previous studies have shown that numerous lipid droplets could be observed after dehydration, which indicates that drought stress also induces the accumulation of lipid droplets in *P. patens* [23]. Lipidome analysis revealed that lipid metabolism plays an important role in *P. patens* to cope with the terrestrial environment stresses, although the composition of monogalactosyldiacylglycerol is different between *P. patens* and vascular plants [24].

RNase H is considered as a class of sequence-nonspecific ribonucleases and was first isolated and identified from calf mammary glands [25]. RNase Hs were found widespread in archaebacteria, prokaryotes, and eukaryotes [26] and are classified into three types (RNase H1, H2, and H3) [27]. Among them, RNase H3 only exists in certain archaebacteria [28,29], while RNase H1 and RNase H2, are present in plants and animals [30,31]. Although the structure of RNase H1s are evolutionarily highly conserved [32], RNase H1s were found to play important roles in diverse biological processes such as replication process, RNA processing process, and development of mitochondrial DNA [33,34]. In addition, they play crucial roles in the maintenance of genomic stability, the repair of DNA damage, and affect the development of organisms as well [35–37]. Three AtRNH1s were found in *Arabidopsis thaliana*, which are localized in the nucleus (AtRNH1A), mitochondria (AtRNH1B), and chloroplasts (AtRNH1C), respectively [38]. The diversification results of the subcellular localization of RNAH1s in *Arabidopsis* facilitate us to hypothesize that RNH1s may have multiple functions in different cellular compartments. In addition, AtRNH1B and AtRNH1C are required for viability in *Arabidopsis*, while AtRNH1A is not [39]. Moreover, studies showed that loss-of-function mutation of AtRNH1C exhibits a distinct growth phenotype of dwarfism and leaf chlorosis [39]. Gene functional mechanism studies showed that AtRNH1C and AtGyrases interact to reduce the damage of DNA [38], and AtRNH1C promotes the repair of DNA damage in chloroplasts mainly by its synergistic action with ssDNA-binding proteins (WHY1/3) and recombinases (RecA1) as well [40]. Interestingly, deletion of AtRNH1B allows AtRNH1C, which is located in chloroplasts, to enter the mitochondria, ensuring the integrity of the mitochondrial genome [41].

Despite functional research of RNH1s in *Arabidopsis* in recent years, it remains unclear how RNH1s function in response to abiotic stresses and how RNH1s evolved within land plant lineage. Mosses, which belong to the early land plants, hold an important position in higher plant evolution. *Physcomitrium patens* (*P. patens*), has been a model plant to study evolutionary developmental and stress tolerance/adaption questions as a non-seed plant [42–44]. Therefore, it is of some far-reaching significance to study the biological functions of the *RNase H1* family in *P. patens*. In our previous research, we found that there are two family members of *RNase H1s* in *P. patens* [45]. PpRNH1A affects shoot growth and branch formation of *P. patens* by controlling the formation of the R-loop and regulating the transcription level of auxin-related genes in the mutant *pprna1a* [45]. However, different from the subcellular location results of AtRNH1A, PpRNH1A was found localized in both nucleus and cytosol, which suggested a species-specific role of RNH1A in *P. patens*. In this research, we further explore the function of PpRNH1A and found that overexpression of *PpRNH1A* (*A*-OE) plants were more sensitive to heat stress by regulation of heat-related gene expressions, such as *DNAJ* and *DNAJC*. Chloroplast development but not pigment synthesis was affected by overexpression of *PpRNH1A*. Ultrastructural and surface structural observations show growth defects in *A*-OE plants, and abnormal accumulation and mobilization of lipid droplets were found in the cytosol, which is associated with the sensitivity of *A*-OE plants to heat stress.

2. Results

2.1. Phylogenetic Analysis of RNases H1s

In our previous study, we identified two family members of *RNase H1s* in *P. patent*, named PpRNH1A (Pp3c4_14290) and PpRNH1B (Pp3c2_25170), which contain the RNase H catalytic domain [45]. To investigate the evolutionary relationship of RNase H1s in plants, we used the full-length protein sequence of PpRNH1A from the Phytozome database to search for the RNase H1s homologs in representative species such as algae, mosses, and higher terrestrial plant groups. Public database Phytozome, Ensembl plants, NCBI, and the *klebsormidium* genome project were used for BLAST analysis. The phylogenetic tree (Figure 1) showed that three main branches were identified according to their full-length protein sequences, and the development process of RNase H1s was consistent with the evolutionary history of organisms. PpRNH1A and PpRNH1B were grouped together and showed a close evolutionary relationship to *Chlamydomonas reinhardii* and *Klebsormidium nitens*. One to three family members of RNase H1s were identified in different plant species, and some of the proteins in the same species were assigned to different sub-branches, which suggested that they may have differentiated functions.

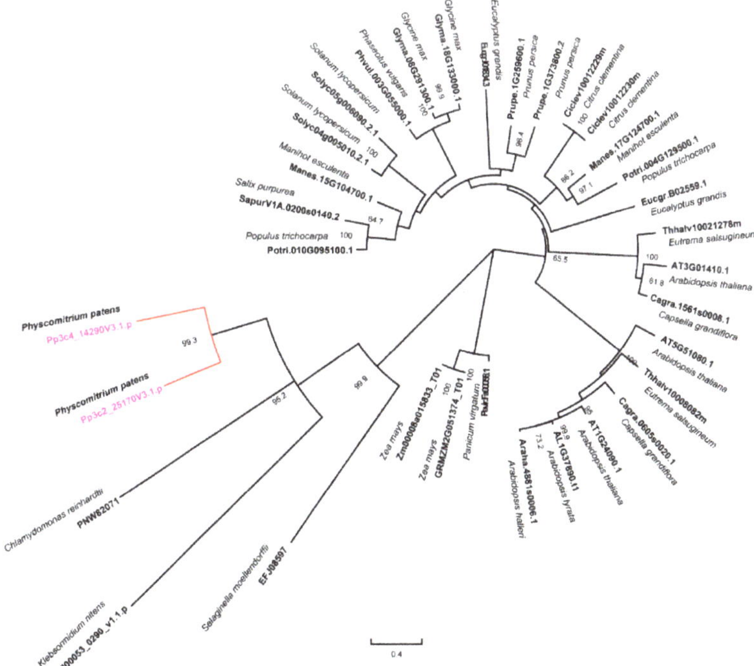

Figure 1. Phylogenetic analyses of RNase H1 in different species. The tree was constructed with MEGA-X using the maximum likelihood (ML) method with 1000 bootstrap replicates. The numbers above the branches represent the bootstrap support values (>50%) from 1000 replications.

2.2. PpRNH1A Is Involved in Growth Development and Stress Tolerance

To explore the features of the *PpRNH1A* (Pp3c4_14290), we first investigated the expression pattern of the *PpRNH1A* in wild type throughout the growth cycle of *P. patens*. Expression data were downloaded from the public database *Physcomitrella* eFP Browser (BAR, http://bar.utoronto.ca/, data shown in Figure 2 were accessed on 19 February 2022). The higher expression levels of *PpRNH1A* were found in spores, rhizoids, archegonia, and at the sporophyte S1 stage, with relatively lower expression levels in chloronema, gametophores, and at the S2, S3, and M stages of sporophyte development (Figure 2A,B). This result suggests that PpRNH1A might be involved in *P. patens* growth and development.

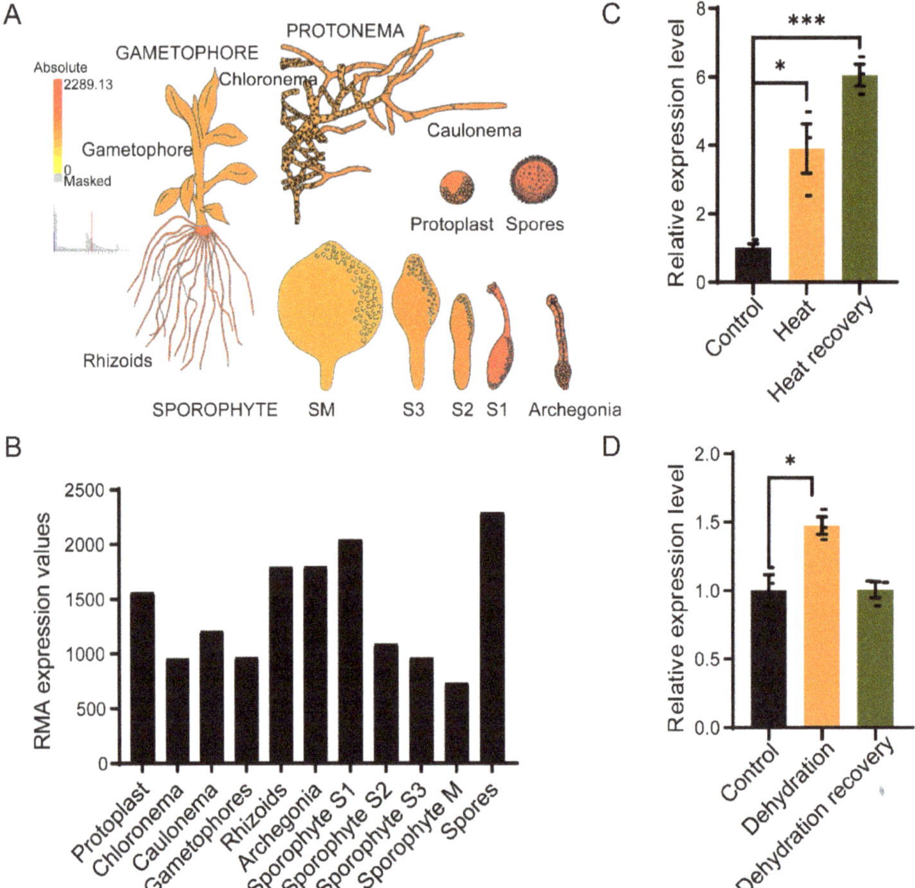

Figure 2. Gene expression patterns of *PpRNH1A* throughout the life cycle of *P. patens* and the expression levels under different stresses. (**A**) Visualization of the expression of *PpRNH1A* at different tissues during *P. patens* growth and development. (**B**) The RMA expression values of *PpRNH1A* at different periods of growth and development. Data were downloaded from *Physcomitrella* eFP Browser. (**C**) Gene expression levels of *PpRNH1A* under heat stress and after heat recovery. (**D**) Gene expression levels of *PpRNH1A* under dehydrating stress and after recovery. For (**C**,**D**), data are presented as means ± SEM of three replicates; *t*-test was used for statistics; asterisks indicate the significant difference between treatment group and the control group, * $p < 0.05$, *** $p < 0.001$.

We next examined the expression levels of *PpRNH1A* under abiotic stresses. The wild-type gametophores of *P. patens* were exposed to heat stress (HS) at 40 °C for 18 h and then recovered for 4 days, and samples at three timepoints (control, heated, and recovered) were taken to determine the relative expression level of the *PpRNH1A*. Our results showed that HS significantly induced the expression of the *PpRNH1A* (3.91 times induction compared to the control) with a continuous induction (6.33 times induction compared to the control) during the recovery period (Figure 2C). Furthermore, *PpRNH1A* expression levels under dehydration stress treatment were extracted from our previous dehydrated transcriptome data [46]. The result showed that *PpRNH1A* was slightly induced (1.47 times induction compared to the control) by dehydration stress treatment and then returned to a normal level after recovery (Figure 2D). Our results showed that *PpRNH1A* participates in both development and stress tolerance in *P. patens*.

2.3. Creation of Stable Transgenic P. patens Line Overexpressing PpRNH1A

To detect the functions of *PpRNH1A*, we obtained the CDS of the *PpRNH1A* gene by PCR amplification. The CDS fragment without the stop codon was integrated into the overexpression vector (Figure 3A) under the promoter of *PpEF1a* followed by a PEG-mediated *P. patens* protoplast transformation. Positive *PpRNH1A*-overexpression plants were selected using the hygromycin resistant marker and confirmed at both DNA and RNA levels (Figure 3B,C). Seven positive transformants at the DNA level were obtained. RNA was taken from each line and followed by quantitative real-time polymerase chain reaction (qRT-PCR) to confirm the expression levels of *PpRNH1A*-overexpressing lines (*A*-OE). One *A*-OE with a three times overexpression level compared to wild-type plants was shown in Figure 3C. The pigment contents, including Chlorophyll A (Chl a), Chl b, total Chls, and carotenoids in WT and *PpRNH1A*-overexpression plants were measured under normal growth conditions, and the result showed that no significant changes were found between WT and *A*-OE in pigment contents (Figure 3D).

2.4. Overexpression of PpRNH1A Affects Lipid Droplets Metabolism and Mobilization

The intracellular ultrastructure of *A*-OE plants was observed by transmission electron microscopy (TEM). We found that the cell walls of the overexpression plants were much thinner than the WT (Figure S1A). Interesting, there was no clear laminal structure of the chloroplast stroma observed in *A*-OE plants (Figure S1B), although the level of pigments in *A*-OE seems not to have been affected. Furthermore, in *A*-OE plants, more plastoglobuli were observed in chloroplasts (Figure S1A,B). The surface of the plant was then observed by scanning electron microscopy (SEM). The results showed that there were protrusions like glandular hairs on the surface of the *A*-OE plants (Figure S1C). These results lead us to hypothesize that the lipid metabolite was affected in *A*-OE plants, since the plant cuticle is a lipid membrane covering plant surfaces, and the plastoglobuli is a kind of lipid droplets within the chloroplasts.

To confirm whether the overexpression plants could affect the accumulation of cytosolic lipid droplets, the dye Nile red was used to stain the *A*-OE plants and wild-type plants to observe the cytosolic lipid droplets under the confocal microscope. As shown in Figure 4, the yellow fluorescence represented the accumulation of cytosolic lipid droplets by Nile red. Chloroplast autofluorescence was marked with red fluorescence in both wild-type and *A*-OE plants (Figure 4). The results showed that the lipid droplets were hardly seen in the gametophore of wild-type plants under normal growth conditions, however, in contrast, abundant of cytosolic lipid droplets around chloroplasts were observed in *A*-OE plants (Figure 4). In addition, a proportion of the lipid droplets observed in the overexpression plants had irregular morphology (Figure 4). Although additional experimental evidence is needed, these results revealed that overexpression of *PpRNH1A* may be associated with the metabolism and mobilization of lipid droplets.

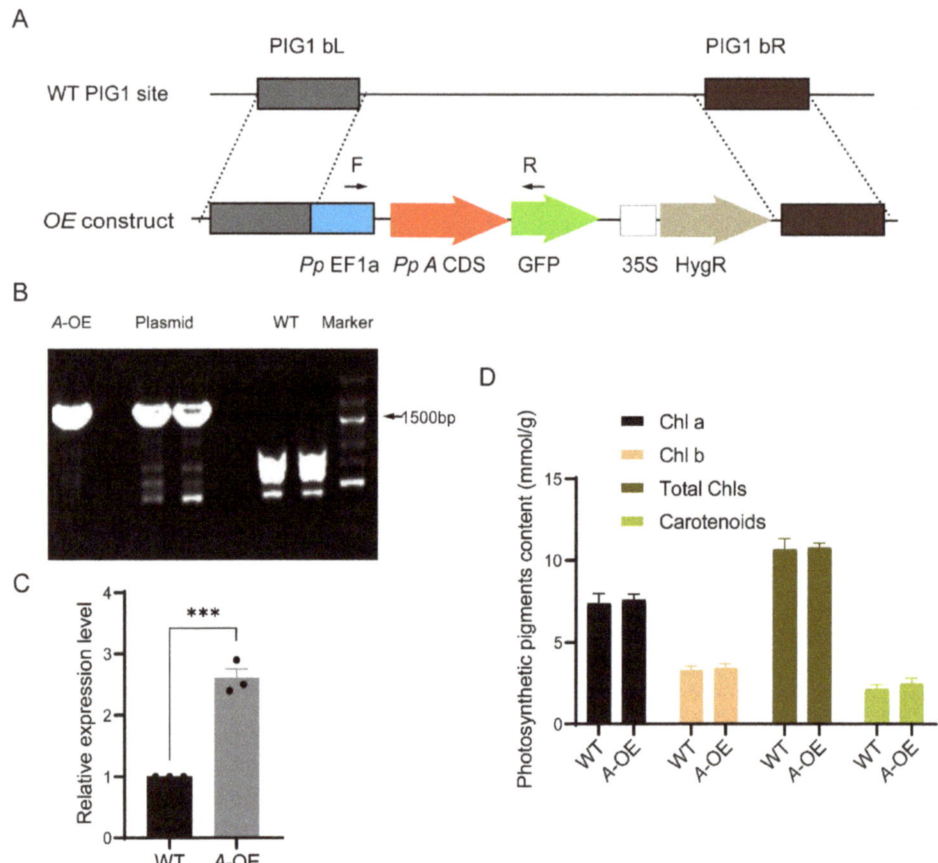

Figure 3. Stable *PpRNH1A*-overexpression lines in *P. patens*. (**A**) Schematic diagram of pPOG1-*A* construct. P, plasmid; PIG1 bL, left targeting site; PIG1 bR, right targeting site; *Pp*EF1a, the EF1a promoter from *P. patens*; HygR, hygromycin selectable marker cassette; GFP, green fluorescent protein. (**B**) Positive *PpRNH1A*-overexpression (*A*-OE) plants were confirmed at DNA level. WT, wild type; *A*-OE, *PpRNH1A*-overexpressing plant. (**C**) Identification of *A*-OE plant by qRT-PCR at RNA level. (**D**) Pigment contents in WT and *A*-OE under normal growth conditions. Chl a: chlorophyll a; Chl b: chlorophyll b; Total Chls: Total chlorophylls. For (**C**,**D**), data are presented as means $\pm$ SEM of three replicates; *t*-test was used for statistics; asterisks indicate the significant difference between WT and *A*-OE, *** $p < 0.001$.

2.5. PpRNH1A-Overexpression Line Is More Sensitive to Heat Stress

To detect the function of the protein deduced by *PpRNH1A*, plants were exposed to heat stress. Thirty days old gametophores of the wild-type plants and *A*-OE plants were treated with 40 °C for 18 h. Followed by a recovery at normal temperature (25 °C) for 5 days. Results showed that *A*-OE plants were more sensitive to heat stress and could not recover from the HS (Figure 5A). Photosynthetic parameters, including the maximal efficiency of PSII photochemistry (Fv/Fm), non-photochemical energy dissipation (NPQ), and electron transport rate (ETR) were measured at the timepoints of before heat stress (the control), after heat stress, and recovered for 4 days. Similar Fv/Fm, NPQ, and ETR levels were observed between *A*-OE plants and WT before HS. Heat stress severely affects the function of the photosystem in both *A*-OE plants and WT, whereas the Fv/Fm and NPQ levels in *A*-OE plants were significantly lower than in WT plants, both under heat stress and after recovery (Figure 5B–D). These results revealed that the overexpression of *PpRNH1A* resulted in decreasing tolerance to heat stress in *P. patens* plants.

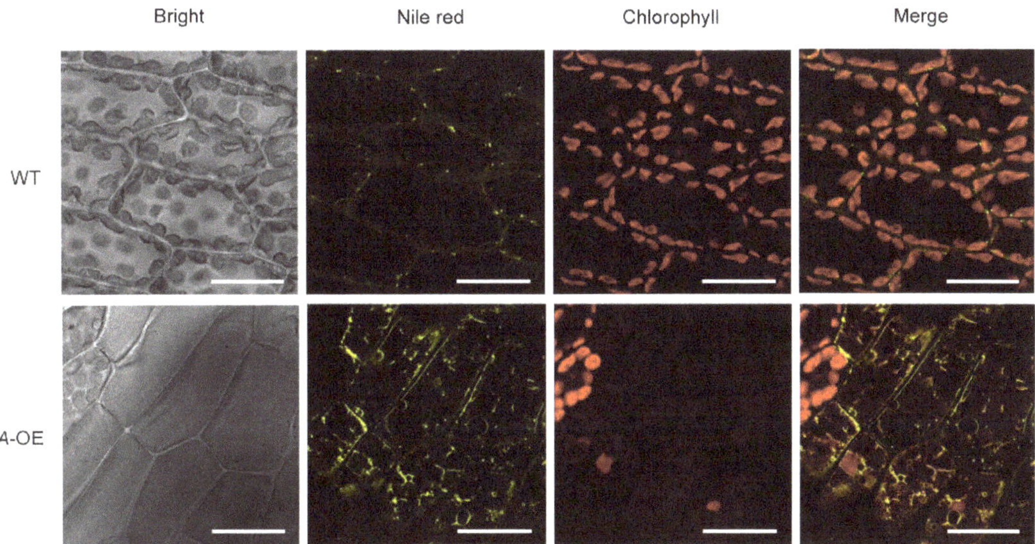

Figure 4. Accumulation of lipid droplets in WT and *A*-OE under normal growth conditions. Scale bar = 20 µm.

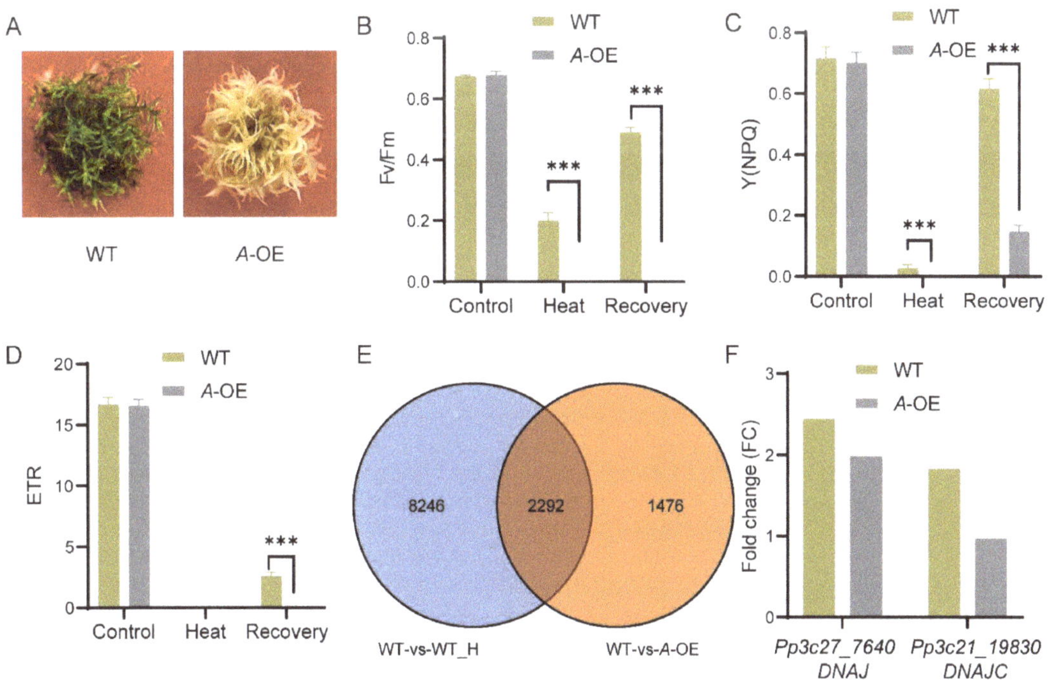

Figure 5. Response of *A*-OE plants to heat stress (HS) treatment. (**A**) *A*-OE plants were more sensitive to heat stress. (**B–D**) Chlorophyll fluorescence (Fv/Fm) (**B**), non-photochemical energy dissipation

(NPQ) (**C**), and electron transport rate (ETR) (**D**) of WT and *A*-OE before HS, after HS, and after recovery from the heat stress. Data are presented as means ± SEM of three replicates; *t*-test was used for statistics; asterisks indicate the significant difference between WT and *A*-OE, *** *p* < 0.001. (**E**) Venn diagram shows the overlapping between HS-regulated DEGs in WT (WT versus WT under heat stress) and *PpRNH1A*-regulated DEGs (WT versus *A*-OE under normal growth conditions). (**F**) Fold change comparison of the expression levels of *DNAJ* and *DNAJC* under HS and normal conditions. Each bar represents the ratio of the average expression value of three biological replicates of HS (after heat stress) vs. control (before heat stress).

2.6. PpRNH1A Regulates the Expression of Heat-Responsive Genes including DNAJ and DNAJC

Considering that overexpression plants are more sensitive to heat stress, we then tested if *PpRNH1A* regulates the heat-related genes. Transcriptomic profiles of WT plants before HS (control) and 18 h after HS were compared to identify the heat-responsive genes. In total, 10,538 DEGs (Fold change > 1.5 or 0 < Fold change < 0.67, *p*-value < 0.05) were found responding to heat stress in wild-type plants (collection: WT-vs-WT_H). To identify the genes regulated by *PpRNH1A*, 3768 DEGs were identified from the comparison between WT and *A*-OE plants under normal growth conditions (collection: WT-vs-*A*-OE). The overlap of these two collections of DEGs showed that 2292 heat-related genes were regulated by *PpRNH1A* (Figure 5E). Representative heat-related genes [47] such as *Pp3c27_7640* (molecular chaperone *DNAJ*, *HSP40*) and *Pp3c21_19830* (DNAJ homolog subfamily C member *DNAJC*) were selected for further analysis. Heat stress significantly induced the expression level of *DNAJ* and *DNAJC* in WT plants. However, in *A*-OE plants, *DNAJC* was not induced by heat (Figure 5F). The fold change of *DNAJ* induced by HS in *A*-OE plants was much less than that in the WT plants. These results suggest that *PpRNH1A* regulates the expression of heat-responsive genes, including *DNAJ* and *DNAJC* (Figure 5F).

3. Discussion

RNase H1s are ribonucleases widely present in organisms [26], and most studies on RNase H1s are related to growth and development. In *Arabidopsis*, AtRNH1B and AtRNH1C play important roles in maintaining the normal development of embryos, and AtRNH1C deletion mutants showed severe growth-defective phenotypes [39]. Our recent progress on *P. patens* revealed that *PpRNH1A* regulates the transcription of auxin-related genes by controlling the formation of R-loops, further regulating the development [45]. In this study, we found novel biological functions of PpRNH1A, which may also participate in heat stress responses, possibly by affecting the numbers and the mobilization of lipid droplets and regulating the expression of heat-related genes.

Phylogeny analysis results in this study suggest that there is functional differentiation of RNase H1s within and among plant species (Figure 1), which confirmed our previous finding that there may be functional differentiation between AtRNH1A and PpRNH1A [45]. Previously, we found that the development of gametophores of the *pprnh1a* mutant was affected through the modulation of R-loop formation on auxin-related genes. Overexpression plants of *PpRNH1A* were obtained (Figure 3A) in this study, and we found that the chloroplasts development was affected and no clear lamellar structure was observed (Figure S1B). Interestingly, although the chloroplast development was impaired, the pigments did not show any difference in *A*-OE plants compared to WT plants (Figure 3D). Abnormal development of lamellar structure and more plastoglobuli were found in the chloroplasts of the *A*-OE plants (Figure S1A,B). Furthermore, protrusions like glandular hairs were found on the surface of the *A*-OE plants as well. In addition, abundant lipid droplets with irregular morphology were accumulated in the cytosol of *A*-OE plants (Figure 4). These results suggest that overexpression of *PpRNH1A* resulted in disordered lipid metabolism and mobilization.

TAGs are the primary constituents of lipid droplets, which do not accumulate in vegetative tissues of plants under normal growth conditions but accumulate significantly under stress conditions, such as drought, high temperature, low temperature, and nutrient starvation, especially in leaves [48,49]. Correspondingly, lipid droplets were reported to be closely associated with plant biotic and abiotic stress as well. *Pseudomonas*-infected leaves caused an accumulation of lipid droplets and induced hypersensitivity reactions in *Arabidopsis* [14]. Seeds with irregular lipid droplets morphology were found susceptible to chilling injury during germination [50,51]. Consistent with the above, we found that *PpRNH1A* is involved in plant responses to heat stress, which provided a novel function of RNase H1s in plants. In our study, the expression levels of *PpRNH1A* were found to be significantly induced by abiotic stresses such as heat (Figure 2C) and drought treatment (Figure 2D). Moreover, *A*-OE plants were sensitive to heat stress (Figure 5A), further confirmed by the levels of photosynthetic parameters (Figure 5B–D). This is coincident with the findings in sunflower (*Helianthus annuus* L.), where stress-sensitive lines showed longer retention of the lipid droplets membrane under salt stress, thus exhibiting higher lipid accumulation and faster mobilization than that of stress-tolerant lines [22]. As a note, the lipid droplets mentioned here in this study are not what had been described as "oil body" in liverworts. Lipid droplets are present in almost all green plants accumulating compounds that are not soluble in the aqueous phase, and oil bodies are present only in liverwort, storing large quantities of toxic sesquiterpenoids [17,52–55]. Lipid droplets are subcellular organelles of monolayer membranes with a diameter of about 0.5–2 μm, while oil bodies are an endocrine structure surrounded by a lipid bilayer membrane [13,55]. The size of small oil bodies is 2–5 × 3–9 μm on average, and the diameter of large oil bodies can reach 70 μm [56,57]. Although both mosses and liverworts are bryophytes, lipid droplets were found in our experiment.

HS affects all aspects of plant growth and elicits responses in a range of genes, including the accumulation of heat shock proteins (HSPs). We found that 60.8% of the genes regulated by overexpression of *PpRNH1A* are responsive to heat stress (Figure 5E), which is one of the explanations for why *A*-OE plants were more sensitive to HS than wild-type plants (Figure 5A–D). Heat responsive genes play critical roles in regulating the resistance/tolerance of plants to heat stress. *GmDNJ1* (a major *HSP40*) was induced by heat stress and was responsible for enhanced heat tolerance in soybean [10]. In tomatoes, the expression levels of *LeCDJ1* of *Lycopersicon esculentum* and *SlDnaJ20* of *Solanum lycopersicum* were induced by heat stress, and overexpression of *LeCDJ1* or *SlDnaJ20* could improve the heat tolerance of plants [9,58]. In alfalfa (*Medicago sativa*), the DnaJ-like protein (*MsDJLP*) gene was induced by heat stress, and ectopic expression of *MsDJLP* in tobacco enhances the heat tolerance of tobacco [59]. All of these indicate that DNAJ could be used as a representative heat responsive marker and its expression level is closely related to heat stress tolerant ability. Our results showed that both *DNAJ* and *DNAJC* were induced by heat stress in WT plants, however, on the contrary, the induction folds in the *A*-OE plants were not as strong as that in the wild type, especially *DNAJC*, whose induction was totally hampered under heat stress. This is corroborated by the result that our *A*-OE plants is more susceptible to heat stress.

4. Materials and Methods

4.1. Phylogenetic Analysis

Amino acid sequences used were obtained from Phytozome (https://phytozome.jgi. doe.gov/pz/portal.html, accessed on 19 February 2022), Ensembl plants (http://plants. ensembl.org/Cyanidioschyzon_merolae/Tools/Blast, accessed on 19 February 2022) and NCBI (https://www.ncbi.nlm.nih.gov, accessed on 19 February 2022) database. And the RNase H1 of the *Klebsormidium nitens* was obtained from the *klebsormidium* genome project website (http://www.plantmorphogenesis.bio.titech.ac.jp/~algae_genome_project/ klebsormidium/index.html, accessed on 19 February 2022) [60]. Multiple sequence alignments of these amino acid sequences were conducted with the ClustalW of MEGA-X, and the phylogenetic tree construction was performed with MEGA-X using the maximum likelihood (ML) method and 1000 bootstrap [61]. The best-fit model we selected was JTT + G + I [62].

4.2. Plant Materials and Growth Conditions

Gransden 2004 (Courtesy Prof. Mitsuyasu Hasebe) was the wild-type (WT) *Physcomitrium (Physcomitrella) patens* genetic material we used. All plant materials were grown on BCD medium supplemented with 5 mM ammonium tartrate and 1 mM $CaCl_2$. Plants were grown at 25 °C under 16 h light photoperiod per 8 h dark photoperiod with light intensity 60–80 μmol photons $m^{-2} s^{-1}$.

4.3. Protoplast Transformation

Protoplast of *P. patens* were prepared from protonema which were continuously disrupted with an electric stirrer every week. The overexpression vector pPOG1-*A* was linearized using the restriction enzyme *Mss*I (Thermo Fisher Scientific, Waltham, MA, USA). The transformants were obtained by transferring the linearized vector into wild-type using polyethylene glycol (PEG)–mediated protoplast transformation [63].

4.4. PCR and Real-Time qRT-PCR Characterization of Overexpression Plants

Stable transgenetic lines were identified by PCR and qRT-PCR at both DNA and RNA level. The transformants were initially screened at the DNA level using primers "F" and "R". Total RNA from *P. patens* tissues was extracted using TRIzol (Invitrogen, Carlsbad, CA, USA) according to the instructions. cDNA synthesis was performed using a PrimeScript™ RT reagent Kit with gDNA Eraser (Takara, Dalian, China) according to the manufacturer's instructions. qRT-PCR was performed using SYBR Premix Ex Taq II (Takara, Dalian, China) and carried out on Bio-Rad CFX96. The primer sequences for PCR and qRT-PCR are listed in Supplementary Table S1.

4.5. Analysis of Gene Expression Patterns and Analysis of Expression by Stress Treatment

Expression pattern data were retrieved and visualized from the public database *Physcomitrella* eFP Browser of BAR (http://bar.utoronto.ca/, accessed on 19 February 2022).

4.6. Observation of Cell Ultrastructure

Leaves were cut into 1 mm^2 with a blade and immediately fixed with 3% glutaraldehyde overnight at 4 °C. Rinse the material by adding 0.1 M phosphate buffer (PBS buffer, pH 7.2) every 30 min for 3 times. Subsequently, 1% osmium tetroxide was added for sample fixation at 4 °C for 2 h. Continue rinsing three times for 20 min each. The materials were continuous rinsed for 3 times and dehydrated in a serial ethanol gradient. After embedded in Epon 812 resin, materials were sectioned and stained with 2% uranyl acetate and lead citrate under EM UC7 ultramicrotome (Leica, Weztlar, Germany). The cellular ultrastructure was observed under JEM-1400Plus transmission electron microscopy (JEOL, Tokyo, Japan).

4.7. Observation of Cell Surface Structure

Leaves of the material were cut into 9 mm^2 with a blade and immediately fixed with 3% glutaraldehyde overnight at 4 °C. The rinsing, fixation, and dehydration procedures are the same as previous mentioned. Samples were dried with a CO_2 critical-point drier. The dried materials were mounted on aluminium stubs using tweezer and sputter-coated with gold. Cell surface structure was observed by Zeiss Sigma 300 scanning electron microscopy (Zeiss, Oberkochen, Germany).

4.8. Nile Red Staining

Gametophytes grown for 30 days were taken and placed in a certain volume of Nile red working solution for staining in dark for 5 mins as previously described, with appropriate modifications [64]. Lipid droplets were observed under Olympus FV1000 confocal microscope (Olympus, Tokyo, Japan). The emission filters for Nile red were 493–636 nm.

4.9. Heat Stress Assay

Thirty-days old gametophytes of *P. patens* were used to perform a heat stress assay. The wild-type plants and overexpression plants were treated at 40 °C for 18 h, followed by a recovery at 25 °C for 4–5 days.

4.10. Measurement of Chlorophyll Fluorescence (Fv/Fm), Non-Photochemical Energy Dissipation (NPQ), Electron Transport Rate (ETR), and Pigments Content

The plants to be tested were placed in the dark for 30 min. Relevant photosynthetic parameters were determined using IMAGING-PAM (Walz, Effeltrich, Germany) and the Imaging Win software (Walz, Effeltrich, Germany) as described previously [65]. Pigment content was determined using the N, N-dimethylformamide (DMF) method, as described previously [65].

4.11. Bioinformatics and Data Analysis

Wild-type and *A*-OE plants before and after heat treatment were used to extract RNA. Samples for RNA sequencing (RNA-seq) were treated as previous description [66]. Differentially expressed genes (DEGs) between two samples were identified with the criteria of "Fold change > 1.5" and "*p* value < 0.05". Differentially expressed genes were visualized with TBtools [67]. The dehydrated transcriptome data used in this study were obtained from Dong et al. [46].

4.12. Statistical Analysis

All experiments were performed with three biological replicates. Student's *t*-test was used for hypothesis testing in statistics between two samples. Significant differences were defined and indicated by asterisks *, **, and ***, corresponding to *p*-values < 0.05, <0.01, and <0.001, respectively.

5. Conclusions

In conclusion, we revealed that *PpRNH1A* not only participates in the regulation of growth development of *P. patens* plants, it also plays a crucial role in plant response and tolerance to abiotic stresses such as heat, possibly by regulating the expression of heat-related genes and causing the abnormal accumulation and the mobilization of lipid droplets in the cytosol. Our data highlights the important role played by *PpRNH1A* in plant heat stress response, providing a novel insight into the function of RNase H1s.

Supplementary Materials: The supporting information can be downloaded at: https://www.mdpi.com/article/10.3390/ijms23169270/s1.

Author Contributions: Z.Y. and T.T. performed experiments. L.L. designed the experiment. Z.Y., L.D., H.Y. and L.L. together contributed to the writing of the original draft. Z.Y., L.D. and L.L. analyzed the data and revised the manuscript. H.L., L.C., K.H. and H.Y. provided experiment support. All authors have read and agreed to the published version of the manuscript.

Funding: This work was supported by the National Natural Science Foundation of China (31971410).

Institutional Review Board Statement: Not applicable.

Informed Consent Statement: Not applicable.

Data Availability Statement: The RNA-seq data reported in this paper have been uploaded in National Genomics Data Center (accession no. PRJCA011244).

Acknowledgments: The authors thank Mitsuyasu Hasebe (National Institute for Basic Biology) for providing Gransden 2004 and plasmid pPOG1.

Conflicts of Interest: The authors declare that they have no conflict of interest.

References

1. Sarwar, M.; Saleem, M.F.; Ullah, N.; Rizwan, M.; Ali, S.; Shahid, M.R.; Alamri, S.A.; Alyemeni, M.N.; Ahmad, P. Exogenously applied growth regulators protect the cotton crop from heat-induced injury by modulating plant defense mechanism. *Sci. Rep.* **2018**, *8*, 17086. [CrossRef] [PubMed]
2. Kaur, H.; Sirhindi, G.; Bhardwaj, R.; Alyemeni, M.N.; Siddique, K.H.M.; Ahmad, P. 28-homobrassinolide regulates antioxidant enzyme activities and gene expression in response to salt- and temperature-induced oxidative stress in *Brassica juncea*. *Sci. Rep.* **2018**, *8*, 8735. [CrossRef] [PubMed]
3. Hatfield, J.L.; Prueger, J.H. Temperature extremes: Effect on plant growth and development. *Weather Clim. Extrem.* **2015**, *10*, 4–10. [CrossRef]
4. Hayes, S.; Schachtschabel, J.; Mishkind, M.; Munnik, T.; Arisz, S.A. Hot topic: Thermosensing in plants. *Plant Cell Environ.* **2021**, *44*, 2018–2033. [CrossRef]
5. Gong, Z.; Xiong, L.; Shi, H.; Yang, S.; Herrera-Estrella, L.R.; Xu, G.; Chao, D.Y.; Li, J.; Wang, P.-Y.; Qin, F.; et al. Plant abiotic stress response and nutrient use efficiency. *Sci. China Life Sci.* **2020**, *63*, 635–674. [CrossRef]
6. Vierling, E. The roles of heat shock proteins in plants. *Annu. Rev. Plant Physiol. Plant Mol. Biol.* **1991**, *42*, 579–620. [CrossRef]
7. Wang, W.; Vinocur, B.; Shoseyov, O.; Altman, A. Role of plant heat-shock proteins and molecular chaperones in the abiotic stress response. *Trends Plant Sci.* **2004**, *9*, 244–252. [CrossRef]
8. Rajan, V.B.V.; D'Silva, P. *Arabidopsis thaliana* J-class heat shock proteins: Cellular stress sensors. *Funct. Integr. Genom.* **2009**, *9*, 433. [CrossRef]
9. Kong, F.; Deng, Y.; Wang, G.; Wang, J.; Liang, X.; Meng, Q. LeCDJ1, a chloroplast DnaJ protein, facilitates heat tolerance in transgenic tomatoes. *J. Integr. Plant Biol.* **2014**, *56*, 63–74. [CrossRef]
10. Li, K.P.; Wong, C.H.; Cheng, C.C.; Cheng, S.S.; Li, M.W.; Mansveld, S.; Bergsma, A.; Huang, T.; van Eijk, M.J.T.; Lam, H.M. GmDNJ1, a type-I heat shock protein 40 (HSP40), is responsible for both Growth and heat tolerance in soybean. *Plant Direct* **2021**, *5*, e00298. [CrossRef]
11. Shao, Q.; Liu, X.; Su, T.; Ma, C.; Wang, P. New insights into the role of seed oil body proteins in metabolism and plant development. *Front. Plant Sci.* **2019**, *10*, 1568. [CrossRef] [PubMed]
12. Kretzschmar, F.K.; Doner, N.M.; Krawczyk, H.E.; Scholz, P.; Schmitt, K.; Valerius, O.; Braus, G.H.; Mullen, R.T.; Ischebeck, T. Identification of low-abundance lipid droplet proteins in seeds and seedlings. *Plant Physiol.* **2020**, *182*, 1326–1345. [CrossRef] [PubMed]
13. Huang, N.-L.; Huang, M.-D.; Chen, T.-L.L.; Huang, A.H.C. Oleosin of subcellular lipid droplets evolved in green algae. *Plant Physiol.* **2013**, *161*, 1862–1874. [CrossRef] [PubMed]
14. Fernández-Santos, R.; Izquierdo, Y.; López, A.; Muñiz, L.; Martínez, M.; Cascón, T.; Hamberg, M.; Castresana, C. Protein profiles of lipid droplets during the hypersensitive defense response of *Arabidopsis* against *Pseudomonas* infection. *Plant Cell Physiol.* **2020**, *61*, 1144–1157. [CrossRef] [PubMed]
15. Feeney, M.; Frigerio, L.; Cui, Y.; Menassa, R. Following vegetative to embryonic cellular changes in leaves of *Arabidopsis* overexpressing LEAFY COTYLEDON2. *Plant Physiol.* **2013**, *162*, 1881–1896. [CrossRef] [PubMed]
16. Lu, J.; Xu, Y.; Wang, J.; Singer, S.D.; Chen, G. The role of triacylglycerol in plant stress response. *Plants* **2020**, *9*, 472. [CrossRef] [PubMed]
17. de Vries, J.; Ischebeck, T. Ties between stress and lipid droplets pre-date seeds. *Trends Plant Sci.* **2020**, *25*, 1203–1214. [CrossRef]
18. Li-Beisson, Y.; Thelen, J.J.; Fedosejevs, E.; Harwood, J.L. The lipid biochemistry of eukaryotic algae. *Prog. Lipid Res.* **2019**, *74*, 31–68. [CrossRef]
19. Légeret, B.; Schulz-Raffelt, M.; Nguyen, H.M.; Auroy, P.; Beisson, F.; Peltier, G.; Blanc, G.; Li-Beisson, Y. Lipidomic and transcriptomic analyses of *Chlamydomonas reinhardtii* under heat stress unveil a direct route for the conversion of membrane lipids into storage lipids. *Plant Cell Environ.* **2016**, *39*, 834–847. [CrossRef]
20. Aizouq, M.; Peisker, H.; Gutbrod, K.; Melzer, M.; Hölzl, G.; Dörmann, P. Triacylglycerol and phytyl ester synthesis in *Synechocystis* sp. PCC6803. *Proc. Natl. Acad. Sci. USA* **2020**, *117*, 6216–6222. [CrossRef]
21. Gogna, M.; Bhatla, S. Biochemical mechanisms regulating salt tolerance in sunflower. *Plant Signal. Behav.* **2019**, *14*, 1670597. [CrossRef] [PubMed]
22. Gogna, M.; Bhatla, S. Salt-tolerant and -sensitive seedlings exhibit noteworthy differences in lipolytic events in response to salt stress. *Plant Signal. Behav.* **2020**, *15*, 1737451. [CrossRef] [PubMed]
23. Guan, Y.; Liu, L.; Wang, Q.; Zhao, J.; Li, P.; Hu, J.; Yang, Z.; Running, M.P.; Sun, H.; Huang, J. Gene refashioning through innovative shifting of reading frames in mosses. *Nat. Commun.* **2018**, *9*, 1555. [CrossRef] [PubMed]
24. Wang, Y.; Zhai, J.; Qi, Z.; Liu, W.; Cui, J.; Zhang, X.; Bai, S.; Li, L.; Shui, G.; Cui, S. The specific glycerolipid composition is responsible for maintaining the membrane stability of *Physcomitrella patens* under dehydration stress. *J. Plant Physiol.* **2021**, *268*, 153590. [CrossRef] [PubMed]
25. Stein, H.; Hausen, P. Enzyme from calf thymus degrading the RNA moiety of DNA-RNA Hybrids: Effect on DNA-dependent RNA polymerase. *Science* **1969**, *166*, 393–395. [CrossRef]
26. Amon, J.D.; Koshland, D. RNase H enables efficient repair of R-loop induced DNA damage. *eLife* **2016**, *5*, e20533. [CrossRef]

27. Ohtani, N.; Haruki, M.; Morikawa, M.; Crouch, R.J.; Itaya, M.; Kanaya, S. Identification of the genes encoding Mn^{2+}-dependent RNase HII and Mg^{2+}-dependent RNase HIII from Bacillus subtilis: Classification of RNases H into three families. *Biochemistry* **1999**, *38*, 605–618. [CrossRef]

28. Chon, H.; Matsumura, H.; Koga, Y.; Takano, K.; Kanaya, S. Crystal Structure and Structure-based Mutational Analyses of RNase HIII from Bacillus stearothermophilus: A New Type 2 RNase H with TBP-like Substrate-binding Domain at the N Terminus. *J. Mol. Biol.* **2006**, *356*, 165–178. [CrossRef]

29. Jongruja, N.; You, D.J.; Angkawidjaja, C.; Kanaya, E.; Koga, Y.; Kanaya, S. Structure and characterization of RNase H3 from *Aquifex aeolicus. FEBS J.* **2012**, *279*, 2737–2753. [CrossRef]

30. Kalhorzadeh, P.; Hu, Z.; Cools, T.; Amiard, S.; Willing, E.-M.; De Winne, N.; Gevaert, K.; De Jaeger, G.; Schneeberger, K.; White, C.I. *Arabidopsis thaliana* RNase H2 deficiency counteracts the needs for the WEE1 checkpoint kinase but triggers genome instability. *Plant Cell* **2014**, *26*, 3680–3692. [CrossRef]

31. Eekhout, T.; Kalhorzadeh, P.; De Veylder, L. Lack of RNase H2 activity rescues HU-sensitivity of WEE1 deficient plants. *Plant Signal. Behav.* **2015**, *10*, e1001226. [PubMed]

32. Hyjek, M.; Figiel, M.; Nowotny, M. RNases H: Structure and mechanism. *DNA Repair* **2019**, *84*, 102672. [CrossRef] [PubMed]

33. Holmes, J.B.; Akman, G.; Wood, S.R.; Sakhuja, K.; Cerritelli, S.M.; Moss, C.; Bowmaker, M.R.; Jacobs, H.T.; Crouch, R.J.; Holt, I.J. Primer retention owing to the absence of RNase H1 is catastrophic for mitochondrial DNA replication. *Proc. Natl. Acad. Sci. USA* **2015**, *112*, 9334–9339. [CrossRef]

34. Akman, G.; Desai, R.; Bailey, L.J.; Yasukawa, T.; Dalla Rosa, I.; Durigon, R.; Holmes, J.B.; Moss, C.F.; Mennuni, M.; Houlden, H.; et al. Pathological ribonuclease H1 causes R-loop depletion and aberrant DNA segregation in mitochondria. *Proc. Natl. Acad. Sci. USA* **2016**, *113*, E4276–E4285. [CrossRef] [PubMed]

35. Cerritelli, S.M.; Frolova, E.G.; Feng, C.; Grinberg, A.; Love, P.E.; Crouch, R.J. Failure to Produce Mitochondrial DNA Results in Embryonic Lethality in Rnaseh1 Null Mice. *Mol. Cell* **2003**, *11*, 807–815. [CrossRef]

36. Kramara, J.; Osia, B.; Malkova, A. Break-Induced Replication: The Where, The Why, and The How. *Trends Genet.* **2018**, *34*, 518–531. [CrossRef] [PubMed]

37. Li, X.; Zhang, J.L.; Lei, Y.N.; Liu, X.Q.; Xue, W.; Zhang, Y.; Nan, F.; Gao, X.; Zhang, J.; Wei, J.; et al. Linking circular intronic RNA degradation and function in transcription by RNase H1. *Sci. China Life Sci.* **2021**, *64*, 1795–1809. [CrossRef]

38. Yang, Z.; Hou, Q.; Cheng, L.; Xu, W.; Hong, Y.; Li, S.; Sun, Q. RNase H1 cooperates with DNA gyrases to restrict R-loops and maintain genome integrity in *Arabidopsis* chloroplasts. *Plant Cell* **2017**, *29*, 2478–2497. [CrossRef]

39. Kuciński, J.; Chamera, S.; Kmera, A.; Rowley, M.J.; Fujii, S.; Khurana, P.; Nowotny, M.; Wierzbicki, A.T. Evolutionary history and activity of RNase H1-like proteins in *Arabidopsis thaliana. Plant Cell Physiol.* **2020**, *61*, 1107–1119. [CrossRef]

40. Wang, W.; Li, K.; Yang, Z.; Hou, Q.; Zhao, W.W.; Sun, Q. RNase H1C collaborates with ssDNA binding proteins WHY1/3 and recombinase RecA1 to fulfill the DNA damage repair in *Arabidopsis* chloroplasts. *Nucleic Acids Res.* **2021**, *49*, 6771–6787. [CrossRef]

41. Cheng, L.; Wang, W.; Yao, Y.; Sun, Q. Mitochondrial RNase H1 activity regulates R-loop homeostasis to maintain genome integrity and enable early embryogenesis in *Arabidopsis. PLoS Biol.* **2021**, *19*, e3001357. [CrossRef] [PubMed]

42. Cove, D. The moss *Physcomitrella patens. Annu. Rev. Genet.* **2005**, *39*, 339–358. [CrossRef] [PubMed]

43. Rensing, S.A.; Lang, D.; Zimmer, A.D.; Terry, A.; Salamov, A.; Shapiro, H.; Nishiyama, T.; Perroud, P.-F.; Lindquist, E.A.; Kamisugi, Y. The *Physcomitrella* genome reveals evolutionary insights into the conquest of land by plants. *Science* **2008**, *319*, 64–69. [CrossRef] [PubMed]

44. Rensing, S.A.; Goffinet, B.; Meyberg, R.; Wu, S.-Z.; Bezanilla, M. The moss *Physcomitrium (Physcomitrella) patens*: A model organism for non-seed plants. *Plant Cell* **2020**, *32*, 1361–1376. [CrossRef] [PubMed]

45. Chen, S.; Dong, X.; Yang, Z.; Hou, X.; Liu, L. Regulation of the Development in *Physcomitrium (Physcomitrella) patens* implicates the functional differentiation of plant RNase H1s. *Plant Sci.* **2021**, *313*, 111070. [CrossRef]

46. Dong, X.M.; Pu, X.J.; Zhou, S.Z.; Li, P.; Luo, T.; Chen, Z.X.; Chen, S.L.; Liu, L. Orphan gene *PpARDT* positively involved in drought tolerance potentially by enhancing ABA response in *Physcomitrium (Physcomitrella) patens. Plant Sci.* **2022**, *319*, 11122. [CrossRef]

47. Ul Haq, S.; Khan, A.; Ali, M.; Khattak, A.M.; Gai, W.X.; Zhang, H.X.; Wei, A.M.; Gong, Z.H. Heat shock proteins: Dynamic biomolecules to counter plant biotic and abiotic stresses. *Int. J. Mol. Sci.* **2019**, *20*, 5321. [CrossRef]

48. Lee, H.G.; Park, M.-E.; Park, B.Y.; Kim, H.U.; Seo, P.J. The *Arabidopsis* MYB96 transcription factor mediates ABA-dependent triacylglycerol accumulation in vegetative tissues under drought stress conditions. *Plants* **2019**, *8*, 296. [CrossRef]

49. Angkawijaya, A.E.; Nguyen, V.C.; Nakamura, Y. LYSOPHOSPHATIDIC ACID ACYLTRANSFERASES 4 and 5 are involved in glycerolipid metabolism and nitrogen starvation response in *Arabidopsis. New Phytol.* **2019**, *224*, 336–351. [CrossRef]

50. Shimada, T.L.; Shimada, T.; Takahashi, H.; Fukao, Y.; Hara-Nishimura, I. A novel role for oleosins in freezing tolerance of oilseeds in *Arabidopsis thaliana. Plant J.* **2008**, *55*, 798–809. [CrossRef]

51. Miquel, M.; Ghassen, T.; d'Andréa, S.; Kelemen, Z.; Baud, S.; Berger, A.; Deruyffelaere, C.; Trubuil, A.; Lepiniec, L.; Dubreucq, B. Specialization of oleosins in oil body dynamics during seed development in *Arabidopsis* Seeds. *Plant Physiol.* **2014**, *164*, 1866–1878. [CrossRef] [PubMed]

52. Lundquist, P.K.; Shivaiah, K.-K.; Espinoza-Corral, R. Lipid droplets throughout the evolutionary tree. *Prog. Lipid Res.* **2020**, *78*, 101029. [CrossRef] [PubMed]

53. Duckett, J.G.; Ligrone, R. The formation of catenate foliar gemmae and the origin of oil bodies in the liverwort *Odontoschisma denudatum* (Mart.) Dum. (Jungermanniales): A light and electron microscope study. *Ann. Bot.* **1995**, *76*, 405–419. [CrossRef]
54. Scholz, P.; Chapman, K.D.; Mullen, R.T.; Ischebeck, T. Finding new friends and revisiting old ones—How plant lipid droplets connect with other subcellular structures. *New Phytol.* **2022**. [CrossRef]
55. Romani, F.; Flores, J.; Tolopka, J.; Suarez, G.; He, X.; Moreno, J. Liverwort oil bodies: Diversity, biochemistry, and molecular cell biology of the earliest secretory structure of land plants. *J. Exp. Bot.* **2022**, *73*, 4427–4439. [CrossRef]
56. Schuster, R.M. The oil-bodies of the Hepaticae. I. Introduction. *J. Hattori Bot. Lab.* **1992**, *72*, 151–162.
57. Schuster, R.M.; Konstantinova, N. Studies on Treubiales, I. On *Apotreubia* Hatt. et al. and *A. hortonae* Schust. & Konstantinova, sp. n. *J. Hattori Bot. Lab.* **1995**, *78*, 41–61.
58. Wang, G.; Cai, G.; Xu, N.; Zhang, L.; Sun, X.; Guan, J.; Meng, Q. Novel DnaJ protein facilitates thermotolerance of transgenic tomatoes. *Int. J. Mol. Sci.* **2019**, *20*, 367. [CrossRef]
59. Lee, K.-W.; Rahman, M.A.; Kim, K.-Y.; Choi, G.J.; Cha, J.-Y.; Cheong, M.S.; Shohael, A.M.; Jones, C.; Lee, S.-H. Overexpression of the alfalfa DnaJ-like protein (*MsDJLP*) gene enhances tolerance to chilling and heat stresses in transgenic tobacco plants. *Turk. J. Biol.* **2018**, *42*, 12–22. [CrossRef]
60. Hori, K.; Maruyama, F.; Fujisawa, T.; Togashi, T.; Yamamoto, N.; Seo, M.; Sato, S.; Yamada, T.; Mori, H.; Tajima, N. *Klebsormidium flaccidum* genome reveals primary factors for plant terrestrial adaptation. *Nat. Commun.* **2014**, *5*, 3978. [CrossRef]
61. Kumar, S.; Stecher, G.; Li, M.; Knyaz, C.; Tamura, K. MEGA X: Molecular evolutionary genetics analysis across computing platforms. *Mol. Biol. Evol.* **2018**, *35*, 1547. [CrossRef] [PubMed]
62. Jones, D.T.; Taylor, W.R.; Thornton, J.M. The rapid generation of mutation data matrices from protein sequences. *Comput. Appl. Biosci.* **1992**, *8*, 275–282. [CrossRef]
63. Liu, L.; McNeilage, R.T.; Shi, L.-x.; Theg, S.M. ATP requirement for chloroplast protein import is set by the Km for ATP hydrolysis of stromal Hsp70 in *Physcomitrella patens*. *Plant Cell* **2014**, *26*, 1246–1255. [CrossRef] [PubMed]
64. Miranda, A.F.; Liu, Z.; Rochfort, S.; Mouradov, A. Lipid production in aquatic plant *Azolla* at vegetative and reproductive stages and in response to abiotic stress. *Plant Physiol. Biochem.* **2018**, *124*, 117–125. [CrossRef] [PubMed]
65. Li, P.; Yang, H.; Liu, G.; Ma, W.; Li, C.; Huo, H.; He, J.; Liu, L. *PpSARK* regulates moss senescence and salt tolerance through ABA related pathway. *Int. J. Mol. Sci.* **2018**, *19*, 2609. [CrossRef]
66. Pu, X.; Yang, L.; Liu, L.; Dong, X.; Chen, S.; Chen, Z.; Liu, G.; Jia, Y.; Yuan, W.; Liu, L. Genome-wide analysis of the MYB transcription factor superfamily in *Physcomitrella patens*. *Int. J. Mol. Sci.* **2020**, *21*, 975. [CrossRef]
67. Chen, C.; Chen, H.; Zhang, Y.; Thomas, H.R.; Frank, M.H.; He, Y.; Xia, R. TBtools: An integrative toolkit developed for interactive analyses of big biological data. *Mol. Plant* **2020**, *13*, 1194–1202. [CrossRef]

International Journal of
Molecular Sciences

Article

Genome-Wide Analysis of the HDAC Gene Family and Its Functional Characterization at Low Temperatures in Tartary Buckwheat (*Fagopyrum tataricum*)

Yukang Hou [1,†] , Qi Lu [1,†], Jianxun Su [1], Xing Jin [1], Changfu Jia [2], Lizhe An [1], Yongke Tian [1,*] and Yuan Song [1,*]

[1]   Ministry of Education Key Laboratory of Cell Activities and Stress Adaptations, School of Life Sciences, Lanzhou University, Lanzhou 730030, China; houyk19@lzu.edu.cn (Y.H.); luq2020@lzu.edu.cn (Q.L.); sujx21@lzu.edu.cn (J.S.); jinx19@lzu.edu.cn (X.J.); lizhean@lzu.edu.cn (L.A.)

[2]   Key Laboratory for Bio-Resources and Eco-Environment of Ministry of Education, College of Life Science, Sichuan University, Chengdu 610017, China; g1020160171@gmail.com

*   Correspondence: tianyk@lzu.edu.cn (Y.T.); songyuan@lzu.edu.cn (Y.S.)

†   These authors have contributed equally to this work.

Abstract: Histone deacetylases (HDACs), widely found in various types of eukaryotic cells, play crucial roles in biological process, including the biotic and abiotic stress responses in plants. However, no research on the HDACs of *Fagopyrum tataricum* has been reported. Here, 14 putative *FtHDAC* genes were identified and annotated in *Fagopyrum tataricum*. Their gene structure, motif composition, *cis*-acting elements, phylogenetic relationships, protein structure, alternative splicing events, subcellular localization and gene expression pattern were investigated. The gene structure showed *FtHDACs* were classified into three subfamilies. The promoter analysis revealed the presence of various *cis*-acting elements responsible for hormone, abiotic stress and developmental regulation for the specific induction of *FtHDACs*. Two duplication events were identified in *FtHDA6-1*, *FtHDA6-2*, and *FtHDA19*. The expression patterns of *FtHDACs* showed their correlation with the flavonoid synthesis pathway genes. In addition, alternative splicing, mRNA enrichment profiles and transgenic analysis showed the potential role of *FtHDACs* in cold responses. Our study characterized *FtHDACs*, providing a candidate gene family for agricultural breeding and crop improvement.

Keywords: *FtHDACs*; genome-wide; low-temperature responses; Tartary buckwheat

Citation: Hou, Y.; Lu, Q.; Su, J.; Jin, X.; Jia, C.; An, L.; Tian, Y.; Song, Y. Genome-Wide Analysis of the HDAC Gene Family and Its Functional Characterization at Low Temperatures in Tartary Buckwheat (*Fagopyrum tataricum*). *Int. J. Mol. Sci.* **2022**, 23, 7622. https://doi.org/10.3390/ijms23147622

Academic Editors: Melvin J. Oliver, Bei Gao and Moxian Chen

Received: 26 May 2022
Accepted: 5 July 2022
Published: 10 July 2022

Publisher's Note: MDPI stays neutral with regard to jurisdictional claims in published maps and institutional affiliations.

1. Introduction

Tartary buckwheat (*Fagopyrum tataricum*) is a pseudocereal that belongs to the genus Fagopyrum within the Polygonaceae family. Tartary buckwheat is not only an essential medicinal and edible crop, but also adapted to growing in adverse environments, such as harsh climates and nutrient-poor soils [1]. Because of the high content of bioactive flavonoids (rutin, anthocyanins, and quercetin), Tartary buckwheat is the preferred healthy food for the "three-highs population" (high blood sugar, high cholesterol, and high blood pressure) [2]. Additionally, flavonoids such as quercetin were found to fight against COVID-19 [3]. In recent years, the research of Tartary buckwheat has become increasingly popular, and the genome is constantly being sequenced and annotated [4–9]. More studies are being conducted on Tartary buckwheat, especially on the synthesis of bioactive flavonoids. The biosynthesis and accumulation of flavonoids are closely related to the living environment in plants. Tartary buckwheat is thought to originate in the mountainous areas of northwest China, and unique phenylalanine pathways have evolved to both respond to and adapt to cold stress [1].

Inducing environmental changes on histone marks at certain loci are important for studying plant stress responses [2]. Moreover, epigenetic editing is a new way of breeding crops [3]. Histone deacetylases (HDACs) are important epigenetic regulators in eukaryotes and are involved in the deacetylation of histone lysine and arginine residues of the H3 and H4 histone. Furthermore, HDACs are highly conserved in many organisms [4,5]. In plants, histone deacetylation is carried out by three HDAC families: RPD3/HDA1, SIR2, and the plant-specific HD2 family [6]. HDACs are associated with transcriptional repression and gene silencing through the deacetylation of lysine residues. The HDACs lack an intrinsic DNA binding domain and are recruited to target genes through their interacting transcription factors and other large multiprotein transcription complexes. The removal of the acetyl groups from histones by HDACs causes tighter chromatin packing, which weakens the combination of the transcription factors and DNA, and is involved in abiotic plant stress responses [7]. The histone deacetylation site, enzyme, and potential function are summarized, and are then associated with transcriptional activation, histone deposition, and DNA repair [8]. HDACs have been characterized in multiple plants, such as in *Marchantia polymorpha* [9], *Arabidopsis* [10], *Zea mays* [11], *Oryza sativa* [12,13], *Camellia sinensis* [14] and *Gossypium* spp. [15]. The *HDAC* gene family is widely involved in plant growth, development, and stress responses. To continue, the function of HDA6 is involved in the morphogenesis of plant roots [16], hypocotyl elongation [17], flowering [18,19], and senescence [20]. However, the basic information and mechanism of *HDACs* in *Fagopyrum tataricum* remain unclear. Alternative splicing events in *FtHDACs* are also an important aspect of understanding Tartary buckwheat adaptation.

In this study, 14 *HDACs* were first identified from *Fagopyrum tataricum*, and they were then comprehensively analyzed through the phylogenetic classification, gene structure and chromosomal location, domain organization, *cis*-acting elements, intraspecific collinearity, protein 3D structure, alternative splicing events, and subcellular localization. In addition, the gene expression patterns of *FtHDACs* were studied during different developmental stages and cold treatments. Here, a fundamental understanding of *FtHDACs* is provided for Tartary buckwheat growth, development, and cold stress responses, even in the flavonoid synthesis pathway. These results provide information on histone deacetylation in *Fagopyrum tataricum*, while providing an essential candidate gene family for crop improvement.

2. Results

2.1. Identification and Classification of FtHDACs Genes

In the present study, 14 *FtHDAC* genes were identified in Tartary buckwheat through the BLASTp methods, including nine *RPD3/HDA1s*, two *SIR2s*, and three *HD2s*. The results showed the basic information of the gene family, including renaming, genome serial number, CDS lengths, protein length, amino acid number, and molecular weight, equipotential point, Aliphatic index, and GRAVY (Table 1). In the *FtHDAC* family, the protein length is between 183aa (FtHDA5) and 485aa (FtHDA6-2), and the MW is between 20.68 (FtHDA5) and 56.18 kDa (FtHDA19). The predicted p*I* is between 4.11 (FtHDT2) and 9.39 (FtHDA5). The aliphatic index was 46.23 (FtHDT3) to 119.34 (FtHDA5). The predicted average hydrophilic coefficient (GRAVY) showed that FtHDA5 and FtHDA14 are hydrophobic proteins, and the others are hydrophilic proteins.

Table 1. Basic information of *HDACs* gene family in *Fagopyrum tataricum.*

Gene Name	Gene ID	CDS bp	Amino Acids aa	Molecular Weight	p*I*	Aliphatic Index	GRAVY
FtHDA6-1	FtPinG000668020	1428	475	53,654.75	5.39	70.78	−0.557
FtHDA6-2	FtPinG000582460	1458	485	54,108.42	5.74	68.72	−0.530
FtHDA19	FtPinG000705950	1509	502	56,189.23	5.08	73.51	−0.492
FtHDA9	FtPinG000324040	1293	430	49,188.5	5.06	79.56	−0.398
FtHDA2	FtPinG000223360	1092	363	39,980.06	8.44	98.87	−0.055
FtHDA5	FtPinG000450700	552	183	20,683.58	9.39	119.34	0.343
FtHDA14	FtPinG000269690	1299	432	46,834.56	6.10	91.23	0.035
FtHDA8-1	FtPinG000633580	849	282	30,216.81	5.34	83.01	−0.191
FtHDA8-2	FtPinG000633540	1149	382	41,256.46	5.28	88.85	−0.128
FtSRT1	FtPinG000031830	1431	476	52,748.52	9.35	91.09	−0.209
FtSRT2	FtPinG000180810	1398	465	51,761.15	8.50	91.96	−0.136
FtHDT1	FtPinG000141150	771	256	27,833.64	5.10	58.67	−0.972
FtHDT2	FtPinG000168350	645	214	23,549.38	4.11	66.07	−0.950
FtHDT3	FtPinG000710910	927	308	33,347.77	4.67	46.23	−1.202

Alignments of the full-length *FtHDAC* sequences were used to generate an unrooted phylogenetic tree by MEGA7.0 software. The conserved structure and conserved domain distribution was analyzed by the SMART and GSDS methods. The results showed that the *FtHDAC* family is divided into three subfamilies with a typical subfamily domain. For the largest subfamily, *RPD3\HDA1*, all members have the conserved histone deacetylation functional domain Hist_deacetyl, and each C terminal has the conserved glycine and histidine aspartic residues. The *FtSIR2* subfamily contains two genes, both of which have functional SIR2 domains. The FtHD2 family contains three members, all of which have conserved functional structures in the SCOP domain (Figure 1A).

The secondary structures of FtHDACs are comprised of an α-helix, extended chain, and random coil. The FtHDT1 and FtHDT3 proteins had a large proportion of random-coiled amino acids (>55%), followed by less α-helix (<20%). However, the α-helix of FtHDA5 accounted for 41.53% of the total. Individual proteins showed different secondary structural properties (Table S4). The predicted 3D structures are shown in (Figure 1B). The structures of FtHDA6-1, FtHDA6-2, and FtHDA19 were similar, suggesting a shared functionality. FtHD1, FtHD2, and FtHD3 were structurally similar, as were FtSRT1 and FtSRT2, illustrating their functional redundancy (Figure 1B).

2.2. Conserved Protein Structure and Cis-Acting Element Prediction

The motifs of the three subfamilies of FtHDAC were analyzed by MEME analysis to identify the putative motifs of the HDAC subfamilies in *Fagopyrum tataricum*, and all members contained one distinct motif for each subfamily, verifying that they belonged to the same subfamily, which further provided evidence for the classification of subfamilies (Figure 2A). The result is the same as the phylogenetic tree analysis, most of the closely related members share common motifs.

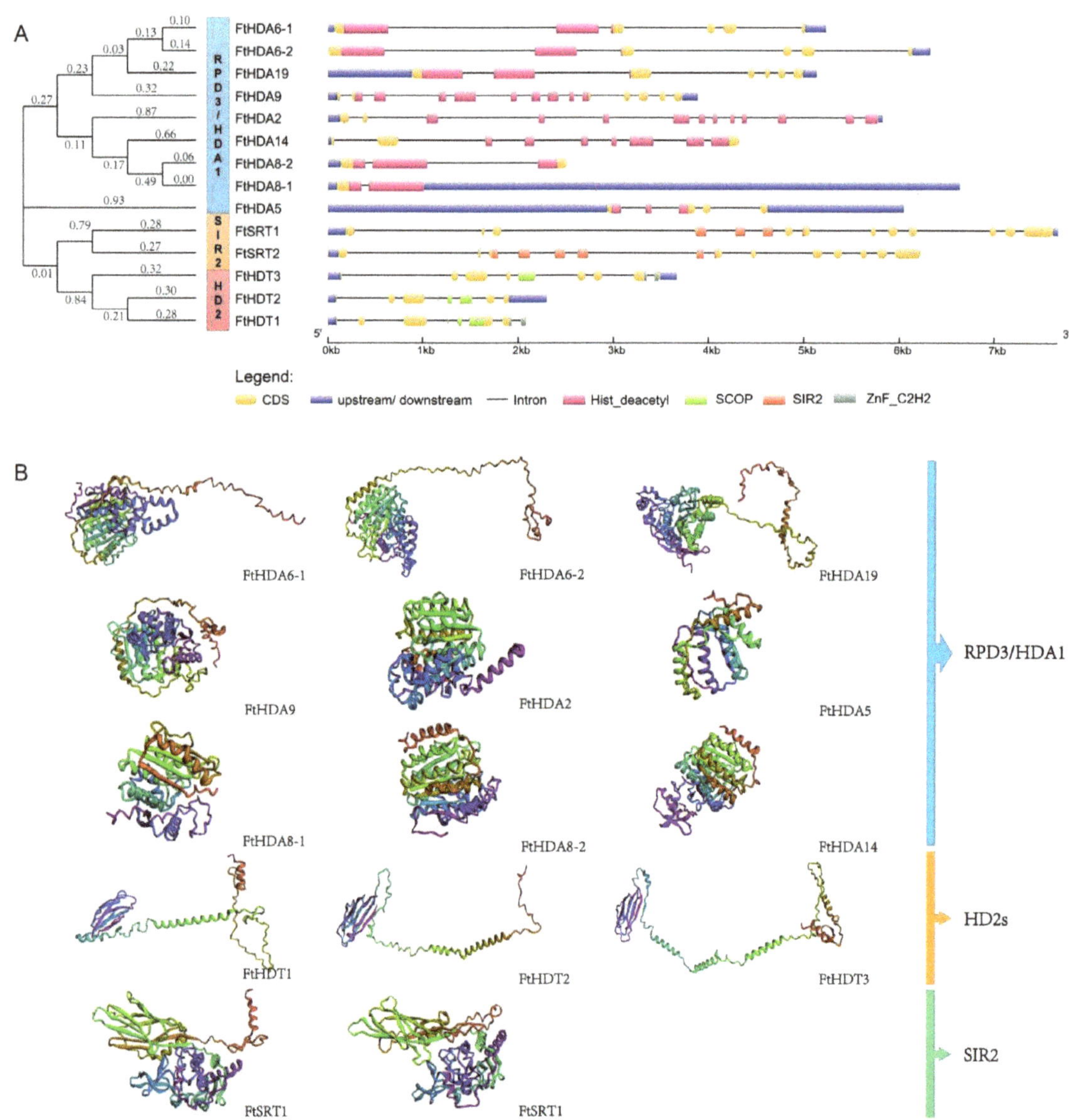

Figure 1. Phylogenetic relationships, motif structure and three-dimensional protein structure of FtHDACs. (**A**) Phylogenetic relationships and exon/intron structures of *FtHDAC* genes. The different structural units are represented by different colors, respectively. (**B**) Models and ribbon diagrams show the 3D domain structure of 14 FtHDACs proteins. Blue to red, N- to C-terminus.

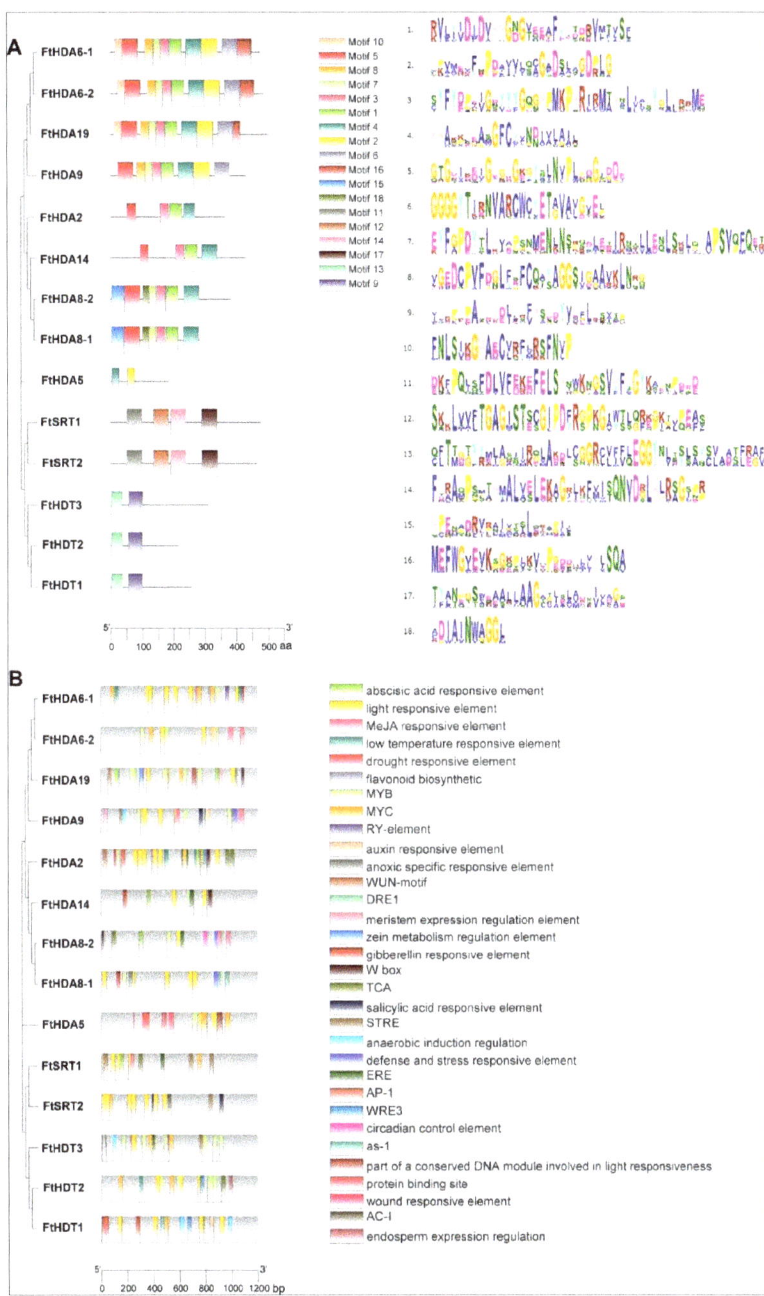

Figure 2. Conserved protein structure and *cis*-acting element prediction of FtHDACs. (**A**) Conserved motifs of FtHDACs proteins using MEME, and details of the 18 conserved motifs shared among the FtHDACs proteins. Each motif is indicated by a colored box numbered on the right. The length of the motifs in each protein is shown as a proportion. Motif symbol and motif consensus also are shown. (**B**) *Cis*-acting element prediction via PLANT CARE. There are 32 types of cis-elements, marked with rectangles of different colors.

Cis-acting elements analysis showed that FtHDACs are mainly involved in hormone response elements (abscisic acid responsive element, MeJA responsive element, auxin responsive element, and salicylic acid responsive element, gibberellin responsive element), stress response elements (light responsive element, low temperature responsive element, drought responsive element, anoxic specific responsive element, defense and stress responsive element), and functional control elements (flavonoid biosynthetic regulation element, meristem expression regulation element, circadian control element, endosperm development regulation element) (Figure 2B). In addition, the MYB and MYC binding sites were also found, suggesting that the *FtHDAC* family might be regulated by specific transcription factors.

2.3. Chromosomal Localization, Phylogenetic Analysis, and Analysis of Gene Duplication Events

The location of the *FtHDACs* gene family in the chromosome and the gene density and gene duplication events among eight chromosomes was explored. Except for the second chromosome, the *FtHDACs* distributed in the other seven chromosomes. Most of the distribution was found on chromosome 6 with four HDAC genes (Figure 3A). The two duplication events were presented in FtHDA6-1, FtHDA6-2, and FtHDA19 to detect the evolutionary relationship between them. The Ka/Ks of FtHDA6-1 was counted by calculating the parameters Ks, Ka, and Ka/Ks ratio in TBtools. The numerical value was 0.11, indicating that the driving force between the evolution of the two genes was mainly purifying selection. To understand the evolutionary relationships of FtHDACs with the HDACs of other plants, a comparative analysis was conducted on the HDAC genes in *Fagopyrum tataricum*, *Arabidopsis*, soybean, tomato, hairy fruit poplar, sweet orange, and rice by MEGA7.0 to construct a phylogenetic tree. The results showed that the HDAC family is highly conserved among the selected species. Most of the genes of the *Fagopyrum tataricum* RPD3\HDA1 subfamily cluster were more closely related with dicots, such as sweet orange and tomato, while the other two subfamilies are basically consistent with monocot dicots, suggesting that *FtHDAC* genes present a different evolutionary history and pathway (Figure 3B).

Furthermore, to explore the underlying evolutionary mechanisms of the *FtHDAC* family, five representative angiosperm species, including some plants from legumes, Chenopodiaceae, cruciferas, Pedaliaceae, and Poaceae, were selected to construct comparative syntenic maps with *Fagopyrum tataricum*. By conducting an interspecific collinearity visualization of major cash crops, such as grasses, maize, and sorghum, the results showed that the *FtHDAC* family exhibited more collinearity with legumes, indicating that *HDACs* are more evolutionarily related between these species. However, it should be noted that they are also closely related to quinoa, roidiaceae, and hemaraceae. The *FtHDAC* family lacked correlation with maize sorghum, and they were not collinear (Figure 4).

2.4. Tartary Buckwheat Multiple Tissue Transcriptome Analysis

Using the Tartary Buckwheat Database (TBD), the relative expression of *Fagopyrum tataricum FtHDACs* and part of the flavonoid synthetic genes in roots, stems, leaves, flowers, and three fruit developmental stages was analyzed (Figure 5A, Table S2). The *FtHDACs* gene family and flavonoid-synthesis-related genes presented different expression patterns. For example, *FtHDT2* showed a high expression level in each tissue but a reduced expression at day 19 of fruit development (Figure 5A). Moreover, *FtHDA19* had a high expression in the roots, stems, and flowers, suggesting some function in these tissues. The flavonoid synthesis-related genes showed different expression patterns during Tartary buckwheat development (Figure S3).

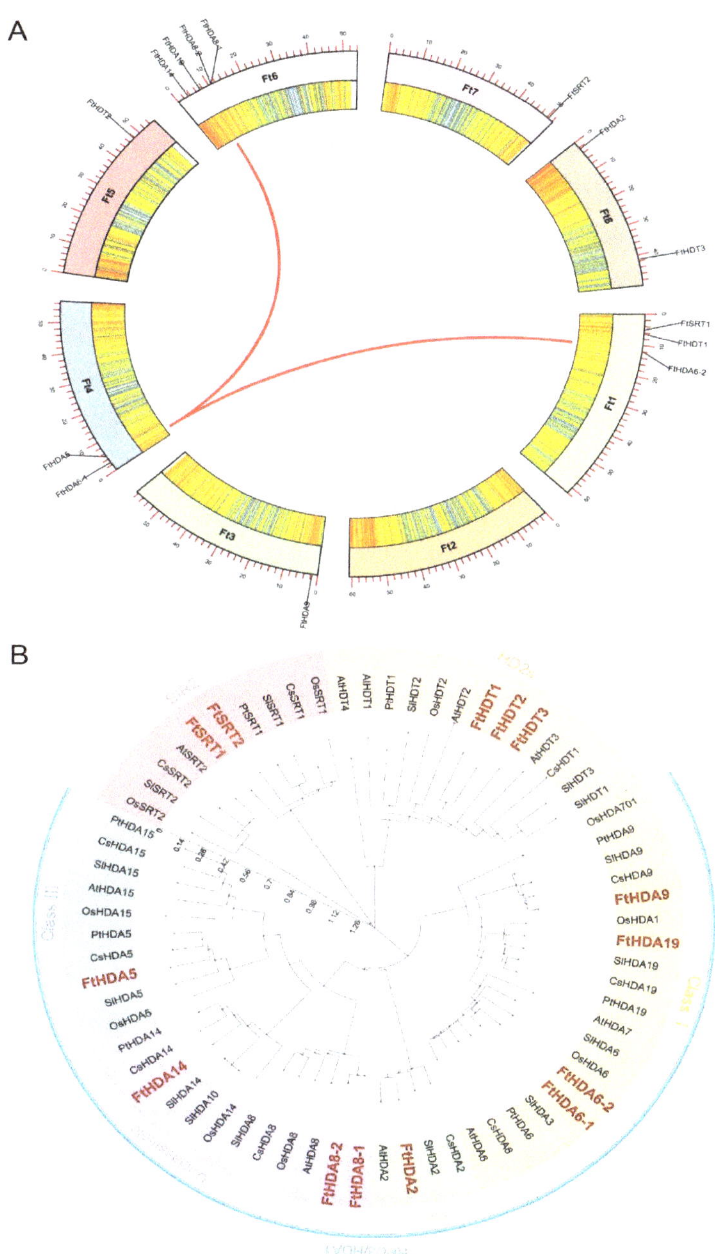

Figure 3. Chromosomal localization and Phylogenetic analysis of *FtHDA6*. (**A**) Distribution of *FtHDACs* gene on *Fagopyrum tataricum* chromosomes, intraspecies collinearity and gene density. (**B**) Phylogenetic tree construction of *HDACs* gene family from *Fagopyrum tataricum* (Ft), *Arabidopsis* (At), Solanum lycopersicum (SI), Oryza sativa (Os), Populus trichocarpa (Pt), Citrus sinensis (Cs). Each subgroup ID number is in the outer circle of the phylogenetic tree and branches with less than 70% bootstrap support are collapsed (replicated 1000 times). The different filling colors indicate different gene subfamilies, and the length of the clade indicates the evolutionary distance.

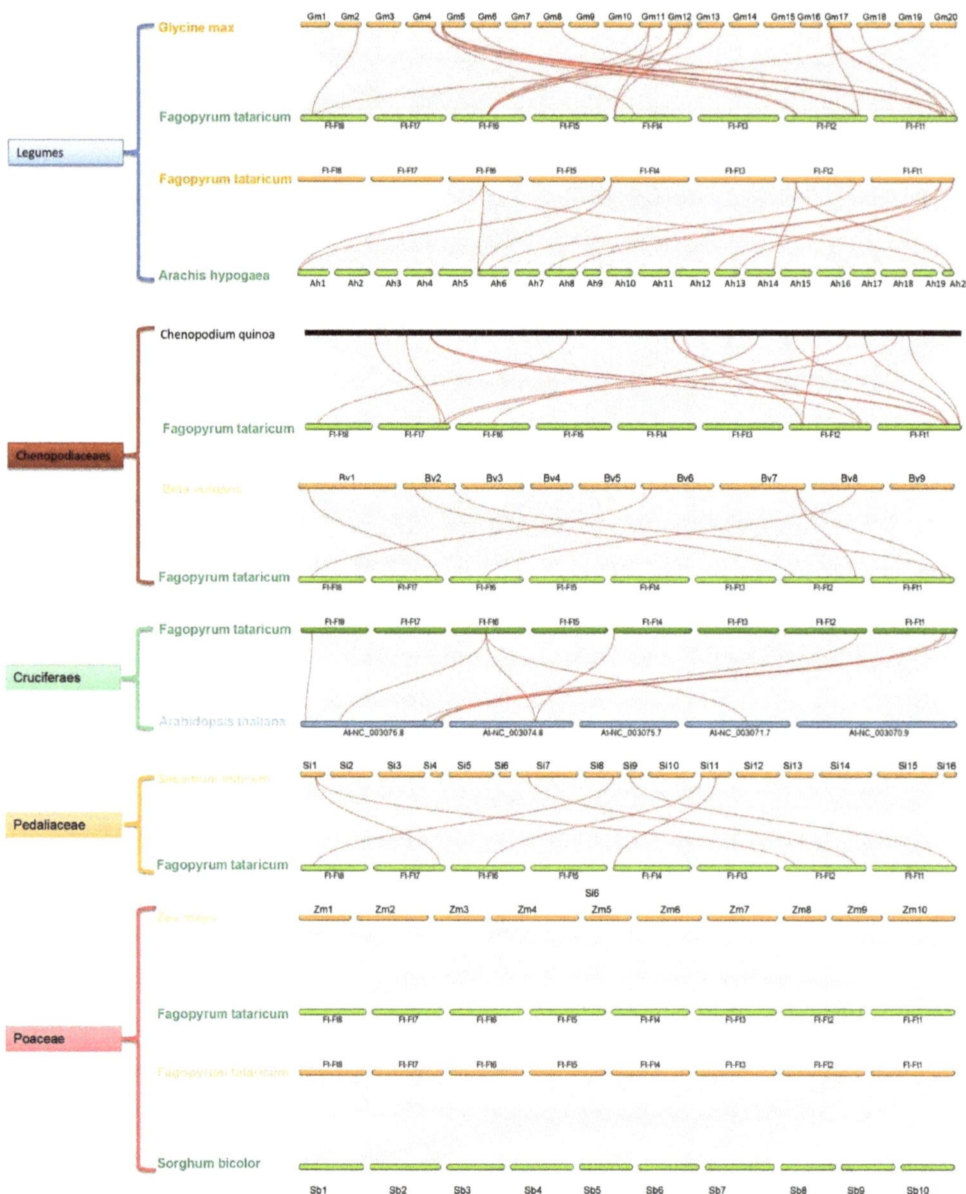

Figure 4. Synteny analyses of *HDACs* between *Fagopyrum tataricum* and other five representative plant species. Gray lines in the background indicate the collinear blocks within *Fagopyrum tataricum* and other plant genomes, while red lines highlight the syntenic *HDACs* gene pairs.

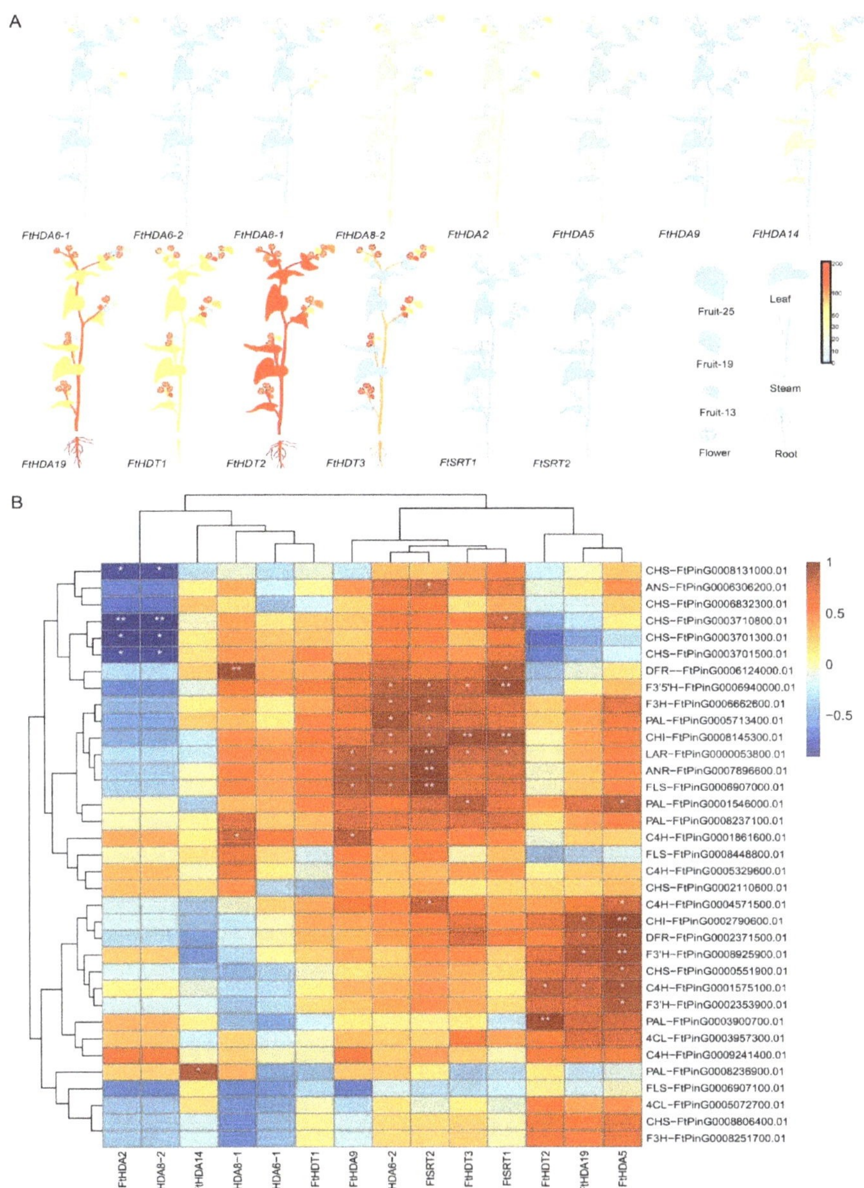

Figure 5. The tissue expression analysis of FtHDACs. (**A**) Expression patterns of *FtHDACs* genes in various Tartary buckwheat tissues. RNA-seq expression data corresponding to *FtHDACs* were retrieved from the Tartary Buckwheat Database (TBD) for further analysis. The RPKM (Reads Per Kilobase of exon model per Million mapped reads) values were transformed to log2(1×10^{-6}). The expression in various Tartary buckwheat tissues is shown, including the root, stem, leaf, flower, and fruit_13, fruit_19, fruit_25. Blue to red, respectively, indicates a high to low expressional level. (**B**) The heatmap of correlation of *FtHDACs* and flavonoid related genes expression level is filtered. Red indicates the positive correlation, and blue indicates a negative correlation. Small white stars indicate significant associations (*: $p \leq 0.05$, **: $p \leq 0.01$).

The correlation analysis between the *FtHDAC* family genes and flavonoid synthesis genes was performed (Figure 5B). There are many key enzymes in the flavonoid synthesis pathway, playing with different roles (Figure S6). For instance, the PAL enzyme catalyzes phenylalanine to produce cinnamic acid and coumaric acid plasma, a key enzyme linking phenylpropane compounds and primary metabolism, and it is important roles for regulation of flavonoid compound synthesis. The C4H enzyme is a single oxidative enzyme of the CYP73 series in plant cytochrome P450, as the second enzyme in the plant phenylalanine metabolism pathway, and it catalyzes the synthesis of coumaric acid by the substrate cinnamic acid, with high catalytic vitality. The 4CL enzyme, the last enzyme in phenylalanine metabolism in plants, catalyzes the generation of coumaric acid to COA ester. The CHS enzyme is a member of the polyketide synthase family, which is the first enzyme in the flavonoids synthesis pathway that is closely related to flavonoids and isoflavone synthesis, and an important rate-limiting enzyme in the synthesis pathway. The CHI enzyme is also a key enzyme in the flavonoids synthesis pathway that catalyzes the stereo-isomerization of the chalcone to synthetize the associated (2S) -flavanones. There are 48 key enzymes encoding genes involved in the flavonoid metabolism pathway in buckwheat [21]. The other enzymes mentioned in Figure 5B function in a branch of the flavone synthesis pathway. For example, DRF, ANS, ANR, LAR, F3'H and F3'5'H are key enzymes regulating anthocyanin synthesis. FLS is involved in the synthesis of quercetin and rutin. The results showed the significant negative correlation between *FtHDA2*, *FtHDA8-2*, and four *CHS* (FtPinG0008131000.01, FtPinG0003701300.01, FtPinG0003701500.01, and FtPinG0003710800. 01), suggesting the role of HDA2/8 in the synthesis of chalcone. Furthermore, there was a significant positive correlation between *FtSRT2* and *FLS* (FtPinG0006907000. 01), *ANR* (FtPinG0007896600. 01), and *LAR* (FtPinG0000053800. 01), indicating FtSRT2 is important for flavonol and flavan synthesis. In addition, *FtHDA15* displayed a significant positive correlation with *F3'H* (FtPinG0008925900. 01), *DFR* (FtPinG0002371500. 01), and *CHI* (FtPinG0002790600. 01). From these results, it can be predicted that the *FtHDAC* gene family is important for regulating both the upstream and downstream effects of the flavonoid synthesis pathway in *Fagopyrum tataricum*.

2.5. Alternative Splicing of FtHDACs at Low Temperature Treatment

To further analyze FtHDAC transcription at low temperatures, alternative splicing events were analyzed. Previous transcriptomic data were used for analysis here [22]. The experiment included plants subjected to cold memory (4 °C for 6 h, then at room temperature for 18 h, repeated four times, and then placed at 0 °C for 6 h), cold stock (not acclimated, directly exposed to 0 °C for 6 h), and control groups with normal growth condition. Qualitative statistics analysis of alternative splicing events was performed separately for each sample using the AS profile software, and gene model was predicted by Stringtie software (transcript.gtf). The results found that many types of alternative splicing were present in the different cold treatment in Tartary buckwheat. Among these, the largest number of alternative splicing events was observed in both the TSS and the TTS (Figure S1). Further statistics of the alternative splicing of *FtHDACs* family showed different types of alternative splicing (Figure S2), and the results showed *FtHDA8-1*, *FtHDA8-2*, *FtHDA2*, and *FtHDA9* differ in Control vs. Memory and Control vs. Shock, while *FtSRT1* and *FtHDA14* differ in Control vs. Memory, but not significantly. Moreover, we analyzed whether the low-temperature responses of *FtHDACs* might instead be occurring through the differential splicing of exons. The results showed that the A5SS (alternative 5' splice site) of *FtHDA8-2* occurred under cold shock treatment (S group) ($p = 0.019$) (Figure 6A), but there was no significant difference ($p = 0.089$) under cold memory (M group) (Figure 6B), suggesting the alternative splicing forms of *FtHDAs* respond to different low-temperature conditions. The alternative splicing was confirmed by PCR analysis with *FtHDA8-2* ORF primers (Figure S8).

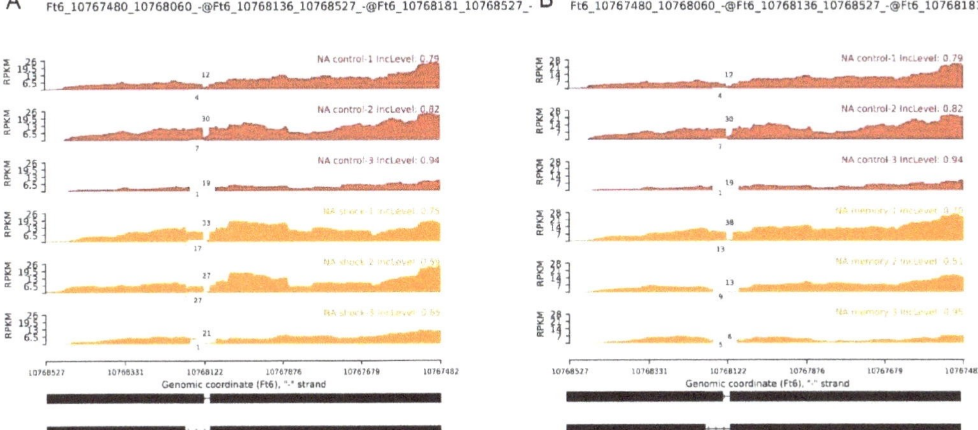

Figure 6. Alternative splicing of *FtHDA8-2* is associated with different low-temperature treatments. Sashimi plot indicating the average RNA-seq read density and splice junction counts for each genotype. (**A**) C group vs. S group. (**B**) C group vs. M group. C: living at room temperature always; S: not acclimated, directly exposed to 0 °C for 6 h; M: 4 °C for 6 h, then at room temperature for 18 h, repeated four times, and then placed at 0 °C for 6 h. C group (red) served as a negative control for S/M group (orange) enrichment.

2.6. Subcellular Localization of the FtHDACs

To determine the location of *FtHDACs* functions, the transient gene expression experiment was performed. We constructed vectors for pCAMBIA1300-FtHDA6-1-eGFP, pCAMBIA1300-FtHDA2-eGFP, and pCAMBIA1300-FtHDT2-eGFP fusion proteins, which were infiltrated into tobacco leaves with the *Agrobacterim tumefaciens* strain GV3101. Three lines of transiently transformed tobacco were obtained, including OE-*FtHDA6-1*, OE-*FtHDA2*, and OE-*FtHDT2*. Through subcellular localization and DAPI staining, the results showed that all five genes were located in the nucleus. Additionally, the light fluorescence signal was found on the cell membrane. (Figure 7).

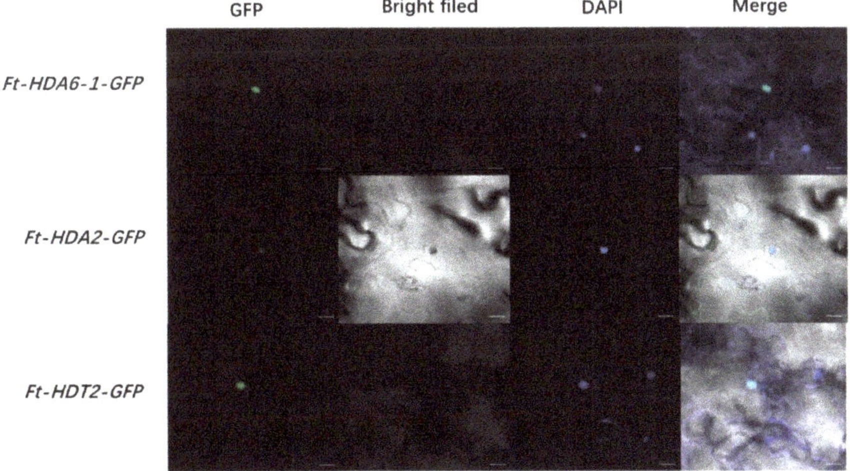

Figure 7. The subcellular localization of FtHDA6-1, FtHDA2 and FtHDT2. Bars = 20 μm.

2.7. Low-Temperature Resistance Analysis of Dingku 1

Our previous studies identified Dingku 1 as a freezing-resistant variety [21], and the role of *FtHDAC* in low-temperature resistance will be further explored here. First, the three cold test groups were established: the control group (23 °C), cold memory group (priming: 4 °C for 6 h, followed by 21 °C for 18 h, repeated for 4 days, then 0 °C for 6 h), and cold shock (0 °C for 6 h directly, without priming). The transcriptome and metabolome data of the three cold test groups were extracted and correlation analyses were performed (Figure S4 and Supplementary Material S2). From the results of the joint analysis, an interesting phenomenon was determined where the different HDACs had completely opposite correlations with the metabolites. For example, FtHDA14 and FtHDA2 exhibited a negative correlation with chloride, vitexin, anthocyanin, apigenin, and kaempferol, which is exactly the opposite of FtSRT1, FtHDA19, FtHDT2, and FtHDA8 (Figure S4A). This is also the same in the analysis of the correlation of FtHDAC with transcription factors. For instance, FtHDA14, FtHDA2, FtHDA6-1, and FtHDA9 were exactly the opposite of FtHDA8-1, FtSRT1, FtHDA19, FtHDT2, and FtHDA8-2 for the correlation with the *MYB*, *BHLH*, *NAC*, and *WRKY* family genes (Figure S4B). These results illustrated the functional differentiation of *FtHDAC* family genes, and the potential role played in cold responses and tolerance.

To further explore the role of *FtHDACs* in low-temperature responses, 2-week *Fagopyrum tataricum* Dingku 1 seedlings were treated at −4, 0, 4, 8, 12, and 16 °C for 3 h, then recovered at 23 °C for 36 h. Their fresh weight, electrical conductivity, SOD, MDA, anthocyanin, and flavonoids were measured (Figure S5). The results showed that seedlings presented more damage in −10 and −4 °C, which was reflected in the higher ion leakage and MDA, lower SOD, and fresh weight. The contents of anthocyanins and flavonoids of the Dingku 1 seedlings were the highest in the 0 °C treatment, indicating that the suitable low-temperature treatment can stimulate and accumulate flavonoid production to protect the activity of plant cells from the cold. Phenotypic analysis showed a greater effect on the growth of Tartary buckwheat below 4 °C (Figure 8A). Moreover, histone H3 acetylation levels responded to low-temperature stress, and the H3 acetylation levels of Dingku1 were measured at different freezing temperatures by Western blotting. The results showed that H3 acetylation levels increased significantly with the fall in temperature (Figure 8B), and under −4 °C treatment, the total protein was ablated and showed less content.

The expression level of *FtHDACs* was tested by RT-qPCR; cold treatment conditions were the same as described above (Figure 8C). The results showed that the expression levels of *FtSRT1*, *FtSRT2*, *FtHDA5*, *FTHDA6-1*, and *FtHDA6-2* increased significantly as the temperature decreased, indicating that these five genes are positive regulators in Dingku 1 cold responses. However, the expression levels of *FtHDA8-2* and *FtHDT1* decreased significantly as the temperature decreased, indicating that these two genes play a negative regulatory role in Dingku 1 when facing low-temperature stress.

To further investigate the function of *FtHDACs*, we constructed the overexpression lines OE-*AtHDA6* and OE-*FtHDA6-1*, and they were identified (Figure 9B,C). Subsequently, the seedlings of four genotypes were frozen at −10 °C for 2 h then recovered in room temperature for 2 days. The seedlings of *axe1-5* showed the worst tolerance in phenotypes and ion leakage (Figure 9A,D); axe1-5 (also called hda6-6) is a hda6 mutant carrying a point mutation on an HDA6 splicing site, and it is a mutant line commonly used to study the function of HDA6 [23]. The levels of mRNA in *AtDREB1A*, *AtDREB1B*, and *AtDREB1C* in the different transgenic lines were assessed. CBF (C-repeat binding transcription factor/dehydrate responsive element binding factor, DREB) is the hub of the plant CBF cold resistance pathway, which mainly regulates the expression of a large number of downstream cold resistance genes, which is extremely important for enhancing plant cold resistance ability. Its expression is also induced by other abiotic stresses, such as, drought, salinity, mechanical injury, and osmotic pressure. The results showed the expression of *CBF* increased dramatically in the cold treatment, especially in

OE-*AtHDA6* and OE-*FtHDA6-1* lines. Interestingly, the expression of *AtDREB1* reduced in *axe1-5* when compared to the other lines (Figure 9E,F), and *AtDREB1C* gave the most significant performance (Figure 9G).

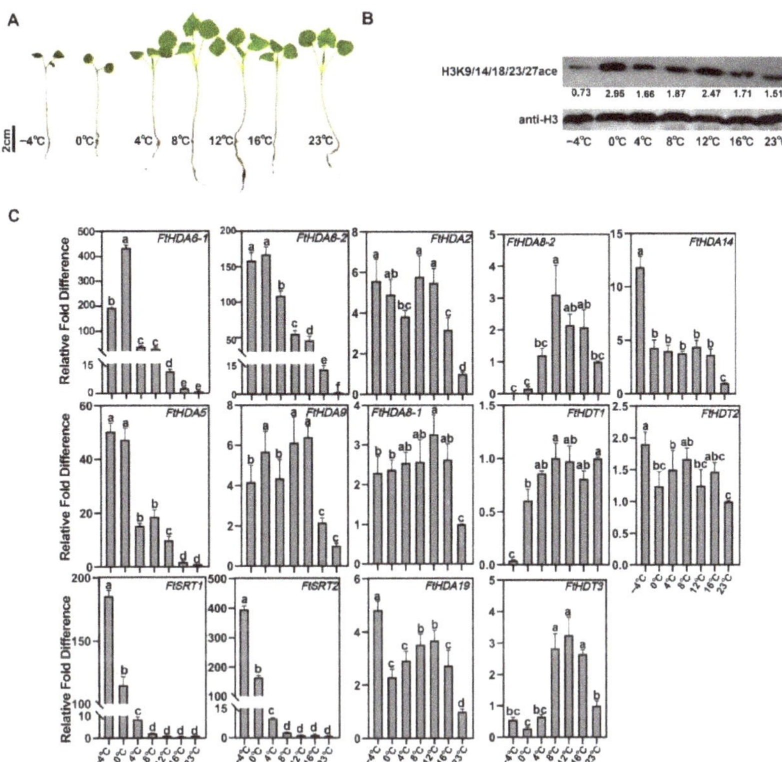

Figure 8. *FtHDACs* response to low temperature. (**A**) Phenotype of Dingku 1 displayed at different low temperatures. (**B**) The seedling of Dingku 1 treated in different low temperatures for 3 h then all recovered in 23 °C at 36 h. Western blot analysis of the change in the global histone 3 acetylation levels after low temperature treatments. Histone H3 was used as an equal loading control. All immunoblots were replicated three times for each sample from three independent experiments. (**C**) mRNA enrichment analysis of Dingku 1 in different cold treatments was performed. Data are presented as the means of three biological replicates (±SD). Different letters indicate significant differences ($p < 0.05$, Tukey's test).

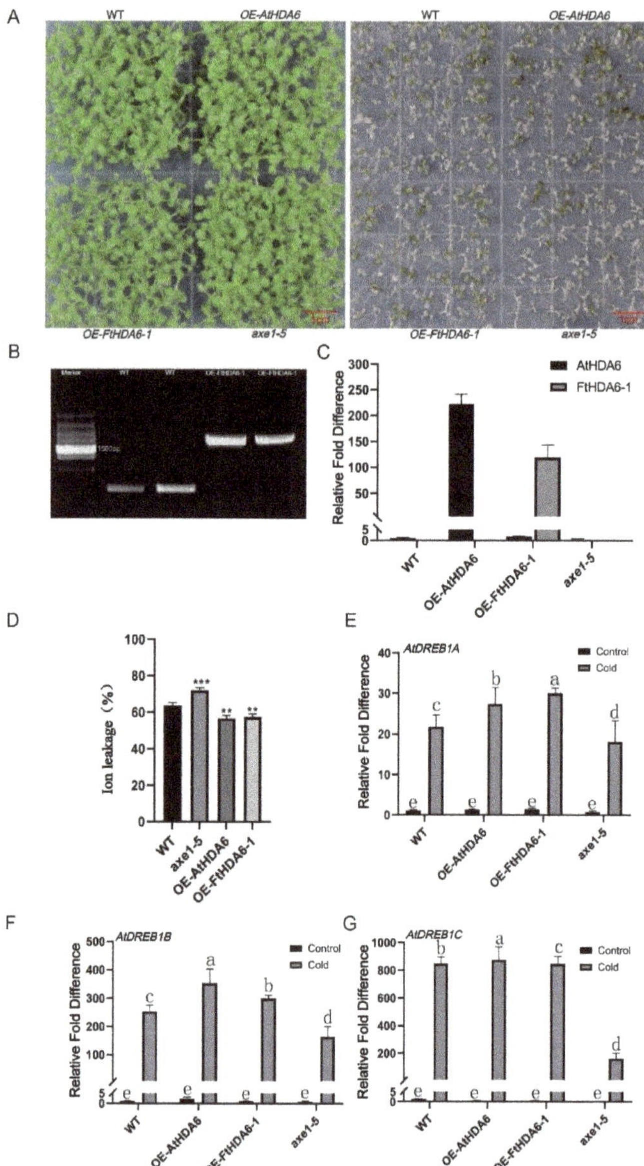

Figure 9. FtHDA6-1 positively regulates cold tolerance. (**A,D**) Phenotypes and ion leakage of seedling of WT, OE-*AtHDA6*, OE-*FtHDA6-1* and *axe1-5* under normal and cold stress conditions ($-10\ °C$ for 2 h, then recovery in $23\ °C$ for 2 days). Data are presented as the means of three biological replicates ($\pm$SD). The asterisks indicate significant differences, one-factor ANOVA (** $p < 0.01$, *** $p < 0.005$). (**B**) Identification of transgenic *Arabidopsis thaliana*. (**C**) qRT-PCR identification the expression level of *HDA6* gene in WT, *axe1-5*, OE-*FtHDA6-1* and OE-*FtHDA6* genotypes. (**E–G**) Expression level of *CBFs* in different transgenic *Arabidopsis* lines under normal (the control) and cold stress conditions ($0\ °C$ for 3 days). Data are presented as the means of three biological replicates ($\pm$SD). Different letters indicate significant differences ($p < 0.05$, Tukey's test).

3. Discussion

Epigenetic studies include DNA methylation, histone acetylation, ubiquitylation, phosphorylation, and intracellular non-coding RNA regulation. These changes in chromatin structure determine the gene expression by activating or silencing, thus adapting to the external growth environment [24]. Histone acetylation modification is jointly regulated by histone acetyltransferase (HAT) and histone deacetylase (HDAC) [25]. Histone lysine acetylation is an important chromatin modification for the epigenetic regulation of gene expression in response to environmental stress [26]. Histone deacetylation affects many growth and developmental events in plants, such as the flowering stage, embryogenesis, root hair development, abscisic acid, and salt reaction [27,28]. All histone modifications are reversible, which may provide a flexible pathway to for regulating gene expression during plant development and in response to environmental stimuli.

HDAC has been isolated from plants including *Arabidopsis*, rice, maize, soybean, cotton and potato. *Arabidopsis* contains 18 HDACs, which can be divided into three families: RPD3/HDA1, SIR2, and plant-specific HD2 [29]. Among them, the RPD Type 3 HDAC serves to maintain chromatin states and regulate housekeeping gene activity in yeast, *Drosophila*, *elegans*, and metazoans [30]. Members of the RPD3/HDA1 family can be further divided into three categories [31]: Class I, including HAD6, HAD7, HAD9, and HAD19; Class II Group HAD2; Class III, which contains HAD5, HAD15, and HAD18. The others were HAD8, HAD14, HAD10, and HAD17. The HD2 subfamily can be divided into HD2A, HD2B, HD2C, and HD2D [32]. The SIR2 family HDAC is a nicotinamide adenine dinucleotide (NAD)-dependent HDAC, and has two members of SIR2-like HDACs, SIR1, and SIR2 [33]. HDACs are found to localize to membranes, nuclei, or nucleoplasmic shuttling, with different functions depending on their localization. RPD3/HDA1 is the largest subfamily in the HDACs family, and the family depends on Zn^{2+}. Members of this family all contain a typical histone deacetylase domain [34]. Its structural analysis found that the HDAC structure of the family members is highly conserved, while the other parts are poorly conserved. Therefore, it may be the main reason for the functional differences between different members of the same family, and the protein activity of this family member can be inhibited by triostatin (TSA).

Here, a total of 14 *FtHDAC* family genes were retrieved from the Tartary buckwheat genome through HDACs conserved domains in *Arabidopsis* (Table 1). It is also composed of three subfamilies, including nine RPD3/HDA1 subfamily genes, two SIR2 subfamily genes, and three HD2 subfamily genes (Figure 1A). The characterization of the MEME motif shows that it conforms to the basic structural compositions of the *HDAC* family genes in plants. The prediction of *cis*-acting elements showed that the *FtHDAC* gene promoters contained hormone responsive elements such as ABA and methyl Jasmonate, light responsive elements, drought responsive elements, and low-temperature responsive elements, suggesting that *FtHDAC* plays an important role in the growth and development of Tartary buckwheat and in coping with various environmental changes (Figure 2). This conclusion is also supported by previous reports in which *HDACs* play a critical role in regulating abiotic stress responses. For example, histone acetylation changes in plant responses to drought stress [35]. *Arabidopsis HDA6* is required for freezing resistance [36] and salt stress [28,37].

The construction of interspecific collinearity can help researchers to better understand the evolutionary relationship between FtHDACs and the HDACs of other plant species (Figures 3 and 4). The secondary structure of FtHDACs is helpful for understanding the mechanism of its actions (Figure 1B). A tissue expression analysis showed that the expression level of the three FtHDT2 subfamily members is relatively higher than other family members during normal development (Figure 5A), especially in the flowers, roots, and stems. It was inferred that these genes in the same clade with similar expression patterns might play similar roles in physiological processes. *AhHDA19* was specifically expressed in the root and stem, and *FtHD19* in Tartary buckwheat is greatly expressed in the root and stem, showing a similar expression pattern, implying they might execute functions dominantly in the root and stem. Furthermore, the *FtHDA2, FtHDA9, FTHDA6-1, FTHDA8-1, FtHDA5,* and *FtRPD3/HDA1* subfamily members were highly expressed in the roots, suggesting that these genes play an important role in root morphogenesis. Correlation analysis revealed the potential relationship between *FtHDACs* and flavonoids synthesis genes (Figure 5B). Seven types of alternative splicing have been observed in both humans and rice (Oryza sativa), including intron retention, exon skipping, mutually exclusive exon, alternative 5′ splicing, alternative 3′ splicing, alternative first exons and alternative last exons [38]. In this study, the type of HDA8-2 alternative splicing is A5SS (alternative 5′ splice site) (Figure 6). Alternative splicing is a possible "molecular temperature" that allows plants to quickly adjust the abundance of functional transcripts to adapt to environmental perturbations [39,40]. A preliminary analysis of subcellular localization was performed, and the results revealed the nucleus localization of FtHDACs (Figure 7). The subcellular localization of target genes occurs through transient transformation. Transient gene expression is an effective experimental tool for research on plant gene function [41]. Figure 7 shows the fluorescent signal that appeared on the membrane, but the function is unknown. *Arabidopsis* HDA6 does have the characteristics of cytoplasmic localization; for example, HDA6 interacts with FLD at the nuclear periphery [42]. BIN2 interacts with HDA6 in the cytoplasm and nucleus [43]. The following low-temperature responses of *FtHDACs* have been discussed. Firstly, phenotype and histone acetylation of the Dingku1 treated at different low temperatures were determined, and the expression of *FtHDACs* was also analyzed in at different low temperatures. Interestingly, the result showed that 16 °C was the optimal growth temperature for Tartary buckwheat. Moreover, the low temperature caused significant changes in the expression of *FtHDACs* when compared to room temperature (Figure 8). A stable transgenic *Arabidopsis* line of *FtHDA6* was constructed to further explore the freezing function of FtHDAC. The results revealed that FtHDA6 showed a trend to improve freezing resistance (Figure 9). Previous studies have conducted a systematic analysis of the cooling tolerance mechanism in Tartary buckwheat [1], and our study fills the gap in the mechanism of epigenetic regulation.

Histone acetylation plays a key role in plant development and the response to various environmental stimuli by regulating gene transcription. It was revealed that Tartary buckwheat HDACs could be classified into three major subgroups: RPD3/HDA1, HD2-like, and SRT, which is similar to *Arabidopsis*. Moreover, FtHDACs also carried the functional catalytic domains and other conserved domains, as well as motifs similar to their counterparts in *Arabidopsis*. The function of FtHDA6 in low-temperature responses and flavonoid synthesis pathways was predicted. In brief, the present study highlights the implication of Tartary buckwheat HDACs in various developmental processes and low temperature stress adaptation. In addition, this study also highlights the potential role of Tartary buckwheat HDACs in flavonoid synthesis pathways. This study provides motivation for the investigation of the biological and cellular functions of histone acetylation, which will eventually lead to the long-term improvement of agronomic characteristics and abiotic stress tolerance in *Fagopyrum tataricum*.

4. Materials and Methods

4.1. Plant Growth, Cold Treatments, and Tissue Collection

Tartary buckwheat seeds (Dingku 1) were provided from the Qinghai Academy of Animal Science and Veterinary Medicine of Qinghai University (Qinghai, Xining, China). Tartary buckwheat and tobacco seedlings were cultured in greenhouses according to [21]. After 4–8 h of soaking the seeds in ddH$_2$O, the seeds were disinfected using a 15% NaClO solution and then placed in a culture dish with two layers of gauze. The culture dish was moved to a greenhouse and cultured until germination. For the cold stress experiment, 2-week-old seedlings were treated in −4, 0, 4, 8, 12, and 16 °C for 3 h, and were all recovered in 23 °C at 36 h. The cold treatment conditions of the transcriptome was mentioned according to [21]. The seedling of the cold memory group (memory) was kept at 4 °C for 6 h, then at room temperature for 18 h, repeated four times, and then placed at 0 °C for 6 h.

The cold stock (not acclimated) was directly exposed to 0 °C for 6 h, and control groups experienced normal growth conditions. The leaves and roots of the samples were collected, immediately frozen in liquid nitrogen, and stored at −80 °C.

4.2. Genome-Wide Identification of Fagopyrum Tataricum HDACs Genes

The genome sequence of the Tartary buckwheat genome was downloaded from the Tartary Buckwheat Genome Project (http://mbkbase.org/Pinku1/) (accessed on 1 March 2020) [44]. Amino acid sequences of the *Arabidopsis HDAC* family genes were downloaded from the TAIR website (https://www.arabidopsis.org/) (accessed on 1 March 2020), then used as queries in local BLASTP against the Tartary buckwheat genome (e-value = 1×10^{-10}). Furthermore, SMART (http://smart.embl-heidelberg.de/) (accessed on 1 March 2020) and HMMER (https://www.ebi.ac.uk/Tools/hmmer/search/phmmer) (accessed on 1 March 2020) were used to confirm the presence of the HDAC domain. The physicochemical properties of the *FtHDACs* genes were predicted by the ExPASy website (https://web.expasy.org/compute_pi/) (accessed on 1 March 2020), including the protein size, molecular weight (MW), isoelectric point (p*I*), and aliphatic and GRAVY index. In addition, the gene structure was visualized by GSDS2. 0 (https://gsds.gao-lab.org/) (accessed on 1 March 2020). The genome sequences and amino acid sequence of the *FtHDACs* are presented in Supplementary Material S1.

4.3. Conserved Protein Structure, Cis-Acting Element Prediction, and Protein 3D Structure Analysis

The conserved motifs of the FtHDACs protein sequences were analyzed via the online Multiple Expectation Maximization for Motif Elicitation (MEME) version 4.11.1 (http://meme-suite.org/tools/meme) (accessed on 1 March 2020) [45], and the maximum number of motifs was set to 18. The 1200 bp upstream genomic DNA sequences were analyzed in the PlantCARE (http://bioinformatics.psb.ugent.be/webtools/plantcare/html/) (accessed on 1 March 2020) database for *cis*-acting element prediction. The data was visualized by TBtools [46]. The secondary protein structure was performed by PRABI (http://www.prabi.fr/) (accessed on 1 March 2020). An automated protein structure building was conducted by the Robetta (https://robetta.bakerlab.org/) (accessed on 1 March 2020) program [47]; the data are showed in Table S3.

4.4. Phylogenetic Analysis, Genome Distribution, and Gene Duplication

The phylogenetic tree of the FtHDAC protein family was constructed in the Neighbor-Joining method via the MEGA7.0 software with 1000 replicated bootstrap values [48,49] and the p-distance and pairwise gap deletion parameters engaged. The chromosomal distribution of *FtHDACs* was built by the TBtools and Itol (https://itol.embl.de/) (accessed on 1 March 2020) [46]. The parameters (Ks-synonymous substitution rate and Ka-nonsynonymous substitution rate) of the duplication events were computed by TBtools. Amino acid sequences used for phylogenetic analysis are listed in Supplementary Material S1.

4.5. Alternative Splicing Analysis

ASprofile software was used to perform qualitative analysis statistics of alternative splicing events for each sample on the gene model (transcript.gtf), predicted by Stringtie. We performed the alternative splicing events analysis based on the gene structure annotation information of Tartary buckwheat [22]. The rMATS is a computational tool for detecting differential alternative splicing events from RNA-Seq data. Based on RNA-Seq data, rMATS can automatically detect and analyze alternative splicing events corresponding to all major types of alternative splicing patterns [50,51].

4.6. Transcriptome and RT-qPCR Analysis

The raw data RNA-seq of *FtHDACs* in different tissues (root, stem, leaf, flower, and three-stage fruit) were retrieved from the Tartary Buckwheat Database (TBD) (http://

shujuku.zuotukeji.net/) (accessed on 1 March 2020) (accessed on 1 March 2020) [52,53]. The correlation analysis was visualized via OmicStudio software and Tbtools; the data are shown in Table S4. The heat map was generated using OmicStudio software (www. omicstudio.cn) (accessed on 1 March 2020). For quantificational real-time PCR, the total RNA was extracted using the RNA prep Pure Plant Plus Kit (Tiangen Biotech, Beijing, China). Then, 2 μg RNA was used for the first strand of cDNA synthesis using reverse transcriptase (Vazyme, R211-02, Nanjing, China). Real-time PCR amplification was carried out with the Bio-Rad CFX96 system using SYBR Green I (Vazyme, Q711-02, Nanjing, China). The reaction system contained 10 μL SYBR Master Mix buffer, 0.4 μL each of the primers (10 μM), 1 μL of template, and 8 μL ddH$_2$O. The thermal profile for qRT-PCR was as follows: pre-denaturation at 95 °C for 5 min; cycling stage at 95 °C for 10 s, 60 °C for 30 s, 72 °C for 15 s, 40 cycles; melting stage at 95 °C for 15 s, 60 °C for 1 min, 95 °C for 15 s. Three independent biological replicates were used in the analysis and the $2^{-(\Delta\Delta Ct)}$ method was applied for the analysis gene expression. Here, *FtH3* was used as a reference gene to normalize the expression level. The RT-qPCR primer sequences used in this paper are all listed in Table S1.

4.7. Western Blot Assays

The tested seedling samples were ground, the total proteins were extracted, and specific proteins were detected as described [54]. Histone H3 was used as an equal loading control. The antibody in this study used in Western blotting was anti-H3K9K14K18K23K27ac (ab47915, Abcam, Lot: GR137984-20, Cambridge, UK).

4.8. Construction of the Arabidopsis Transgenic Plants and Low Temperature Treatments

The sequence of HDA6-1 CDs was amplified and inserted into the pCAMBIA1300-eGFP vector using a ClonExpress II One Step Cloning kit (Vazyme, C112-02) to generate pCAMBIA1300-FtHDA6-1-eGFP constructs. The true recombinant plasmid was transformed into *Agrobacterium tumefaciens* GV3101, then it was transferred into wild-type *Arabidopsis* by the floral-dip method. The positive transgenic plants were obtained by resistance screening (the selectable marker gene is hygromycin) and quantificational real-time PCR identification. The target protein has been proven in transgenic lines by confocal fluorescence microscopy (ZEISS/LSM880) (Figure S7) [55,56]. The genomic DNA of *Arabidopsis* was isolated from leaves using the Plant Genomic DNA Extraction Kit (TiangenP5008, China). For phenotypes and ion leakage analysis of low-temperature stress experiments in transgenic plants, 2-week-old transgenic seedlings were treated in −10 °C for 2 h, and were recovered in 23 °C at 3 days. For *CBF* gene expressional analysis of low-temperature stress experiments in transgenic plants, 2-week-old transgenic seedlings were treated in 0 °C for 3 days, and total RNA was extracted, cDNA was reverse-transcribed, and subjected to the qPCR assay.

5. Conclusions

We performed a comprehensive analysis of the HDAC gene family in Tartary buckwheat. A total of 14 *FtHDAC* genes, including nine *FtRPD3/HDA1*, three *FtHD2s*, and two *FtSIR2s*, were genome-wide identified. The gene structure, chromosomal distribution, motif prediction, *cis*-acting element prediction, phylogenetic relationship, 3D structure, expression patterns, alternative splicing, subcellular localization, and heterologous expression were conducted and analyzed. Moreover, the expression pattern of *FtHDACs* was shown in various tissues/organs to be involved in the developmental process. The *cis*-acting element prediction and RNA-seq data indicated that the *FtHDAC* is involved in low-temperature stress responses and flavonoids' synthesis. This conclusion was also validated in FtHDA6 transgenic *Arabidopsis thaliana*, which indeed affects plant freezing tolerance. According to the above results, the potential mechanism for the roles of *FtHDACs* in the regulation of flavonoids at low temperatures in Tartary buckwheat was proposed (Figure 10). We first collated and analyzed the *HDACs* in Tartary buckwheat and attempts to discover the biological function of *FtHDACs*. This result implies that chromatin regulations are important for low-temperature tolerance and the flavonoid synthesis of Tartary buckwheat.

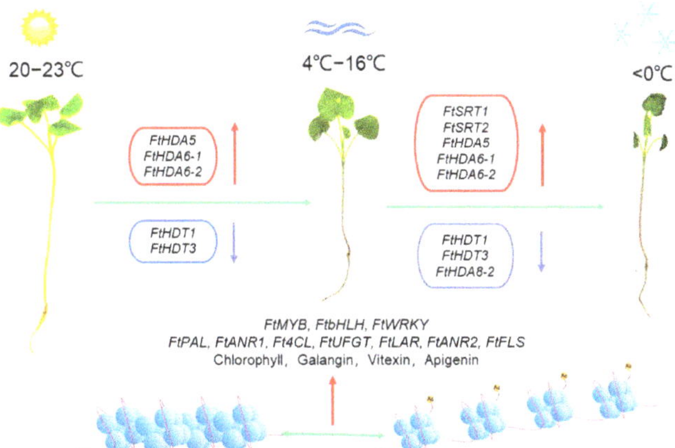

Figure 10. The proposed mechanism for the roles of *FtHDACs* in the regulation of low temperature and flavonoids biosynthesis in Tartary buckwheat. The level of histone acetylation affects chromatin conformation, and chromatin contraction and loosening regulate gene expression in response to low temperature and accumulation of flavonoids in Tartary buckwheat.

Supplementary Materials: The following supporting information can be downloaded at: https://www.mdpi.com/article/10.3390/ijms23147622/s1.

Author Contributions: Y.H.: Conceptualization, Methodology, Data curation, Writing—original draft. Q.L.: Conceptualization, Methodology. J.S.: Methodology. X.J.: Methodology. C.J.: Methodology. L.A.: Supervision. Y.T.: Supervision, Project administration. Y.S.: Conceptualization, Writing—review & editing, Supervision, Project administration. All authors have read and agreed to the published version of the manuscript.

Funding: This work was supported by Key Laboratory of Superior Forage Germplasm in the Qinghai-Tibetan Plateau, Qinghai (2020-ZJ-Y03); The National Natural Science Foundation of China (No. 31872682, No. 32101244); Earmarked Fund for China Agriculture Research System (CARS-07-G14).

Data Availability Statement: The transcriptome data have been deposited in National Center for Biotechnology Information's Gene Expression Omnibus and are accessible through GSE138546.

Acknowledgments: We thank Core Facility of School of Life Science (Lanzhou University) for technical support.

Conflicts of Interest: The authors have no conflict of interest to declare.

References

1. Song, Y.J.; Feng, J.C.; Liu, D.M.; Long, C.L. Different Phenylalanine Pathway Responses to Cold Stress Based on Metabolomics and Transcriptomics in Tartary Buckwheat Landraces. *J. Agric. Food Chem.* **2022**, *70*, 687–698. [CrossRef] [PubMed]
2. He, Y.H.; Li, Z.C. Epigenetic Environmental Memories in Plants: Establishment, Maintenance, and Reprogramming. *Trends Genet.* **2018**, *34*, 856–866. [CrossRef]
3. Selma, S.; Orzaez, D. Perspectives for epigenetic editing in crops. *Transgenic Res.* **2021**, *30*, 381–400. [CrossRef]
4. Tahir, M.S.; Tian, L. HD2-type histone deacetylases: Unique regulators of plant development and stress responses. *Plant Cell Rep.* **2021**, *40*, 1603–1615. [CrossRef]
5. Ekwall, K. Genome-wide analysis of HDAC function. *Trends Genet.* **2005**, *21*, 608–615. [CrossRef] [PubMed]
6. Bourque, S.; Jeandroz, S.; Grandperret, V.; Lehotai, N.; Aime, S.; Soltis, D.E.; Miles, N.W.; Melkonian, M.; Deyholos, M.K.; Leebens-Mack, J.H.; et al. The Evolution of HD2 Proteins in Green Plants. *Trends Plant Sci.* **2016**, *21*, 1008–1016. [CrossRef]

7. Luo, M.; Cheng, K.; Xu, Y.; Yang, S.; Wu, K. Plant Responses to Abiotic Stress Regulated by Histone Deacetylases. *Front. Plant Sci.* **2017**, *8*, 2147. [CrossRef]
8. Peterson, C.L.; Laniel, M.A. Histones and histone modifications. *Curr. Biol.* **2004**, *14*, R546–R551. [CrossRef] [PubMed]
9. Chu, J.; Chen, Z. Molecular identification of histone acetyltransferases and deacetylases in lower plant Marchantia polymorpha. *Plant Physiol. Biochem.* **2018**, *132*, 612–622. [CrossRef]
10. Hollender, C.; Liu, Z. Histone deacetylase genes in Arabidopsis development. *J. Integr. Plant Biol.* **2008**, *50*, 875–885. [CrossRef]
11. Zhang, K.; Yu, L.; Pang, X.; Cao, H.; Si, H.; Zang, J.; Xing, J.; Dong, J. In silico analysis of maize HDACs with an emphasis on their response to biotic and abiotic stresses. *PeerJ* **2020**, *8*, e8539. [CrossRef] [PubMed]
12. Hu, Y.; Qin, F.; Huang, L.; Sun, Q.; Li, C.; Zhao, Y.; Zhou, D.X. Rice histone deacetylase genes display specific expression patterns and developmental functions. *Biochem. Biophys. Res. Commun.* **2009**, *388*, 266–271. [CrossRef] [PubMed]
13. Hou, J.; Ren, R.; Xiao, H.; Chen, Z.; Yu, J.; Zhang, H.; Shi, Q.; Hou, H.; He, S.; Li, L. Characteristic and evolution of HAT and HDAC genes in Gramineae genomes and their expression analysis under diverse stress in Oryza sativa. *Planta* **2021**, *253*, 72. [CrossRef] [PubMed]
14. Yuan, L.; Dai, H.; Zheng, S.; Huang, R.; Tong, H. Genome-wide identification of the HDAC family proteins and functional characterization of CsHD2C, a HD2-type histone deacetylase gene in tea plant (*Camellia sinensis* L. O. Kuntze). *Plant Physiol. Biochem.* **2020**, *155*, 898–913. [CrossRef]
15. Imran, M.; Shafiq, S.; Naeem, M.K.; Widemann, E.; Munir, M.Z.; Jensen, K.B.; Wang, R.R. Histone Deacetylase (HDAC) Gene Family in Allotetraploid Cotton and Its Diploid Progenitors: In Silico Identification, Molecular Characterization, and Gene Expression Analysis under Multiple Abiotic Stresses, DNA Damage and Phytohormone Treatments. *Int. J. Mol. Sci.* **2020**, *21*, 321. [CrossRef]
16. Liu, C.; Li, L.C.; Chen, W.Q.; Chen, X.; Xu, Z.H.; Bai, S.N. HDA18 affects cell fate in Arabidopsis root epidermis via histone acetylation at four kinase genes. *Plant Cell* **2013**, *25*, 257–269. [CrossRef]
17. Tang, Y.; Liu, X.; Liu, X.; Li, Y.; Wu, K.; Hou, X. Arabidopsis NF-YCs Mediate the Light-Controlled Hypocotyl Elongation via Modulating Histone Acetylation. *Mol. Plant* **2017**, *10*, 260–273. [CrossRef]
18. Liew, L.C.; Singh, M.B.; Bhalla, P.L. An RNA-seq transcriptome analysis of histone modifiers and RNA silencing genes in soybean during floral initiation process. *PLoS ONE* **2013**, *8*, e77502. [CrossRef]
19. Luo, M.; Tai, R.; Yu, C.W.; Yang, S.; Chen, C.Y.; Lin, W.D.; Schmidt, W.; Wu, K. Regulation of flowering time by the histone deacetylase HDA5 in Arabidopsis. *Plant J.* **2015**, *82*, 925–936. [CrossRef]
20. Chen, X.; Lu, L.; Mayer, K.S.; Scalf, M.; Qian, S.; Lomax, A.; Smith, L.M.; Zhong, X. POWERDRESS interacts with HISTONE DEACETYLASE 9 to promote aging in Arabidopsis. *eLife* **2016**, *5*, e17214. [CrossRef]
21. Yao, Y.J.; Sun, L.; Wu, W.J.; Wang, S.; Xiao, X.; Hu, M.L.; Li, C.L.; Zhao, H.X.; Chen, H.; Wu, Q. Genome-Wide Investigation of Major Enzyme-Encoding Genes in the Flavonoid Metabolic Pathway in Tartary Buckwheat (Fagopyrum tataricum). *J. Mol. Evol.* **2021**, *89*, 269–286. [CrossRef] [PubMed]
22. Song, Y.; Jia, Z.; Hou, Y.; Ma, X.; Li, L.; Jin, X.; An, L. Roles of DNA Methylation in Cold Priming in Tartary Buckwheat. *Front Plant Sci.* **2020**, *11*, 608540. [CrossRef] [PubMed]
23. Murfett, J.; Wang, X.J.; Hagen, G.; Guilfoyle, T.J. Identification of arabidopsis histone deacetylase HDA6 mutants that affect transgene expression. *Plant Cell* **2001**, *13*, 1047–1061. [CrossRef] [PubMed]
24. Chang, Y.N.; Zhu, C.; Jiang, J.; Zhang, H.M.; Zhu, J.K.; Duan, C.G. Epigenetic regulation in plant abiotic stress responses. *J. Integr. Plant Biol.* **2020**, *62*, 563–580. [CrossRef] [PubMed]
25. Wakeel, A.; Ali, I.; Khan, A.R.; Wu, M.J.; Upreti, S.; Liu, D.D.; Liu, B.H.; Gan, Y.B. Involvement of histone acetylation and deacetylation in regulating auxin responses and associated phenotypic changes in plants. *Plant Cell Rep.* **2018**, *37*, 51–59. [CrossRef] [PubMed]
26. Hu, Y.F.; Lu, Y.; Zhao, Y.; Zhou, D.X. Histone Acetylation Dynamics Integrates Metabolic Activity to Regulate Plant Response to Stress. *Front. Plant Sci.* **2019**, *10*, 1236. [CrossRef] [PubMed]
27. Liu, X.C.; Yang, S.G.; Zhao, M.L.; Luo, M.; Yu, C.W.; Chen, C.Y.; Tai, R.; Wu, K.Q. Transcriptional Repression by Histone Deacetylases in Plants. *Mol. Plant* **2014**, *7*, 764–772. [CrossRef]
28. Chen, L.T.; Luo, M.; Wang, Y.Y.; Wu, K.Q. Involvement of Arabidopsis histone deacetylase HDA6 in ABA and salt stress response. *J. Exp. Bot.* **2010**, *61*, 3345–3353. [CrossRef]
29. Pandey, R.; Muller, A.; Napoli, C.A.; Selinger, D.A.; Pikaard, C.S.; Richards, E.J.; Bender, J.; Mount, D.W.; Jorgensen, R.A. Analysis of histone acetyltransferase and histone deacetylase families of Arabidopsis thaliana suggests functional diversification of chromatin modification among multicellular eukaryotes. *Nucleic Acids Res.* **2002**, *30*, 5036–5055. [CrossRef]
30. Yang, X.J.; Seto, E. The Rpd3/Hda1 family of lysine deacetylases: From bacteria and yeast to mice and men. *Nat. Rev. Mol. Cell Biol.* **2008**, *9*, 206–218. [CrossRef]
31. Alinsug, M.V.; Yu, C.W.; Wu, K.Q. Phylogenetic analysis, subcellular localization, and expression patterns of RPD3/HDA1 family histone deacetylases in plants. *BMC Plant Biol.* **2009**, *9*, 1–15. [CrossRef] [PubMed]
32. Han, Z.F.; Yu, H.M.; Zhao, Z.; Hunter, D.; Luo, X.J.; Duan, J.; Tian, L.N. AtHD2D Gene Plays a Role in Plant Growth, Development, and Response to Abiotic Stresses in Arabidopsis thaliana. *Front. Plant Sci.* **2016**, *7*, 310. [CrossRef] [PubMed]

33. Yu, C.W.; Tai, R.; Wang, S.C.; Yang, P.; Luo, M.; Yang, S.G.; Cheng, K.; Wang, W.C.; Cheng, Y.S.; Wu, K.Q. HISTONE DEACETY-LASE6 Acts in Concert with Histone Methyltransferases SUVH4, SUVH5, and SUVH6 to Regulate Transposon Silencing. *Plant Cell* **2017**, *29*, 1970–1983. [CrossRef] [PubMed]

34. De Ruijter, A.J.M.; Van Gennip, A.H.; Caron, H.N.; Kemp, S.; Van Kuilenburg, A.B.P. Histone deacetylases (HDACs): Characterization of the classical HDAC family. *Biochem. J.* **2003**, *370*, 737–749. [CrossRef] [PubMed]

35. Li, S.; He, X.; Gao, Y.; Zhou, C.G.; Chiang, V.L.; Li, W. Histone Acetylation Changes in Plant Response to Drought Stress. *Genes* **2021**, *12*, 1409. [CrossRef]

36. To, T.K.; Nakaminami, K.; Kim, J.M.; Morosawa, T.; Ishida, J.; Tanaka, M.; Yokoyama, S.; Shinozaki, K.; Seki, M. Arabidopsis HDA6 is required for freezing tolerance. *Biochem. Bioph. Res. Commun.* **2011**, *406*, 414–419. [CrossRef]

37. Luo, M.; Wang, Y.Y.; Liu, X.C.; Yang, S.G.; Lu, Q.; Cui, Y.H.; Wu, K.Q. HD2C interacts with HDA6 and is involved in ABA and salt stress response in Arabidopsis. *J. Exp. Bot.* **2012**, *63*, 3297–3306. [CrossRef]

38. Barbazuk, W.B.; Fu, Y.; McGinnis, K.M. Genome-wide analyses of alternative splicing in plants: Opportunities and challenges. *Genome Res.* **2008**, *18*, 1381–1392. [CrossRef]

39. Capovilla, G.; Delhomme, N.; Collani, S.; Shutava, I.; Bezrukov, I.; Symeonidi, E.; Amorim, M.D.; Laubinger, S.; Schmid, M. PORCUPINE regulates development in response to temperature through alternative splicing. *Nat. Plants* **2018**, *4*, 534–539. [CrossRef]

40. Liu, X.X.; Guo, Q.H.; Xu, W.B.; Liu, P.; Yan, K. Rapid Regulation of Alternative Splicing in Response to Environmental Stresses. *Front. Plant Sci.* **2022**, *13*, 832177. [CrossRef]

41. Tyurin, A.A.; Suhorukova, A.V.; Kabardaeva, K.V.; Goldenkova-Pavlova, I.V. Transient Gene Expression is an Effective Experimental Tool for the Research into the Fine Mechanisms of Plant Gene Function: Advantages, Limitations, and Solutions. *Plants* **2020**, *9*, 1187. [CrossRef] [PubMed]

42. Yu, C.W.; Liu, X.C.; Luo, M.; Chen, C.Y.; Lin, X.D.; Tian, G.; Lu, Q.; Cui, Y.H.; Wu, K.Q. HISTONE DEACETYLASE6 Interacts with FLOWERING LOCUS D and Regulates Flowering in Arabidopsis. *Plant Physiol.* **2011**, *156*, 173–184. [CrossRef] [PubMed]

43. Hao, Y.H.; Wang, H.J.; Qiao, S.L.; Leng, L.N.; Wang, X.L. Histone deacetylase HDA6 enhances brassinosteroid signaling by inhibiting the BIN2 kinase. *Proc. Natl. Acad. Sci. USA* **2016**, *113*, 10418–10423. [CrossRef] [PubMed]

44. Zhang, L.J.; Li, X.X.; Ma, B.; Gao, Q.; Du, H.L.; Han, Y.H.; Li, Y.; Cao, Y.H.; Qi, M.; Zhu, Y.X.; et al. The Tartary Buckwheat Genome Provides Insights into Rutin Biosynthesis and Abiotic Stress Tolerance. *Mol. Plant* **2017**, *10*, 1224–1237. [CrossRef] [PubMed]

45. Bailey, T.L.; Williams, N.; Misleh, C.; Li, W.W. MEME: Discovering and analyzing DNA and protein sequence motifs. *Nucleic Acids Res.* **2006**, *34*, W369–W373. [CrossRef] [PubMed]

46. Chen, C.J.; Chen, H.; Zhang, Y.; Thomas, H.R.; Frank, M.H.; He, Y.H.; Xia, R. TBtools: An Integrative Toolkit Developed for Interactive Analyses of Big Biological Data. *Mol. Plant* **2020**, *13*, 1194–1202. [CrossRef]

47. Pennisi, E. Protein structure prediction now easier, faster. *Science* **2021**, *373*, 262–263. [CrossRef]

48. Ma, X.J.; Zhang, C.; Zhang, B.; Yang, C.P.; Li, S.J. Identification of genes regulated by histone acetylation during root development in Populus trichocarpa. *BMC Genom.* **2016**, *17*, 1–17. [CrossRef]

49. Zhao, L.M.; Lu, J.X.; Zhang, J.X.; Wu, P.Y.; Yang, S.G.; Wu, K.Q. Identification and characterization of histone deacetylases in tomato (Solanum lycopersicum). *Front. Plant Sci.* **2015**, *5*, 760. [CrossRef]

50. Fei, T.; Chen, Y.W.; Xiao, T.F.; Li, W.; Cato, L.; Zhang, P.; Cotter, M.B.; Bowden, M.; Lis, R.T.; Zhao, S.G.; et al. Genome-wide CRISPR screen identifies HNRNPL as a prostate cancer dependency regulating RNA splicing. *Proc. Natl. Acad. Sci. USA* **2017**, *114*, E5207–E5215. [CrossRef]

51. Park, E.; Pan, Z.C.; Zhang, Z.J.; Lin, L.; Xing, Y. The Expanding Landscape of Alternative Splicing Variation in Human Populations. *Am. J. Hum. Genet.* **2018**, *102*, 11–26. [CrossRef] [PubMed]

52. Liu, M.Y.; Sun, W.J.; Ma, Z.T.; Hu, Y.; Chen, H. Tartary buckwheat database (TBD): An integrative platform for gene analysis of and biological information on Tartary buckwheat. *J. Zhejiang Univ.-Sic. B* **2021**, *22*, 954–958. [CrossRef] [PubMed]

53. Liu, M.Y.; Ma, Z.T.; Zheng, T.R.; Sun, W.J.; Zhang, Y.J.; Jin, W.Q.; Zhan, J.Y.; Cai, Y.T.; Tang, Y.J.; Wu, Q.; et al. Insights into the correlation between Physiological changes in and seed development of tartary buckwheat (Fagopyrum tataricum Gaertn.). *BMC Genom.* **2018**, *19*, 1–20. [CrossRef] [PubMed]

54. Zhao, L.; Wang, P.; Yan, S.H.; Gao, F.; Li, H.; Hou, H.L.; Zhang, Q.; Tan, J.J.; Li, L.J. Promoter-associated histone acetylation is involved in the osmotic stress-induced transcriptional regulation of the maize ZmDREB2A gene. *Physiol. Plant.* **2014**, *151*, 459–467. [CrossRef]

55. Clough, S.J.; Bent, A.F. Floral dip: A simplified method for Agrobacterium-mediated transformation of Arabidopsis thaliana. *Plant J.* **1998**, *16*, 735–743. [CrossRef]

56. Wang, H.M.; Li, K.P.; Sun, X.M.; Xie, Y.H.; Han, X.M.; Zhang, S.G. Isolation and characterization of larch BABY BOOM2 and its regulation of adventitious root development. *Gene* **2019**, *690*, 90–98. [CrossRef]

International Journal of
Molecular Sciences

Overexpression of the *Arabidopsis* MACPF Protein AtMACP2 Promotes Pathogen Resistance by Activating SA Signaling

Xue Zhang [1], Yang-Shuo Dai [1,2], Yu-Xin Wang [1], Ze-Zhuo Su [1,3,4], Lu-Jun Yu [1], Zhen-Fei Zhang [2], Shi Xiao [1] and Qin-Fang Chen [1,*]

[1] State Key Laboratory of Biocontrol, Guangdong Provincial Key Laboratory of Plant Resources, School of Life Sciences, Sun Yat-sen University, Guangzhou 510275, China
[2] Plant Protection Research Institute Guangdong Academy of Agricultural Sciences, Guangzhou 510640, China
[3] School of Biomedical Sciences, Li Ka Shing Faculty of Medicine, The University of Hong Kong, Pokfulam, Hong Kong SAR, China
[4] Laboratory of Data Discovery for Health Limited (D24H), Hong Kong Science Park, Hong Kong SAR, China
[*] Correspondence: chenqf3@mail.sysu.edu.cn

Abstract: Immune response in plants is tightly regulated by the coordination of the cell surface and intracellular receptors. In animals, the membrane attack complex/perforin-like (MACPF) protein superfamily creates oligomeric pore structures on the cell surface during pathogen infection. However, the function and molecular mechanism of MACPF proteins in plant pathogen responses remain largely unclear. In this study, we identified an Arabidopsis MACP2 and investigated the responsiveness of this protein during both bacterial and fungal pathogens. We suggest that MACP2 induces programmed cell death, bacterial pathogen resistance, and necrotrophic fungal pathogen sensitivity by activating the biosynthesis of tryptophan-derived indole glucosinolates and the salicylic acid signaling pathway dependent on the activity of enhanced disease susceptibility 1 (EDS1). Moreover, the response of MACP2 mRNA isoforms upon pathogen attack is differentially regulated by a posttranscriptional mechanism: alternative splicing. In comparison to previously reported MACPFs in Arabidopsis, MACP2 shares a redundant but nonoverlapping role in plant immunity. Thus, our findings provide novel insights and genetic tools for the MACPF family in maintaining SA accumulation in response to pathogens in Arabidopsis.

Keywords: MACP2; membrane attack complex/perforin-like protein; pathogen infection; salicylic acid signaling; indole glucosinolates

Citation: Zhang, X.; Dai, Y.-S.; Wang, Y.-X.; Su, Z.-Z.; Yu, L.-J.; Zhang, Z.-F.; Xiao, S.; Chen, Q.-F. Overexpression of the *Arabidopsis* MACPF Protein AtMACP2 Promotes Pathogen Resistance by Activating SA Signaling. *Int. J. Mol. Sci.* **2022**, 23, 8784. https://doi.org/10.3390/ijms23158784

Academic Editors: Melvin J. Oliver, Bei Gao and Moxian Chen

Received: 7 July 2022
Accepted: 3 August 2022
Published: 7 August 2022

Publisher's Note: MDPI stays neutral with regard to jurisdictional claims in published maps and institutional affiliations.

1. Introduction

As sessile organisms, plants have evolved sophisticated mechanisms to communicate with surrounding microorganisms, including beneficial and pathogenic interaction. Plant pathogens secrete effector proteins to suppress host immune responses during their colonization [1,2]. Thus, plant cells are equipped with a variety of cell surface or intracellularly localized receptor proteins that can recognize microorganisms and initiate downstream immune responses to restrict pathogen proliferation [3]. In particular, cell surface receptors, also known as pattern recognition receptors (PRRs), function in the recognition of conserved pathogen-associated molecular patterns (PAMPs) and are able to trigger PAMP-triggered immunity (PTI) responses during plant interactions with either nonpathogenic or pathogenic microbes [3,4]. For example, PAMP receptors such as PRR FLAGELLIN SENSING 2 (FLS2), chitin elicitor receptor kinase 1 (CERK1), and EF-Tu receptor (EFR) recognize conserved microbial effectors, flagellin (or the minimal epitope flg22), fungal chitin, and elongation factor thermo unstable (EF-Tu), respectively, in *Arabidopsis* [3,5]. Furthermore, PAMP receptors form heterocomplexes with coreceptors such as Brassinosteroid insensitive 1-associated kinase 1 (BAK1) to activate downstream responses under microbial infection [3]. In addition to PTI, plants exhibit counter-defense strategies to initiate a locally rapid immune response to microbial-derived effectors through intracellular nucleotide-binding domains and leucine-rich repeat-containing receptors (NLRs). Specifically, a typical NLR consists of a Toll/interleukin 1-receptor (TIR) or coiled-coil domain at the N-terminus, leucine-rich repeats (LRR) at the C-terminus, and an internal nucleotide-binding region [6,7]. The NLR-mediated strategy is characterized as effector-triggered immunity (ETI) and is frequently linked with the hypersensitive response (HR) [8], a form of programmed cell death (PCD) [9]. To date, HR has been considered an efficient and immediate immune reaction in response to pathogen invasion, leading to rapid cell death to limit pathogen proliferation at the entry site [10]. However, little is known about the underlying mechanism of this response in plant immunity. Recent studies have suggested that the plant defense hormone salicylic acid (SA) emerges as a pivotal signal to mediate immunity-related HR, linking the activation of pathogenesis-related (PR) genes and HR-induced PCD to confer resistance to pathogens [11–13]. In particular, plant mutants with misregulation of the cell death pathway result in lesion-mimicking phenotypes and the constant activation of SA signaling, H_2O_2 accumulation, and PR genes [7,14,15]; these materials are powerful tools for studying the underlying connection between PCD and plant immunity [15].

While the HR can be a rapid response for containing disease progress at the site of pathogen entry, mechanisms that cause microbial death remain to be investigated. In recent years, mammalian pore-forming proteins have been found to target and lyse infected microorganisms by forming the membrane attack complex (MAC), an oligomeric cylindrical ring at the surface of target membranes [16]. Since then, phylogenetic analysis has indicated that the conserved signature of the membrane attack complex and perforin (MACPF) proteins, i.e., Y/S-G-T/S-H-X7-G-G (X), is present in both eukaryotic and prokaryotic organisms to form the MACPF superfamily [17,18]. The members of this family have been documented to play essential roles in diverse developmental processes and immune responses [19]. MACPF proteins—which are related to cholesterol-dependent cytolysins and are structurally similar to C6, C7, C8α, C8β, and C9 complement system proteins [20]—can form cellular membrane pores and perform biological functions [21,22]. The formation of pore structures causes a breach of cellular integrity, ultimately inducing cell death by allowing the free passage of molecules in/out of the corresponding cell [19]. Several MACPF proteins have been structurally resolved with conserved MACPF oligomers and varied C-terminal domains [23].

However, there is little information related to the biological function of plant MACPF domain-containing proteins. In particular, several reports have indicated that MACPF proteins play important roles in viral and bacterial infections and plant PAMP-triggered immunity [1,4,23]. Many plant *MACPF* genes have been identified in *Arabidopsis* [15], *Poaceae* species [18], and cotton [24] and have been divided into four groups using phylogenetic analysis and domain organization [18]. Several MACPF genes (such as C6, C7, C8α, C8β, and C9) have been experimentally identified in animals with functions in growth and immunity [19,21,25–27]. The published transcriptomic data indicate that *MACPF* genes are involved in plant growth, development, and response to biotic and abiotic stresses [18]. In contrast, few genes have been experimentally confirmed in plants. For example, in *Gossypium*, silencing of the *GhMACPF26* gene enhanced tolerance of cotton plants to cold stress [24]. *Arabidopsis* constitutively activated cell death 1 (CAD1), localized to both the plasma membrane and cytosol, is a salicylic acid (SA)-responsive protein, which plays important roles in immunity-induced PCD [10,15,28] and could influence endophytic phyllosphere microbiota [29]. The *cad1* mutant shows a lesion phenotype that mimics HR and is also regulated by chitin elicitors independent of the SA-mediated pathway [10]. Similarly, another *Arabidopsis* membrane-localized necrotic spotted lesion 1 (NSL1) took part in SA-mediated defense responses and PCD. The *nsl1* mutant showed spotted necrotic lesions, retarded plant growth, and high accumulation of SA, thus activating the SA signaling pathway and linking PAMP-induced PCD to antimicrobial metabolism upon pathogen attack [4,30].

A total of four MACPF proteins have been identified in Arabidopsis; however, in addition to CAD1 and NSL1, the other two MACPFs have not been studied well. To this end, we isolated two independent mutants of *AtMACP2* (At4g24290) from *Arabidopsis*. Phenotypic and genetic analysis suggested that MACP2 is involved in the SA-mediated PCD response during pathogen infection and that the activation of SA signaling may result from altered metabolism of tryptophan (Trp)-derived indole glucosinolates. Further analysis indicated that MACP2 undergoes posttranscriptional regulation by alternative splicing (AS), and the three spliced isoforms detected based on the database information differentially respond to treatment with bacterial and fungal pathogens, implying distinct responsive pathways derived from plant immunity.

2. Results

2.1. Characterization of T-DNA Insertional Mutants and Transgenic Overexpression Lines in MACP2

In *Arabidopsis* thaliana, besides the two reported proteins, NSL1 and CAD1, the function of the other two MACPF-containing proteins, MACP1 (encoded by At1g14780) and MACP2 (encoded by At4g24290), remain unknown. To explore the role of the MACP2 in *Arabidopsis thaliana*, we bought two T-DNA insertional mutants (Figure 1A, B) of this gene from TAIR and identified homozygotes. Amplification with primer pairs containing the T-DNA fragment LBa1 showed obvious bands but showed a blank with primer pairs for full-length MACP2 in mutants, which indicated that the mutants were exactly homozygous T-DNA insertion mutants. RT-PCR showed that no full-length transcripts were present in the corresponding *KO-1* and *KO-2* mutants (Figure 1C), indicating that these lines are knockout mutants. Subsequently, transgenic lines introducing the MACP2-YFP construct into the wild-type were generated at the same time. Clear bands of the vector contained fragment amplification results (Figure 1D) and higher relative expression levels of *MACP2* (Figure 1E); specific YFP-tagged MACP2 detection in *MACP2-YFP(OE)* plants via Western blot analysis (Figure 1F) showed the correct *MACP2-overexpression* transgenic plants we obtained.

2.2. Overexpression of MACP2-Accelerated Cell Death in Rosettes

To investigate whether disruption or overexpression of MACP2 shows linkage to SA, ROS accumulation, and constitutive cell death, we examined the rosettes of 4-, 5-, and 6-week-old wild-type, *KO* mutants, and *OE* transgenic plants using diaminobenzidine (DAB) and trypan blue staining. Interestingly, as shown in Figure 2, trypan blue-stained lesions displayed no significant differences at the fourth week and were apparently more severe in the leaves of *OE* transgenic plants than in the wild-type, whereas this was reversed in the *KO* mutants at the fifth and sixth weeks (Figure 2A). Meanwhile, *OE* leaves generated higher levels of H_2O_2 at the fifth and sixth weeks. This was indicated by the brown color upon DAB staining compared with the wild-type control rather than the lower level of H_2O_2 in the *KO* mutants than wild-type plants in the same period (Figure 2B). These results suggest that MACP2 promotes natural continuous cell death and ROS eruptions during leaf senescence.

2.3. MACP2-Strengthened Plant Resistance to Bacterial Pathogens Relying on the SA Pathway

To address whether the overexpression of MACP2 affects the plant defense response to bacterial pathogens, we conducted *Pst* DC3000 inoculation assays on four-week-old wild-type, *MACP2-KO* mutants, and *MACP2-OE* plants. According to the results, the *OE* plants showed a more tolerant phenotype than the wild-type when responding to *Pst* DC3000 infection (Figure 3A) and significantly repressed the bacterial population (Figure 3B), while the *KO* mutants showed a more sensitive phenotype than the wild-type (Figure 3A,B).

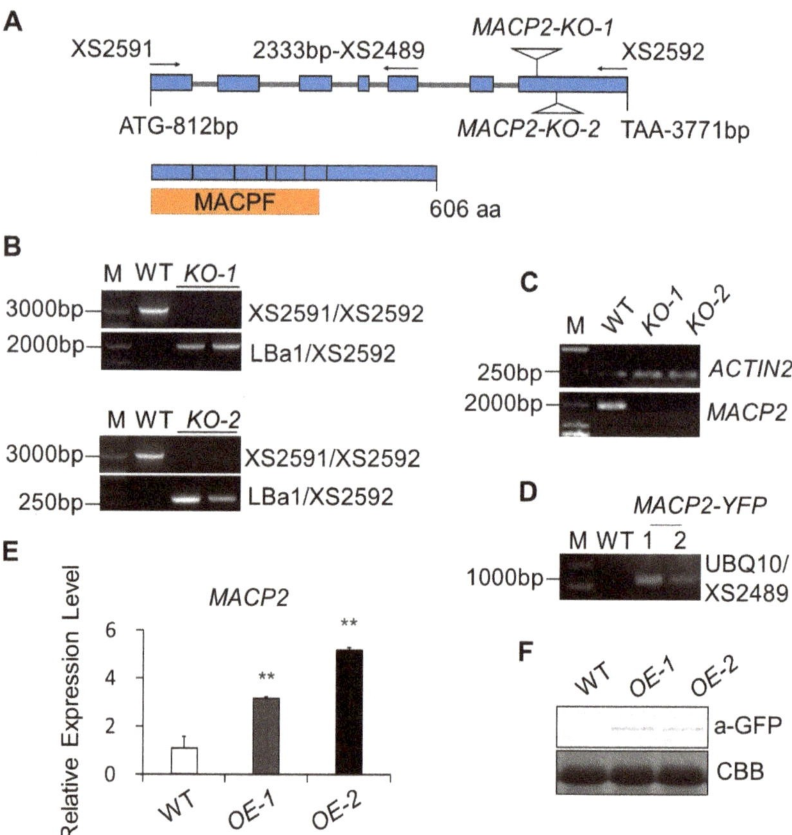

Figure 1. Characterization of T-DNA insertional mutants and transgenic overexpression lines in *MACP2*. (**A**) Schematic depicting the MACP2 gene, indicating the T-DNA insertion site of *MACP2-KO-1* and *MACP2-KO-2*. (**B**) Identification of *MACP2-KOs* via PCR. The full-length *MACP2* was amplified with primer pair XS2591 and XS2592. The length-contained T-DNA sequence was amplified via primer pair LBa1 and XS2592 in *KO-1*, and *KO-2*. (**C**) Semiquantitative PCR of *MACP2* in WT and *MACP2-KOs*. The full-length *MACP2* was amplified with primer pair XS2591 and XS2592. The *ACTIN2* was amplified with primer pair ACTIN2-F and ACTIN2-R. (**D**) Identification in DNA level of *MACP2-YFP* transgenic plants. MACP2 CDS was cloned into pFGC-RCS binary vector then the expression cassette of MACP2-YFP was inserted into the Arabidopsis genome. UBQ10 and XS2489 were derived from the pFGC-RCS plasmid and MACP2 CDS, respectively. (**E**) Identification in RNA level of *MACP2-YFP* transgenic plants. Transcriptional level of *MACP2* in *MACP2-OE-1* and *MACP2-OE-2* upregulated 3–5 times as that in wild-type. The data represent means from three independent repeats. Statistical differences were identified using Student's *t* test. ** $p < 0.01$. (**F**) Identification in protein level of *MACP2-YFP* transgenic plants. Anti-GFP was used to recognize the specific YFP tag. CBB represented Coomassie blue staining.

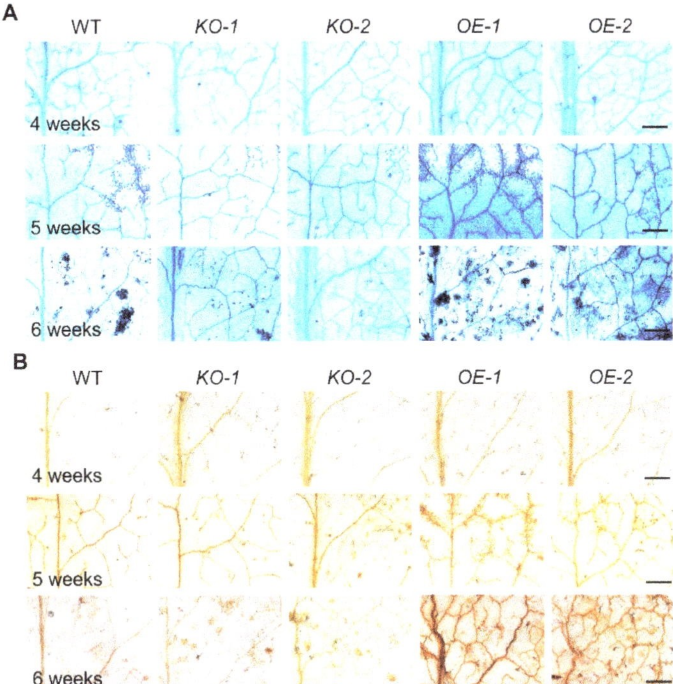

Figure 2. Overexpression of MACP2 showed accelerated cell death in rosettes. Trypan blue staining (**A**) and DAB staining (**B**) of wild-type, *MACP2-KO* mutants, and *MACP2-OE* rosettes after 4, 5, and 6-week development. *MACP2-OEs* obtained more cell death lesions and higher levels of H_2O_2, indicated by the brown color, than wild-type in the 5th and 6th weeks, while *MACP2-KOs* obtained less cell death and lower levels of H_2O_2 than wild-type. Bar = 1 mm.

In addition, we measured the endogenous SA levels in wild-type, *KO* mutants, and *OE* transgenic lines during pathogen infection using liquid chromatography-mass spectrometry. As shown in Figure 3C, in the CK group, the contents of SA and SAG were higher in the *OE* plants but lower in the *KO* plants than in the wild-type (WT) plants. After *Pst* DC3000 treatment, although the SA content increased sharply in general, the SA and SAG contents in *OEs* were significantly higher than those in the wild-type. In contrast, a reverse trend of their content variation in response to *Pst* DC3000 treatment was found in the *KOs*.

On the basis of MACP2 promoting SA accumulation in response to bacterial pathogens, we selected enhanced disease susceptibility 1 (EDS1), forming heterodimers with phytoalexin deficient 4 (PAD4) to promote SA accumulation [31], and to generate *OE eds1-22* plants to further define the connection of the pathogen response and SA accumulation in *MACP2-OE* plants. Then, we conducted *Pst* DC3000 inoculation assays in four-week-old rosettes among different genotypes, including wild-type, *OE*, *eds1-22*, and *OE eds1-22*. Disrupting the SA signaling pathway EDS1 suppressed the resistance phenotype to *Pst* DC3000 in *OE* plants (Figure 3D,E). These results suggest that MACP2 strengthened plant resistance to bacterial pathogens depending on the SA pathway in Arabidopsis.

2.4. MACP2-Weakened Plant Resistance Depending on the SA Pathway to Necrotrophic Fungal Pathogens

To explore whether the accumulation of SA in *OE* plants affected the response to necrotrophic fungal pathogens, we conducted *B. cinerea* inoculation assays on four-week-old wild-type, *KO* mutant, and *OE* plants. *OE* plants were hypersensitive to *B. cinerea* infection compared to wild-type plants (Figure 4A,B), as confirmed by the larger lesion size in *OE* plants instead of the resistant phenotype, and the smaller lesion size in *KO* plants (Figure 4A, B). Naturally, we also tested the contents of SA and SAG that were also induced during fungal infection and accumulated in *OEs* (Figure 4C). *B. cinerea* infection experiments were also carried out on *OE eds1-22* plants and revealed that the absence of the SA signal pathway EDS1 crippled the sensitive phenotype to *B. cinerea* in *OE* plants (Figure 4D,E). These results suggest that MACP2 operates differentially in response to bacterial and necrotrophic fungal pathogens.

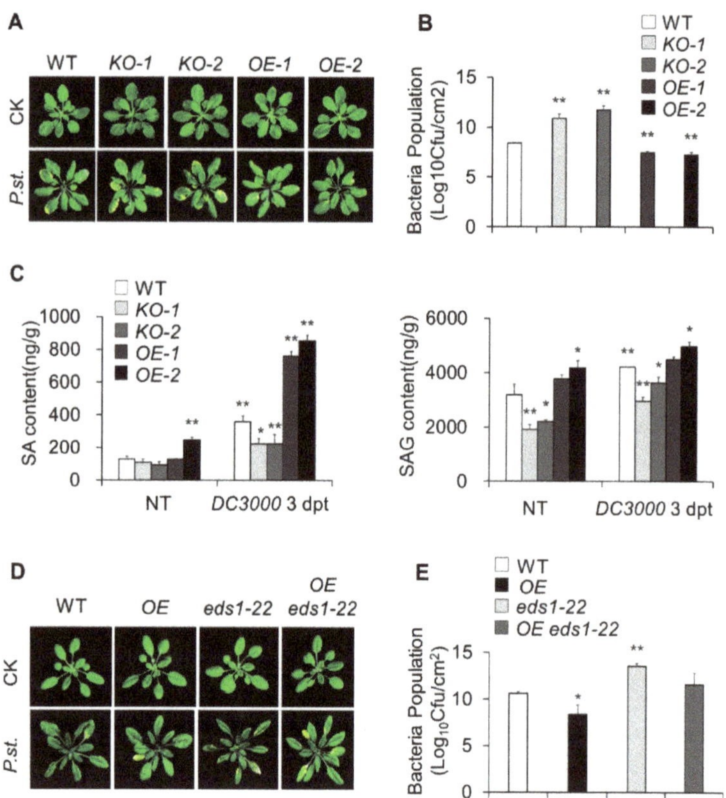

Figure 3. MACP2-strengthened plant resistance relying on SA pathway to the bacterial pathogen. (**A**) Phenotypes of wild-type, *MACP2-KOs*, and *MACP2-OEs* in response to *Pst* DC3000 infection. Four-week-old wild-type, *MACP2-KOs*, and *MACP2-OEs* were infected with *Pst* DC3000 on leaf surface and photographed 5 days after treatment. (**B**) Bacterial populations at 5 days postinoculation in wild-type, *MACP2-KOs*, and *MACP2-OEs* leaves. The data represent means from three independent repeats. Statistical differences were identified using Student's *t* test. ** $p < 0.01$. (**C**) SA contents detection of wild-type, *MACP2-KO* mutants, and *MACP2-OEs* adult plants during *Pst* DC3000 infection. The contents of SA and SAG were measured by LC-MS. The "g" in "ng/g" represents the fresh weight. The experiments were biologically repeated three times with similar results. Error bars represent SD ($n = 3$ biological replicates). * $p < 0.05$, ** $p < 0.01$ by Student's *t* test. (**D**) Phenotypes of leaves from 4-week-old wild-type, *MACP2-OE, eds1-22,* and *MACP2-OE eds1-22* leaves in response to *Pst* DC3000 infection. (**E**) Bacterial populations at 5 days postinoculation in wild-type, *MACP2-OE, eds1-22,* and *MACP2-OE eds1-22* leaves. The data represent means from three independent repeats. Statistically significant differences were identified using Student's *t* test. * $p < 0.05$, ** $p < 0.01$.

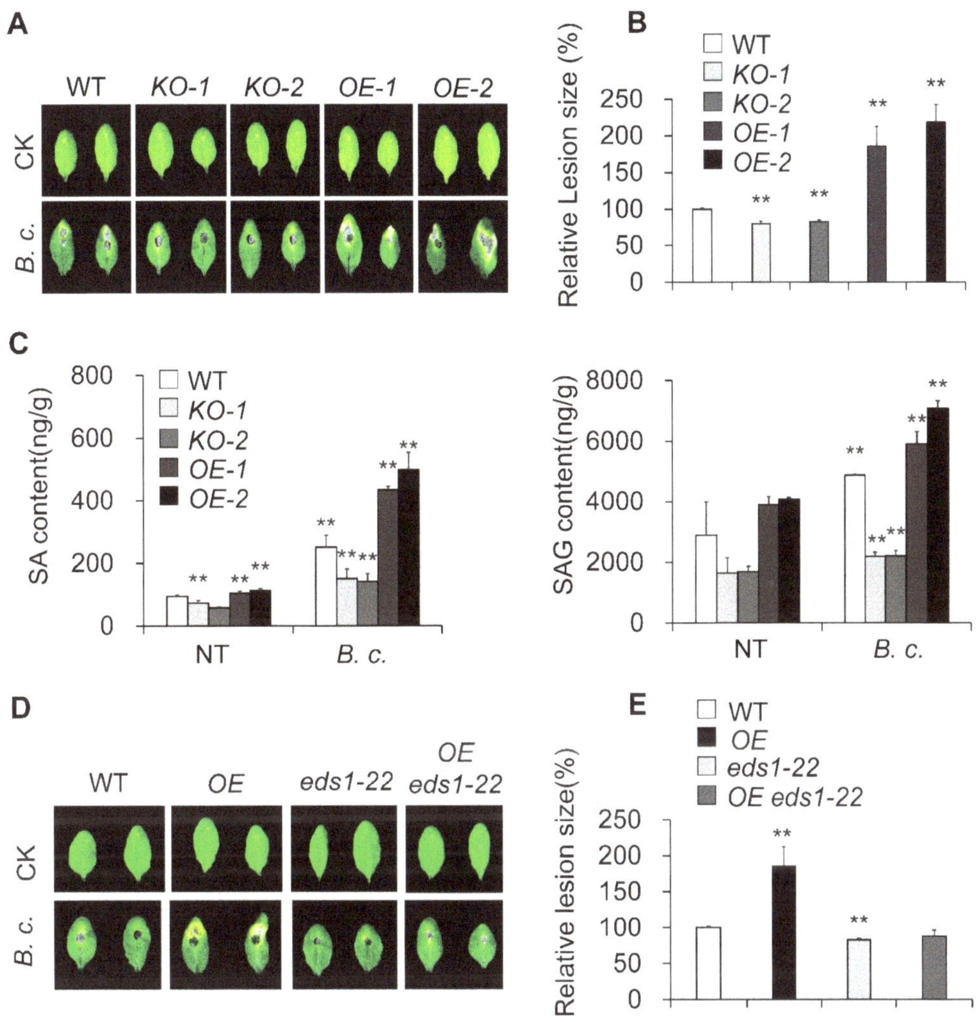

Figure 4. MACP2-weakened plant resistance depending on SA pathway to fungal pathogen. (**A**) Phenotypes of leaves from 4-week-old wild-type, *MACP2-KO* mutants, and *MACP2-OEs* plants in response to *B. cinerea* infection. Added *B.c.* on leaf surface and photographed 3 days after treatment. (**B**) Relative lesion size of wild-type, *MACP2-KOs*, and *MACP2-OEs* leaves after 3 days of *B. cinerea* infection. The lesion size was calculated by ImageJ and relative lesion size was calculated by comparing the values from treated leaves versus mock leaves. Asterisks indicate significant differences from the wild-type. ** $p < 0.01$ by Student's t test. (**C**) SA contents detection of wild-type, *MACP2-KO* mutants, and *MACP2-OEs* adult plants during *B. cinerea* infection. The contents of SA and SAG were measured by LC-MS. The "g" in "ng/g" represented the fresh weight. The experiments were biologically repeated three times with similar results. Error bars represent SD. $n = 3$ biological replicates. ** $p < 0.01$ by Student's t test. (**D**) Phenotypes of leaves from 4-week-old wild-type, *MACP2-OE, eds1-22*, and *MACP2-OE eds1-22* leaves in response to *B. cinerea* infection. (**E**) Relative lesion size of wild-type, *MACP2-OE, eds1-22*, and *MACP2-OE eds1-22* leaves after 3 days of *B. cinerea* infection. Asterisks indicate significant differences from the wild-type. ** $p < 0.01$ by Student's t test.2.5. MACP2 Differentially Modulates Plant Sensitivities to Fungal and Bacterial Pathogens via the SA Signaling Pathway.

To further investigate the difference between bacterial and fungal immunity caused by *MACP2* overexpression, we detected the expression of genes related to SA-associated defense responses, including *PR1*, *PR5*, *ST1*, and *EDR2* [32–34], and JA-associated defense responses, including *PDF1.2a*, *PDF1.2b*, *VPS1*, and *VPS2* [35] in rosettes. Consistent with the phenotype resistant to bacterial pathogen in *OE* plants, the transcript levels of SA-associated responsive genes in *OE* plants (with the exception of *EDR2*) showed a sharply upregulated trend compared with the wild-type but displayed a significantly downregulated trend in *KO* mutants. On the contrary, an inverse trend was observed for JA-related responsive genes (Figure 5A), manifesting in significant upregulation in *KO* mutants but downregulation in *OEs*. With the necrotrophic fungal pathogen *(B. cinerea)*, the *OEs* displayed downregulation of JA-responsive genes, and *KOs* showed fiercely higher expression than the wild-type. In addition, SA-responsive genes were also induced slightly by *B. cinerea* and maintained a higher expression level in *OEs,* resulting in a fungus-sensitive phenotype based on the antagonizing role of SA and JA. Our results suggest that overexpression of MACP2 may contribute to accelerated SA accumulation, thus activating the SA signaling pathway in response to pathogen invasion.

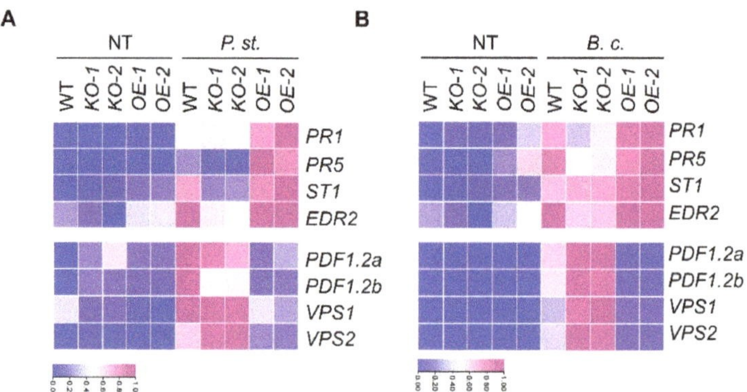

Figure 5. MACP2 differentially modulate plant sensitivities to fungal and bacterial pathogens via SA signaling pathway. Heatmaps show the fold change of key regulators in SA and JA signaling pathways in wild-type, *MACP2-KO* mutants, and *MACP2-OEs* plants after infection with *Pst* DC3000 (**A**) and *B. cinerea* (**B**). The transcriptional profiles of relative gene expression values were analyzed using the TB tools.

2.5. Alternatively Spliced Isoforms of MACP2 Are Differentially Expressed under Pathogen Treatment

To investigate the relationship between AS regulation of MACP2 and pathogen resistance, we designed isoform-specific primers to explore the expression levels of these isoforms either under normal conditions or treatment with *Pst* DC3000 or *B. cinerea* (Figure 6A). A total of three mRNAs can be detected for MACP2, named *MACP2-1*, *MACP2-2*, and *MACP2-3*. Expression analysis using semiquantitative and real-time quantitative PCR indicated that *MACP2-1* and *MACP2-3* were highly expressed in four-week-old rosettes of the wild-type before any treatments (Figure 6B, C). Interestingly, *Pst* DC3000 induced the expression of the *MACP2* locus, particularly through the transcription of *MACP2-1*, but not the other two mRNA isoforms. In contrast, the transcript abundance of *MACP2* was reduced in response to *B. cinerea* inoculation. Isoform-specific expression analysis suggested that *MACP2-1* and *MACP2-2* contribute to this reduction, whereas *MACP2-3* was elevated in comparison to untreated controls. Thus, the underlying mechanism of this differential expression of MACP2 isoforms in response to different pathogens remains to be further investigated.

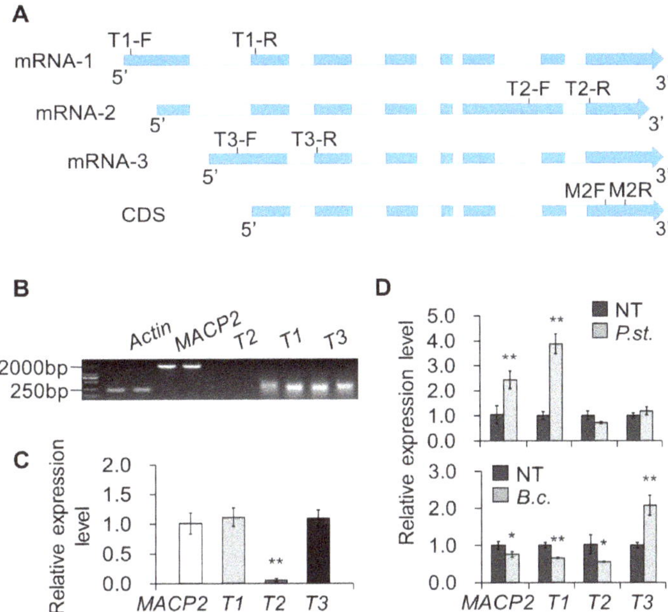

Figure 6. AS of MACP2 responded to fungal and bacterial pathogens. (**A**) Classification of three AS of MACP2 gene. The blue represents exons and white represents introns. (**B**) Semiquantitative PCR detection of full-length MACP2 and specific sequence of each AS (T1, T2, T3) in 4-week-old rosettes of wild-type. The full-length of MACP2 was amplified with primer pair XS2591 and XS2592. The specific sequences of AS were amplified with primer pairs MACP2-1 F/MACP2-1 R, MACP2-2 F/MACP2-2 R, and MACP2-3 F/MACP2-3 R, respectively. (**C**) qRT-PCR detection of common sequence (MACP2) and specific sequence of each AS (T1, T2, T3) in 4-week-old rosettes of wild-type. The common sequence of three AS was amplified with primer pair MACP2-F and MACP2-R. The specific sequences of AS were amplified with primer pair mentioned in (**B**). The *ACTIN2* was amplified with primer pair ACTIN2-F and ACTIN2-R. Asterisks indicate significant differences from the wild-type. ** $p < 0.01$ by Student's *t* test. (**D**) qRT-PCR detection of AS responding to *Pst* DC3000 and *B. cinerea* after 3 days of infection on 4-week-old rosettes of wild-type. Asterisks indicate significant differences from the wild-type. * $p < 0.05$, ** $p < 0.01$ by Student's *t* test.

2.6. Indolic GS Contributed to Bacterial Resistance to MACP2-OE

To explore whether the increased SA contents in *MACP2-OEs* are related to tryptophan (Trp)-derived glucosinolates, we examined the transcription level of six vital regulators in GS biosynthesis in *Arabidopsis* plants, including MYB34, MYB51, MYB122 (involved in the synthesis of indolic GSs), MYB28, MYB29, and MYB76 (related to the synthesis of aliphatic GSs), and the level of GS contents, including the indolic GSs indol-3-ylmethyl-GS (I3M) and 1-methoxyindol-3-ylmethyl-GS (1-MOI3M), and the aliphatic GSs 4-methylsulfinylbutyl-GS (4-MOSB), 5-methylsulfinylpentyl-GS (5-MSOP), and 8-methylsulfinyloctyl-GS (8-MSOO) in wild-type, *MACP2-KO* mutants, and *MACP2-OE* plants rosettes after 3 days of *Pst* DC3000 infection. The levels of indolic GS synthesis regulators (MYB34, MYB51, and MYB122) and indolic GS species (I3M and 1-MOI3M) in response to *Pst* DC3000 infection were significantly elevated in *OE* plants but significantly reduced in *KO* mutants (with the exception of *MYB51*) compared to the wild-type (Figure 7A, B). In contrast, the transcriptional level of aliphatic GS synthesis regulators (MYB28, MYB29, and MYB76) and the levels of aliphatic GSs, including 4-MOSB, 5-MSOP, and 8-MSOO, were not significantly altered in *OEs*, while those related to aliphatic GSs and the levels of aliphatic GSs increased in response to *Pst* DC3000 infection in general (Supplementary Figure S1). These findings indicate that MACP2 responds to pathogen infection in a tryptophan (Trp)-derived indole glucosinolate-activated SA-dependent manner.

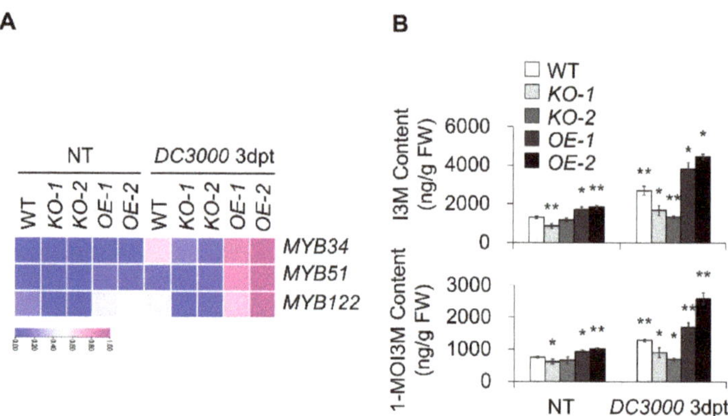

Figure 7. Indolic GS contributed to bacteria resistance of *MACP2-OE*. (**A**) Heatmaps show the fold change of key regulators in GS biosynthesis, containing *MYB34*, *MYB51*, and *MYB122* in wild-type, *MACP2-KO* mutants, and *MACP2-OEs* plants after *Pst* DC3000 infection. The transcriptional profiles of relative gene expression values were analyzed using the TB tools. (**B**) Indolic GS contents of wild-type, *MACP2-KO* mutants, and *MACP2-OEs* plants after *Pst* DC3000 infection. The contents of indolic GS were measured by LC-MS. The experiments were biologically repeated three times with similar results. Error bars represent SD. n = 3 biological replicates. Asterisks indicate significant differences from the wild-type. * $p < 0.05$, ** $p < 0.01$ by Student's t test.

3. Discussion

3.1. Pleiotropic Function of MACPF Proteins in Plant Immunity and Programmed Cell Death

An orthologue search indicated that there are four MACPF proteins in Arabidopsis, and two of them have been extensively studied in the past twenty years [19]. In this study, we demonstrated that *macp2* knockout mutants and *MACP2-OEs* display altered sensitivity to bacterial and fungal pathogens (Figures 3 and 4), suggesting that MACP2 participates in plant immunity responses to external pathogens. Interestingly, the knockout mutant of *CAD1* showed similar lesion mimic phenotypes [10,15,36], suggesting the activation of immune responses in this knockout mutant. Previous genetic and physiological studies have demonstrated that the *cad1-1* mutant is resistant to the virulent bacterial pathogen *Pst* DC3000 [15]. Similarly, the *MACP2-OEs* had a similar phenotype (Figure 3), showing fewer lesions in comparison to the wild-type. The MACP2 shared 52.1% and 43.9% identity with NSL1 and CAD1, respectively [15]. The underlying mechanism of phenotypic variation among these three MACPFs remains to be further investigated. One hypothesis that has been proposed previously is that nonoverlapping functions of NSL1 and CAD1 may be related to downstream defense-related R proteins [30]. At the molecular level, the *cad1-5* mutant has been found to elevate *PR1* gene expression, the marker of plant immunity [15]. However, this hypothesis needs direct experimental evidence for further investigation.

Furthermore, the expression of NSL1 and CAD1 is not induced by biotic stress treatments but is altered under abiotic stress conditions, suggesting that constitutive defense responses of these Arabidopsis mutant lines may not be the primary function of these MACPF proteins [30]. Indeed, the massive production of reactive oxygen species (ROS) through oxidative bursts during plant-pathogen interactions will trigger PCD in plants [37]. NSL1 has been proposed to disturb ROS production, thus impairing PCD during plant–disease responses [30], whereas overexpression of MACP2 caused higher levels of H_2O_2 and profound cell death in rosette leaves (Figure 2).

In mammals, to form a transmembrane pore structure, MACPF domain-containing proteins require the assisted assembly of other complement proteins [38,39]. However, no secretory peptide signal could be detected in the protein sequence of *Arabidopsis* MACPFs [15]. A previous study of NSL1 suggested that the metabolic imbalance detected in the *nsl1* mutant may be the result of improper assembly of these pore structures [30]. By using 35S- and native promoter-driven constructs, NSL1 was found to localize at the plasma membrane in Arabidopsis. Similarly, the subcellular localization of CAD1 has been confirmed by fractionation and confocal microscopy approaches [15]. In this study, MACP2 is a membrane-localized protein (Supplemental Figure S2), indicating that all three MACPFs can be deployed by the plant immune system to the entry site as a defense mechanism during host–microbe interactions. Unfortunately, there is little hard experimental evidence to prove the formation of protein complexes by these MACPFs in plants. Our previous study characterized *MACPF* genes in plants and revealed that several of those in Poaceae participated in plant vegetative growth and environmental stress adaptation [18]. In addition, nonredundant phenotypes of CAD1 and NSL1 suggested that plant MACPFs may function differently from their animal counterparts by assembling heteromeric complexes themselves to create pore structures on cell membranes [15]. Furthermore, although NSL1 is localized at the cell membrane, it did not kill pathogens at the entry site, suggesting that NSL1 has a differential mechanism in comparison to their animal orthologues. However, further molecular and biochemical experiments are required to unravel the underlying mechanism of plant MACPFs.

3.2. Plant Hormonal Signaling Is Critical to Influence MACP2-Mediated Disease Resistance

Plant hormones are important for all aspects of plant growth and physiology [40–43]. To unravel the molecular mechanism of MACP2-mediated PCD in plant immune responses, the relationship between plant defense hormones and MACP2 were evaluated. Genetic analysis of transgenic plants suggested that MACP2-mediated PCD is dependent on the plant hormone SA (Figure 3), which is similar to the molecular mechanism of CAD1 [15]. It has long been reported that the elimination of SA content could inhibit the expression of *PR* genes and thus lower resistance to pathogen infection [44]. The *cad1-1* mutant has a higher level of SA content than the wild-types, and the introduction of the bacterial enzyme NahG for SA degradation could rescue the PCD phenotype of *cad1-1* [15]. Meanwhile, the SA content increased significantly after treatment with both bacterial (*Pst DC3000*) and fungal pathogens (*B. cinerea*).

Furthermore, NSL1-mediated PCD triggered by flg22 has been considered a potential PAMP response and is characterized by the accumulation of SA and ROS, which are typical MTI outputs in response to pathogen attacks [4,45]. In contrast, CAD1 has been proposed to induce HR-related cell death by activating NLR signaling [36,46]. In this article, subsequent analysis indicated that EDS1 is downstream of MACP2 to confer plant immune responses (Figures 3 and 4). Similarly, most of the phenotypic and biochemical changes among CAD1 transgenic lines are proposed to be dependent on EDS1-mediated signaling [15], and approximately 90% of SA biosynthesis in plants is affected by EDS1-PAD4 signaling in the cytosol and nucleus [47,48]. EDS1 is a nucleocytoplasmic lipase-like protein that is classified as a member of the NLR-TNL signaling pathway by forming heterodimers with either phytoalexin deficient 4 (PAD4) or SAG101 [12,13]. Furthermore, the *nsl1-3 pad4* double mutant did not show a hyperactive immunity phenotype, indicating that NSL1 is guarded by NLR-TNL signaling [4]. In addition to SA biosynthesis activation, the EDS1-PAD4 complex is able to induce the expression of genes involved in the cell death response, such as *PR1*. In this study, the expression of *PR1* and *PR5* among transgenic Arabidopsis *MACP2-OEs* correlated with SA levels, further validating that EDS1 is responsible for MACP2-mediated PCD. Nevertheless, except for EDS1, downstream signaling of MACPFs in response to plant pathogens remains to be further investigated.

Intriguingly, a recent report demonstrated that the EDS1-PAD4 pair participates in sphingolipid metabolism to trigger cell death in response to the fungal pathogen *B. cinerea* [13]. The involvement of sphingolipids, especially long-chain ceramides, in the MACP2-mediated PCD response is valuable for study. Furthermore, SA resists the biotrophic pathogens living and reproducing on live host cells, whereas jasmonic acid (JA) acts on necrotrophic pathogens that kill host cells for nutrition and reproduction. Both of them play important but antagonistic signaling roles in pathogen responses [49]. EDS1-PAD4 signaling has been reported to play a negative role in response to *B. cinerea*, a necrotrophic fungal pathogen that can activate the JA pathway in plants [50]. Thus, the phenotypes of MACP2 in response to *B. cinerea* could be explained (Figure 4), suggesting that the repression of cell death in Arabidopsis effectively confers plant resistance to *B. cinerea*. Similarly, JA accumulated in the *cad1* mutant, and the JA/ethylene-induced gene *PDF1.2* was altered compared to wild-types [10]. Specifically, *PDF1.2a* and *PDF1.2b* were differentially expressed in MACP2 transgenic lines in response to bacterial and fungal pathogens (Figure 5), suggesting crosstalk between multiple plant hormonal signaling pathways downstream of MACP2. Finally, different splice isoforms responded to bacterial or fungal inoculation (Figure 6), indicating that the distinct response mechanism of MACP2 to bacterial and fungal pathogens can be controlled by posttranscriptional regulation, i.e., alternative splicing [51–55].

3.3. Glucosinolates Are Crucial Signal Messengers That Transduce Immunity-Triggered PCD Downstream of MACPF Proteins

Previously identified EDS1-PAD4 signaling has been documented as a universal regulator of plant immunity, which also regulates multiple metabolic pathways of plant hormones, phytoalexins (camalexin), and other secondary metabolisms (tocopherols and N-hydroxypipecolic acid) [56]. In the study of NSL1, glucosinolates (GSs), an unsuspected role of tryptophan-derived secondary metabolites, are pivotal messengers to initiate PCD by activating SA biosynthesis in Arabidopsis [4]. Glucosinolates (sulfur- and nitrogen-containing thioglucosides) show broad activity against insect herbivores and plant pathogens [57] and are classified into three subcategories: aromatic GSs, methionine-derived aliphatic GSs (AGSs), and tryptophan-derived indole GSs (IGSs) [58].

A pathogen-inducible myrosinase, penetration 2 (PEN2) involved in the bioconversion of indole glucosinolates (IGSs) [59], plays an important role in PAMP-triggered PCD in the absence of NSL1 [4]. PEN2 is responsible for releasing bioactive molecules (e.g., isothiocyanates) with a wide range of toxicity to insects and plant pathogens [60]. In particular, 4-methoxyindol-3-ylmethylglucosinolate (4MI3G, IGS against a broad spectrum of fungal pathogens) is accumulated [61] under pathogen infection via PEN2 activity. In our study, indolic GS species (I3M and 1-MOI3M) were highly accumulated in *MACP2-OEs* and were less accumulated in *MACP2-KOs* in comparison to the levels of these compounds in the wild-type plants (Figure 7), implying that IGs may function similarly as signal molecules to connect MACP2 and downstream PCD responses. Furthermore, the conversion of I3G to 4MI3G has been proposed to be tightly regulated by the mitogen-activated protein kinase (MAPK)-transcription factor (TF) cascade. The MPK3/MPK6-MYB34/51/122 cascade has been suggested to participate in this regulation [62]. Here, the transcript abundance of three R2R3-MYB TFs, *MYB34*, *MYB51*, and *MYB122*, was tested, showing a high correlation with the content of I3M and 1-MOI3M in *MACP2-KOs* and *MACP2-OEs* (Figure 7). However, the mechanism by which the EDS1-PAD4 pair triggers IGS biosynthesis remains elusive. Further study of signal transduction downstream and the assembly mechanism of MACP2 will be informative because this is the general defense mechanism that plants possess to restrict pathogen infection.

4. Materials and Methods

4.1. Plant Material, Growth Condition, and Treatment

Arabidopsis thaliana accession Columbia-0 (Col-0) was used as the wild-type line in this study. *MACP2-KO-1* (SALK_040186) and *MACP2-KO-2* (SALK_052845C) were obtained from The Arabidopsis Biological Resource Center (ABRC, USA, http://www.arabidopsis. org, accessed on 24 April 2015). The *eds1-22* mutant used in this study has been described previously [63]. For genetic analysis, the *eds1-22* mutant was crossed with *MACP2-OE* to generate *OE eds1-22*. For the seed germination assay, the Arabidopsis seeds were surface sterilized with 20% bleach containing 0.1% Tween 20 (Sigma, P2287, St. Louis, MI, USA) for 15 min, washed with distilled water 6 times, and then plated on 1/2 MS (Sigma, M5519, USA) agar with 1% sucrose. The plates were incubated at 4 °C for 2 days and then transferred to a greenhouse under a 16 h light/8 h dark photoperiod at 20 °C for 7 days according to a previous study [64].

4.2. Plasmid Construction and Transgenic Plant Generation

All constructs were generated using the ClonExpress II One Step Cloning Kit (Vazyme, C112, China). The gene-specific primers with 15 bp extensions homologous to the corresponding vectors are listed in Supplemental Table S1.

To generate stable transgenic plants, MACP2 CDS was cloned into pUC119-YFP to construct the expression cassette MACP2-YFP, which was cloned into the binary vector pFGC-RCS via the same *AscI* digestion site between the two vectors [65,66]. The expression cassettes were subsequently introduced into wild-type Arabidopsis (Col-0) by Agrobacterium tumefaciens-medium transformation via the floral dip method [67] to generate *MACP2-YFP* transgenic plants.

4.3. DAB and Trypan Blue Staining

Trypan blue staining and DAB staining were performed according to Xiao and Chye [68]. For trypan blue staining, rosettes of 4-, 5-, and 6-week-old wild-type, *MACP2-Kos*, and *MACP2-OE* plants were collected in 10 mL tubes and boiled for 5 min in trypan blue staining buffer composed of 12.5% phenol (Thermo, K2599312, Waltham, MA, USA), 12.5% glycerol (Guangzhou Chemical Reagent Factory, Guangzhou, China), 12.5% lactic acid, 48% ethanol (Guangzhou Chemical Reagent Factory, China) and 0.025% trypan blue. The rosettes were incubated for 10 min at room temperature and then decolorized five times in 70% chloral hydrate. For DAB staining, rosettes of 4-, 5-, and 6-week-old wild-type, *MACP2-Kos*, and *MACP2-OE* plants were collected and incubated in 1 mg/mL DAB (Sigma, D8001, USA) solution of 10 mM PBS (Vazyme, G101, pH 7.0, Nanjing, China) and 0.05% Tween 20 at 37 °C in darkness overnight and subsequently decolorized in 95% ethanol at 65 °C 3 times every 2 h.

4.4. Pathogen Infection

Pathogen inoculation was carried out as previously described [69–71] with minor modifications. The fungal pathogen *Botrytis cinerea* was maintained on V8 juice agar medium at 25 °C in the dark for 10 days. Spore masses were collected and suspended in Vogel buffer composed of 50 mM sucrose (Guangzhou Chemical Reagent Factory, China), 20 mM K_2HPO_4 (Damao Chemical Reagent Factory, China), 10 mM sodium citrate, 20 mM $(NH_4)_2SO_4$, 1 mM $MgSO_4$ and 10 mM $CaCl_2$ (pH 5.0). More than 9 mature rosettes per genotype from different 4-week-old plants were placed in petri dishes containing 0.6% (w/v) agar. Each leaf was inoculated with 5 μL droplets containing 1.6×10^6 spores/mL of *B. cinerea* suspension, incubated in the dark for 36 h, and then cultivated in a greenhouse with a 16 h light/8 h dark photoperiod at 20 °C. The lesion diameter (mm) was calculated using ImageJ software.

The bacterial pathogen *P. syringae* pv. *tomato* (*Pst* DC3000) was cultivated at 28 °C and 200 rpm in King's medium B [70] containing rifampicin (New Probe, 50 mg/L, China). Then, *Pst* DC3000 was collected by centrifugation and resuspended in 10 mM $MgCl_2$ at $A_{600} = 0.2$. Bacteria were then diluted 10 times to approximately 107 colony-forming units/mL in 10 mM $MgCl_2$ and 0.02% Silwet L-77 (New Probe, P001374, China) for inoculation. After inoculation, the plants were kept in high humidity. To calculate the bacterial populations, leaf discs (0.6 cm diameter) were collected from infected leaves, washed three times with sterile water, and homogenized in 10 mM $MgCl_2$, followed by applying appropriate dilutions on solid King's B medium with rifampicin. All experiments were repeated three times with similar results.

4.5. SA Measurements

SA was extracted and measured as described previously [72,73]. Approximately 150 mg powdered tissue was weighed in a 2 mL centrifuge tube and extracted with 800 µL of extraction buffer of 2-propanol/water/concentrated HCl (2:1:0.005, $v/v/v$) with internal standards of 10 ng d4-SA (Sigma-Aldrich, USA). The mixtures were shaken mildly for 30 min at 4 °C, followed by adding 1 mL dichloromethane and shaking for an additional 30 min at 4 °C. The samples were then centrifuged at 13,000× g and 4 °C for 10 min. Solvent (1 mL) from the lower phase was collected and dried using a nitrogen evaporator with nitrogen flow. The samples were dissolved in a 200 µL mixture of 60% methanol (Mreda, M042749, China) and 40% sterile ultrapure water. Quantitative analysis of SA was performed via a chromatography (Shimadzu, Japan)–mass spectrometry (Triple TOF 5600, AB SCIEX, USA) system according to Chen et al. [72].

4.6. RNA Extraction and RT-qPCR

Total RNA was extracted from 5- and 6-week-old *Arabidopsis* leaves referring to a previous study [35]. Two milligrams of total RNA were extracted by HiPure Plant RNA Mini kit (Magen, China) and converted into cDNA with the HiScript II QRT Super Mix kit with gDNA Wiper (Vazyme). RT-qPCR assays (10 µL reaction volumes with gene-specific primers, Supplemental Materials Table S1) were performed on a StepOne Plus Real-time PCR System (Applied Biosystems) using ChamQ SYBR Color qPCR Master Mix (Vazyme, China) and the following protocol: 95 °C for 5 min followed by 40 cycles of 95 °C for 15 s, 55 °C for 15 s, and 72 °C for 30 s.

Primers for RT-PCR were described in both a previous publication and qPrimer DB (https://biodb.swu.edu.cn/qprimerdb/, accessed on 24 April 2015, [74]). Primers for specific AS were designed in exon–exon junction, for which specificity was verified via Primer-Blast software and amplified in restricted extension time to tule out genomic DNA contribution. The efficiency of each primer pair was not evaluated and only comparisons for each particular mRNA isoform under normal conditions or pathogen treatment were compared to draw further conclusions. For calculation of relative transcription levels, the delta of threshold cycle (ΔCt) values was calculated by subtracting the arithmetic mean Ct values of the target genes from the normalizing *ACTIN2*. The relative transcription level ($2^{\hat{}}\Delta\Delta$Ct) was calculated from three independent experiments. The fold change values were visualized, illustrated, and standardized in a heatmap generated by the TBtools package [75]. In the heatmap, the color represents the fold change value. The closer it is to pink, the greater the fold change value.

4.7. GS Measurements

GSs were extracted and detected as described previously [35,76] with minor modifications. Frozen leaf samples (120 mg) were ground with a glass rod in 1.2 mL ice-cold MeOH/H$_2$O (70:30, v/v) and incubated at 80 °C for 15 min. The homogenate was centrifuged at 3500× g and 4 °C for 10 min, and the supernatant was filtered through a 0.22 µm filter for analysis. Chromatography (Shimadzu, Japan)–mass spectrometry (Triple TOF 5600, AB SCIEX, USA) was used to detect and analyze the GS contents according to Liao et al. [35]. Quantification was performed using three technical replicates. Experiments were repeated three times with similar results, and five plants of each genotype were collected for one technical replicate.

4.8. Protein Isolation and Immunoblot Analysis

For total protein extraction, 4-week-old *Arabidopsis* seedlings grown in soil were ground in liquid nitrogen and homogenized in ice-cold protein extraction buffer of 50 mM sodium phosphate (pH 7.0), 200 mM NaCl, 10 mM MgCl$_2$, 0.2% β-mercaptoethanol (Westgene, WG0482, China), and 10% glycerol, and supplemented with protease inhibitor cocktail (Roche, 04693132001) according to Xia et al. [77]. The samples were placed on ice for 30 min and centrifuged at 4 °C at 12,000× *g* for 10 min. The supernatant was transferred to a new microfuge tube before electrophoresis.

For immunoblot analysis, clarified extracts were subjected to SDS-PAGE and transferred to a Hybond-C membrane (Cytiva, 10600002, USA). Specific anti-GFP (Abmart, M20004S, 1:5000, China) antibody was used in the protein blotting analysis.

4.9. Statistical Analysis

The significance of the difference between 2 groups was determined using Student's *t* test. The level of statistical significance is indicated by asterisks (* $p < 0.05$ and ** $p < 0.01$). The numbers of samples are indicated in the figure legends.

5. Conclusions

Collectively, this study reveals the molecular mechanism of the Arabidopsis MACPF domain-containing protein MACP2 in the plant immune response. The natural PCD, bacterial pathogen resistance and necrotrophic fungal pathogen sensitivity observed in *MACP2-OEs* is possibly mediated by the activation of IGSs and endogenous SA biosynthesis through the EDS1 signaling pathway. These findings provide a genetic framework and knowledge base to study the biochemical function of plant MACPF proteins in future works.

Supplementary Materials: The following supporting information can be downloaded at: https://www.mdpi.com/article/10.3390/ijms23158784/s1.

Author Contributions: Data curation X.Z. and Z.-Z.S.; formal analysis, X.Z. and Y.-S.D.; experiments conduction, X.Z.; materials generation, Z.-Z.S.; pathogen cultivation, Y.-X.W.; funding acquisition, L.-J.Y., Z.-F.Z., S.X., and Q.-F.C.; methodology, X.Z. and Y.-S.D.; supervision, S.X; writing—original draft, X.Z.; writing–review and editing, L.-J.Y., S.X., and Q.-F.C. All authors have read and agreed to the published version of the manuscript.

Funding: This work was supported by the Key Realm R&D Program of Guangdong Province (Project 2020B0202090001), the National Natural Science Foundation of China (Project 31725004, 31870237), and the Natural Science Foundation of Guangdong Province (2022A1515012402).

Institutional Review Board Statement: Not applicable.

Informed Consent Statement: Not applicable.

Data Availability Statement: Not applicable.

Acknowledgments: We thank the ABRC (www.arabidopsis.org, accessed on 24 April 2015) for providing the *macp2* mutant seed pools.

Conflicts of Interest: The authors declare no conflict of interest.

References

1. Dou, D.; Zhou, J.M. Phytopathogen effectors subverting host immunity: Different foes, similar battleground. *Cell Host Microbe* **2012**, *12*, 484–495. [CrossRef] [PubMed]
2. Wang, Y.; Pruitt, R.N.; Nürnberger, T.; Wang, Y.C. Evasion of plant immunity by microbial pathogens. *Nat. Rev. Microbiol.* **2022**, *20*, 449–464. [CrossRef] [PubMed]
3. Couto, D.; Zipfel, C. Regulation of pattern recognition receptor signalling in plants. *Nat. Rev. Immunol.* **2016**, *16*, 537–552. [CrossRef] [PubMed]
4. Fukunaga, S.; Sogame, M.; Hata, M.; Singkaravanit-Ogawa, S.; Pislewska-Bednarek, M.; Onozawa-Komori, M.; Nishiuchi, T.; Hiruma, K.; Saitoh, H.; Terauchi, R.; et al. Dysfunction of Arabidopsis MACPF domain protein activates programmed cell death via tryptophan metabolism in MAMP-triggered immunity. *Plant J.* **2017**, *89*, 381–393. [CrossRef] [PubMed]
5. Boller, T.; Felix, G. A renaissance of elicitors: Perception of microbe-associated molecular patterns and danger signals by pattern-recognition receptors. *Annu. Rev. Plant Biol.* **2009**, *60*, 379–406. [CrossRef]
6. Jones, J.D.G.; Dangl, J.L. The plant immune system. *Nature* **2006**, *444*, 323–329. [CrossRef]
7. Wu, F.; Chi, Y.; Jiang, Z.; Xu, Y.; Xie, L.; Huang, F.; Wan, D.; Ni, J.; Yuan, F.; Wu, X.; et al. Hydrogen peroxide sensor HPCA1 is an LRR receptor kinase in *Arabidopsis*. *Nature* **2020**, *578*, 577–581. [CrossRef]
8. Heath, M.C. Hypersensitive response-related death. *Plant. Mol. Biol.* **2000**, *44*, 321–334. [CrossRef]

9. Tsuda, K.; Katagiri, F. Comparing signaling mechanisms engaged in pattern-triggered and effector-triggered immunity. *Curr. Opin. Plant Biol.* **2010**, *13*, 459–465. [CrossRef]

10. Tsutsui, T.; Morita-Yamamuro, C.; Asada, Y.; Minami, E.; Shibuya, N.; Ikeda, A.; Yamaguchi, J. Salicylic acid and a chitin elicitor both control expression of the *CAD1* gene involved in the plant immunity of *Arabidopsis*. *Biosci. Biotech. Biochem.* **2006**, *70*, 2042–2048. [CrossRef]

11. Heath, M.C. Nonhost resistance and nonspecific plant defenses. *Curr. Opin. Plant Biol.* **2000**, *3*, 315–319. [CrossRef]

12. Lapin, D.; Bhandari, D.D.; Parker, J.E. Origins and immunity networking functions of EDS1 family proteins. *Annu. Rev. Phytopathol.* **2020**, *58*, 253–276. [CrossRef] [PubMed]

13. Zeng, H.Y.; Liu, Y.; Chen, D.K.; Bao, H.N.; Huang, L.Q.; Yin, J.; Chen, Y.L.; Xiao, S.; Yao, N. The immune components ENHANCED DISEASE SUSCEPTIBILITY 1 and PHYTOALEXIN DEFICIENT 4 are required for cell death caused by overaccumulation of ceramides in Arabidopsis. *Plant J.* **2021**, *107*, 1447–1465. [CrossRef] [PubMed]

14. Zhao, X.; Song, L.; Jiang, L.; Zhu, Y.; Gao, Q.; Wang, D.; Xie, J.; Lv, M.; Liu, P.; Li, M. The integration of transcriptomic and transgenic analyses reveals the involvement of the SA response pathway in the defense of chrysanthemum against the necrotrophic fungus *Alternaria* sp. *Hortic. Res.* **2020**, *7*, 80. [CrossRef] [PubMed]

15. Morita-Yamamuro, C.; Tsutsui, T.; Sato, M.; Yoshioka, H.; Tamaoki, M.; Ogawa, D.; Matsuura, H.; Yoshihara, T.; Ikeda, A.; Uyeda, I.; et al. The *Arabidopsis* gene *CAD1* controls programmed cell death in the plant immune system and encodes a protein containing a MACPF domain. *Plant Cell Physiol.* **2005**, *46*, 902–912. [CrossRef]

16. Bayly-Jones, C.; Bubeck, D.; Dunstone, M.A. The mystery behind membrane insertion: A review of the complement membrane attack complex. *Philos. Trans. R. Soc. Lond. B Biol. Sci.* **2017**, *372*, 20160221. [CrossRef]

17. Ni, T.; Gilbert, R.J.C. Repurposing a pore: Highly conserved perforin-like proteins with alternative mechanisms. *Philos. Trans. R. Soc. Lond. B Biol. Sci.* **2017**, *372*, 20160212. [CrossRef]

18. Yu, L.J.; Liu, D.; Chen, S.Y.; Dai, Y.S.; Guo, W.X.; Zhang, X.; Wang, L.N.; Ma, S.R.; Xiao, M.; Qi, H.; et al. Evolution and expression of the membrane attack complex and perforin gene family in the Poaceae. *Int. J. Mol. Sci.* **2020**, *21*, 5736. [CrossRef]

19. Bakopoulos, D.; Whisstock, J.C.; Johnson, T.K. Control of growth factor signalling by MACPF Proteins. *Biochem. Soc. Trans.* **2019**, *47*, 801–810. [CrossRef]

20. Rosado, C.J.; Buckle, A.M.; Law, R.H.; Butcher, R.E.; Kan, W.T.; Bird, C.H.; Ung, K.; Browne, K.A.; Baran, K.; Bashtannyk-Puhalovich, T.A.; et al. A common fold mediates vertebrate defense and bacterial attack. *Science* **2007**, *317*, 1548–1551. [CrossRef]

21. Lukoyanova, N.; Hoogenboom, B.W.; Saibil, H.R. The membrane attack complex, perforin and cholesteroldependent cytolysin superfamily of pore-forming proteins. *J. Cell Sci.* **2016**, *129*, 2125–2133. [PubMed]

22. Moreno-Hagelsieb, G.; Vitug, B.; Medrano-Soto, A.; Saier, M.H., Jr. The Membrane Attack Complex/Perforin Superfamily. *J. Mol. Microb. Biotech.* **2017**, *27*, 252–267. [CrossRef] [PubMed]

23. Rosado, C.J.; Kondos, S.; Bull, T.E.; Kuiper, M.J.; Law, R.H.; Buckle, A.M.; Voskoboinik, I.; Bird, P.I.; Trapani, J.A.; Whisstock, J.C.; et al. The MACPF/CDC family of pore-forming toxins. *Cell Microbiol.* **2008**, *10*, 1765–1774. [CrossRef] [PubMed]

24. Chen, P.; Jian, H.; Wei, F.; Gu, L.; Hu, T.; Lv, X.; Guo, X.; Lu, J.; Ma, L.; Wang, H.; et al. Phylogenetic Analysis of the Membrane Attack Complex/Perforin Domain-Containing Proteins in Gossypium and the Role of GhMACPF26 in Cotton Under Cold Stress. *Front. Plant. Sci.* **2021**, *12*, 684227. [CrossRef] [PubMed]

25. Esser, A.F. The membrane attack complex of complement. Assembly, structure and cytotoxic activity. *Toxicology* **1994**, *87*, 229–247. [CrossRef]

26. Hadders, M.A.; Beringer, D.X.; Gros, P. Structure of C8α-MACPF reveals mechanism of membrane attack in complement immune defense. *Science* **2007**, *317*, 1552–1554. [CrossRef]

27. Dudkina, N.V.; Spicer, B.A.; Reboul, C.F.; Conroy, P.J.; Lukoyanova, N.; Elmlund, H.; Law, R.H.; Ekkel, S.M.; Kondos, S.C.; Goode, R.J.; et al. A Structure of the poly-C9 component of the complement membrane attack complex. *Nat. Commun.* **2016**, *7*, 10588. [CrossRef]

28. Asada, Y.; Yamamoto, M.; Tsutsui, T.; Yamaguchi, J. The Arabidopsis NSL2 negatively controls systemic acquired resistance via hypersensitive response. *Plant Biotechnol.* **2011**, *28*, 9–15. [CrossRef]

29. Chen, T.; Nomura, K.; Wang, X.; Sohrabi, R.; Xu, J.; Yao, L.; Paasch, B.C.; Ma, L.; Kremer, J.; Cheng, Y.; et al. A plant genetic network for preventing dysbiosis in the phyllosphere. *Nature* **2020**, *580*, 653–657. [CrossRef]

30. Noutoshi, Y.; Kuromori, T.; Wada, T.; Hirayama, T.; Kamiya, A.; Imura, Y.; Yasuda, M.; Nakashita, H.; Shirasu, K.; Shinozaki, K. Loss of NECROTIC SPOTTED LESIONS 1 associates with cell death and defense responses in *Arabidopsis thaliana*. *Plant Mol. Biol.* **2006**, *62*, 29–42. [CrossRef]

31. Bhandari, D.D.; Lapin, D.; Kracher, B.; von Born, P.; Bautor, J.; Niefind, K.; Parker, J.E. An EDS1 heterodimer signalling surface enforces timely reprogramming of immunity genes in *Arabidopsis*. *Nat. Commun.* **2019**, *10*, 772. [CrossRef] [PubMed]

32. Ascencio-Ibáñez, J.T.; Sozzani, R.; Lee, T.J.; Chu, T.M.; Wolfinger, R.D.; Cella, R.; Hanley-Bowdoin, L. Global Analysis of Arabidopsis Gene Expression Uncovers a Complex Array of Changes Impacting Pathogen Response and Cell Cycle during Geminivirus Infection. *Plant Physiol.* **2008**, *148*, 436–454. [CrossRef]

33. Wu, Z.J.; Han, S.M.; Zhou, H.D.; Tuang, Z.K.; Wang, Y.Z.; Jin, Y.; Shi, H.Z.; Yang, W.N. Cold stress activates disease resistance in *Arabidopsis thaliana* through a salicylic acid dependent pathway. *Plant Cell Environ.* **2019**, *42*, 2645–2663. [CrossRef] [PubMed]

34. Tang, D.Z.; Ade, J.; Frye, C.A.; Innes, R.W. Regulation of plant defense responses in Arabidopsis by EDR2, a PH and START domain-containing protein. *Plant J.* **2005**, *44*, 245–257. [CrossRef]

35. Liao, K.; Peng, Y.-J.; Yuan, L.-B.; Dai, Y.-S.; Chen, Q.-F.; Yu, L.-J.; Bai, M.-Y.; Zhang, W.-Q.; Xie, L.-J.; Xiao, S. Brassinosteroids Antagonize Jasmonate-Activated Plant Defense Responses through BRI1-EMS-SUPPRESSOR1 (BES1). *Plant Physiol.* **2020**, *182*, 1066–1082. [CrossRef] [PubMed]

36. Holmes, D.R.; Bredow, M.; Thor, K.; Pascetta, S.A.; Sementchoukova, I.; Siegel, K.R.; Zipfel, C.; Monaghan, J. A novel allele of the *Arabidopsis thaliana* MACPF protein CAD1 results in deregulated immune signaling. *Genetics* **2021**, *217*, iyab022. [CrossRef] [PubMed]

37. Torres, M.A.; Dangl, J.L.; Jones, J.D. *Arabidopsis* gp91phox homologues *AtrbohD* and *AtrbohF* are required for accumulation of reactive oxygen intermediates in the plant defense response. *Proc. Natl. Acad. Sci. USA* **2002**, *99*, 517–522. [CrossRef]

38. Musingarimi, P.; Plumb, M.E.; Sodetz, J.M. Interaction between the C8 alpha-gamma and C8 beta subunits of human complement C8: Role of the C8 beta *N*-terminal thrombo-spondin type 1 module and membrane attack complex/perforin domain. *Biochemistry* **2002**, *41*, 11255–11260. [CrossRef]

39. Ishino, T.; Chinzei, Y.; Yuda, M. A Plasmodium sporozoite protein with a membrancade attack complex domain is required for breaching the liver sinusoidal cell layer prior to hepatocyte infection. *Cell Microbiol.* **2005**, *7*, 199–208. [CrossRef]

40. Robert-Seilaniantz, A.; Grant, M.; Jones, J.D. Hormone Crosstalk in Plant Disease and Defense: More Than Just JASMONATE-SALICYLATE Antagonism. *Annu. Rev. Phytopathol.* **2011**, *49*, 317–343. [CrossRef]

41. Qi, H.; Xia, F.N.; Xiao, S.; Li, J. Traf proteins as key regulators of plant development and stress responses. *J. Integr. Plant Biol.* **2022**, *64*, 431–448. [CrossRef] [PubMed]

42. Xie, L.J.; Zhou, Y.; Chen, Q.F.; Xiao, S. New insights into the role of lipids in plant hypoxia responses. *Prog. Lipid Res.* **2021**, *81*, 101072. [CrossRef] [PubMed]

43. Qi, H.; Xia, F.N.; Xiao, S. Autophagy in plants: Physiological roles and post-translational regulation. *J. Integr. Plant Biol.* **2021**, *63*, 161–179. [CrossRef] [PubMed]

44. Gaffney, T.; Friedrich, L.; Vernooji, B.; Negrotto, D.; Nye, G.; Uknes, S.; Ward, E.; Kessmann, H.; Ryals, J. Requirement of salicylic acid for the induction of systemic acquired resistance. *Science* **1993**, *261*, 754–756. [CrossRef]

45. Tsuda, K.; Sato, M.; Glazebrook, J.; Cohen, J.D.; Katagiri, F. Interplay between MAMP-triggered and SA-mediated defence responses. *Plant J.* **2008**, *53*, 763–775. [CrossRef]

46. Rodriguez, E.; Ghoul, H.E.; Mundy, J.; Petersen, M. Making sense of plant autoimmunity and "negative regulators". *FEBS J.* **2016**, *283*, 1385–1391. [CrossRef]

47. Rekhter, D.; Lüdke, D.; Ding, Y.; Feussner, K.; Zienkiewicz, K.; Lipka, V.; Wiermer, M.; Zhang, Y.; Feussner, I. Isochorismate-derived biosynthesis of the plant stress hormone salicylic acid. *Science* **2019**, *365*, 498–502. [CrossRef]

48. Wagner, S.; Stuttmann, J.; Rietz, S.; Guerois, R.; Brunstein, E.; Bautor, J.; Niefind, K.; Parker, J.E. Structural basis for signaling by exclusive EDS1 heteromeric complexes with SAG101 or PAD4 in plant innate immunity. *Cell Host Microbe* **2013**, *14*, 619–630. [CrossRef]

49. Spoel, S.H.; Johnson, J.; Dong, X.N. Regulation of tradeoffs between plant defenses against pathogens with different lifestyles. *Proc. Natl. Acad. Sci. USA* **2007**, *104*, 18842–18847. [CrossRef]

50. Cui, H.; Qiu, J.; Zhou, Y.; Bhandari, D.D.; Zhao, C.; Bautor, J.; Parker, J.E. Antagonism of transcription factor MYC2 by EDS1/PAD4 complexes bolsters salicylic acid defense in Arabidopsis effector-triggered immunity. *Mol. Plant* **2018**, *11*, 1053–1066. [CrossRef]

51. Chen, M.X.; Zhang, K.L.; Yang, J.F.; Chen, C.; Gao, B.; Tian, Y.; Hao, G.F.; Yang, G.F.; Zhang, J.H.; Zhu, F.Y.; et al. Phylogenetic comparison of plant U1-70K gene family, central regulators on 5′ splice site determination, in response to developmental cues and stress conditions. *Plant J.* **2020**, *103*, 357–378. [CrossRef] [PubMed]

52. Zhang, D.; Chen, M.X.; Zhu, F.Y.; Zhang, J.H.; Liu, Y.G. Emerging Functions of Plant Serine/Arginine-Rich (SR) Proteins: Lessons from Animals. *Crit. Rev. Plant Sci.* **2020**, *39*, 173–194. [CrossRef]

53. Chen, M.X.; Zhang, Y.J.; Alisdair, F.R.; Liu, Y.G.; Zhu, F.Y. SWATH-MS based proteomics: Strategies and Applications in Plants. *Trends Biotechnol.* **2021**, *39*, 433–437. [CrossRef]

54. Chen, M.X.; Mei, L.C.; Wang, F.; Wijethunge, I.K.; Yang, J.F.; Dai, L.; Yang, G.F.; Gao, B.; Cheng, C.L.; Liu, Y.G.; et al. PlantSPEAD: A web resource towards comparatively analyzing stress-responsive expression of splicing-related proteins in plant. *Plant Biotechnol. J.* **2021**, *19*, 227–229. [CrossRef] [PubMed]

55. Chen, M.X.; Zhang, K.L.; Zhang, M.; Das, D.; Fang, Y.M.; Zhang, J.H.; Zhu, F.Y. Alternative splicing and its regulatory role in woody plants. *Tree Physiol.* **2020**, *40*, 1475–1486. [CrossRef]

56. Hartmann, M.; Zeier, T.; Bernsdorff, F.; Reichel-Deland, V.; Kim, D.; Hohmann, M.; Scholten, N.; Schuck, S.; Brautigam, A.; Holzel, T.; et al. Flavin monooxygenase-generated N-hydroxypipecolic acid is a critical element of plant systemic immunity. *Cell* **2018**, *173*, 456–469.e416. [CrossRef]

57. Fan, J.; Doerner, P. Genetic and molecular basis of nonhost disease resistance: Complex, yes; silver bullet, no. *Curr. Opin. Plant Biol.* **2012**, *15*, 400–406. [CrossRef]

58. Wittstock, U.; Halkier, B.A. Glucosinolate research in the *Arabidopsis* era. *Trends Plant Sci.* **2002**, *7*, 263–270. [CrossRef]

59. Hiruma, K.; Onozawa-Komori, M.; Takahashi, F.; Asakura, M.; Bednarek, P.; Okuno, T.; Schulze-Lefert, P.; Takano, Y. Entry mode-dependent function of an indole glucosinolate pathway in Arabidopsis for non-host resistance against anthracnose pathogens. *Plant Cell* **2010**, *22*, 2429–2443. [CrossRef]

60. Bednarek, P.; Pislewska-Bednarek, M.; Svatos, A.; Schneider, B.; Doubsky, J.; Mansurova, M.; Humphry, M.; Consonni, C.; Panstruga, R.; Sanchez-Vallet, A.; et al. A glucosinolate metabolism pathway in living plant cells mediates broad-spectrum antifungal defense. *Science* **2009**, *323*, 101–106. [CrossRef]
61. Stotz, H.U.; Sawada, Y.; Shimada, Y.; Hirai, M.Y.; Sasaki, E.; Krischke, M.; Brown, P.D.; Saito, K.; Kamiya, Y. Role of camalexin, indole glucosinolates, and side chain modification of glucosinolate-derived isothiocyanates in defense of Arabidopsis against *Sclerotinia sclerotiorum*. *Plant J.* **2011**, *67*, 81–93. [CrossRef] [PubMed]
62. Xu, J.; Meng, J.; Meng, X.; Zhao, Y.; Liu, J.; Sun, T.; Liu, Y.; Wang, Q.; Zhang, S. Pathogen-Responsive MPK3 and MPK6 Reprogram the Biosynthesis of Indole Glucosinolates and Their Derivatives in Arabidopsis Immunity. *Plant Cell* **2016**, *28*, 1144–1162. [CrossRef] [PubMed]
63. Chen, Q.F.; Xu, L.; Tan, W.J.; Chen, L.; Qi, H.; Xie, L.J.; Chen, M.X.; Liu, B.Y.; Yu, L.J.; Yao, N.; et al. Disruption of the *Arabidopsis* Defense Regulator Genes SAG101, EDS1, and PAD4 Confers Enhanced Freezing Tolerance. *Mol. Plant* **2015**, *8*, 1536–1549. [CrossRef] [PubMed]
64. Zhou, Y.; Zhou, D.M.; Yu, W.W.; Shi, L.L.; Zhang, Y.; Lai, Y.X.; Huang, L.P.; Qi, H.; Chen, Q.F.; Yao, N.; et al. Phosphatidic acid modulates MPK3- and MPK6-mediated hypoxia signaling in Arabidopsis. *Plant Cell* **2022**, *34*, 889–909. [CrossRef]
65. Li, J.F.; Chung, H.S.; Niu, Y.; Bush, J.; McCormack, M.; Sheen, J. Comprehensive protein-based artificial microRNA screens for effective gene silencing in plants. *Plant Cell* **2013**, *25*, 1507–1522. [CrossRef]
66. Qi, H.; Li, J.; Xia, F.-N.; Chen, J.-Y.; Lei, X.; Han, M.-Q.; Xie, L.-J.; Zhou, Q.-M.; Xiao, S. Arabidopsis SINAT Proteins Control Autophagy by Mediating Ubiquitylation and Degradation of ATG13. *Plant Cell* **2020**, *32*, 263–284. [CrossRef]
67. Clough, S.; Bent, A. Floral dip: A simplified method for *Agrobacterium*-mediated transformation of *Arabidopsis thaliana*. *Plant J.* **1998**, *16*, 735–743. [CrossRef]
68. Xiao, S.; Chye, M.L. Overexpression of Arabidopsis ACBP3 enhances NPR1-dependent plant resistance to *Pseudomonas syringe* pv *tomato* DC3000. *Plant Physiol.* **2011**, *156*, 2069–2081. [CrossRef]
69. Gong, B.Q.; Guo, J.; Zhang, N.; Yao, X.; Wang, H.B.; Li, J.F. Cross-Microbial Protection via Priming a Conserved Immune Co-Receptor through Juxtamembrane Phosphorylation in Plants. *Cell Host Microbe* **2019**, *26*, 810–822.e7. [CrossRef]
70. Hsu, F.C.; Chou, M.Y.; Chou, S.J.; Li, Y.R.; Peng, H.P.; Shih, M.C. Submergence confers immunity mediated by the WRKY22 transcription factor in *Arabidopsis*. *Plant Cell* **2013**, *25*, 2699–2713. [CrossRef]
71. Hu, P.; Zhou, W.; Cheng, Z.W.; Fan, M.; Wang, L.; Xie, D.X. JAV1 controls jasmonate-regulated plant defense. *Mol. Cell* **2013**, *50*, 504–515. [CrossRef]
72. Chen, L.; Liao, B.; Qi, H.; Xie, L.J.; Huang, L.; Tan, W.J.; Zhai, N.; Yuan, L.B.; Zhou, Y.; Yu, L.J.; et al. Autophagy contributes to regulation of the hypoxia response during submergence in *Arabidopsis thaliana*. *Autophagy* **2015**, *11*, 2233–2246. [CrossRef] [PubMed]
73. Pan, X.; Welti, R.; Wang, X. Quantitative analysis of major plant hormones in crude plant extracts by high-performance liquid chromatography-mass spectrometry. *Nat Protoc.* **2010**, *5*, 986–992. [CrossRef] [PubMed]
74. Lu, K.; Li, T.; He, J.; Chang, W.; Zhang, R.; Liu, M.; Yu, M.; Fan, Y.; Ma, J.; Sun, W.; et al. qPrimerDB: A thermodynamics-based gene-specific qPCR primer database for 147 organisms. *Nucleic Acids Res.* **2018**, *46*, D1229–D1236. [CrossRef] [PubMed]
75. Chen, C.; Chen, H.; Zhang, Y.; Thomas, H.R.; Frank, M.H.; He, Y.; Xia, R. TBtools: An Integrative Toolkit Developed for Interactive Analyses of Big Biological Data. *Mol. Plant* **2020**, *13*, 1194–1202. [CrossRef]
76. Glauser, G.; Schweizer, F.; Turlings, T.C.; Reymond, P. Rapid profiling of intact glucosinolates in *Arabidopsis* leaves by UHPLC-QTOFMS using a charged surface hybrid column. *Phytochem. Anal.* **2012**, *23*, 520–528. [CrossRef]
77. Xia, F.-N.; Zeng, B.; Liu, H.-S.; Qi, H.; Xie, L.-J.; Yu, L.-J.; Chen, Q.-F.; Li, J.-F.; Chen, Y.-Q.; Jiang, L.; et al. SINAT E3 Ubiquitin Ligases Mediate FREE1 and VPS23A Degradation to Modulate Abscisic Acid Signaling. *Plant Cell* **2020**, *32*, 3290–3310. [CrossRef]

International Journal of
Molecular Sciences

Article

Identifying QTLs for Grain Size in a Colossal Grain Rice (*Oryza sativa* L.) Line, and Analysis of Additive Effects of QTLs

Xuanxuan Hou [1,†], Moxian Chen [2,3,4,†], Yinke Chen [1], Xin Hou [3], Zichang Jia [3], Xue Yang [3], Jianhua Zhang [5,6], Yinggao Liu [3,*] and Nenghui Ye [1,*]

1   College of Agriculture, Hunan Agricultural University, Changsha 410128, China; 18854806295@163.com (X.H.); chenyinkede@foxmail.com (Y.C.)
2   Co-Innovation Center for Sustainable Forestry in Southern China & Key Laboratory of National Forestry and Grassland Administration on Subtropical Forest Biodiversity Conservation, College of Biology and the Environment, Nanjing Forestry University, Nanjing 210037, China; cmx2009920734@gmail.com
3   State Key Laboratory of Crop Biology, College of Life Science, Shandong Agricultural University, Tai'an 271018, China; h15621523656@163.com (X.H.); jiazc973@163.com (Z.J.); xueyang202001@163.com (X.Y.)
4   CAS Key Laboratory of Quantitative Engineering Biology, Shenzhen Institute of Synthetic Biology, Shenzhen Institutes of Advanced Technology, Chinese Academy of Sciences, Shenzhen 518055, China
5   Department of Biology, Hong Kong Baptist University, Hong Kong 999077, China; jzhang@hkbu.edu.hk
6   State Key Laboratory of Agrobiotechnology, The Chinese University of Hong Kong, Hong Kong 999077, China
*   Correspondence: liuyg@sdau.edu.cn (Y.L.); nye@hunau.edu.cn (N.Y.)
†   These authors contributed equally to this work.

Citation: Hou, X.; Chen, M.; Chen, Y.; Hou, X.; Jia, Z.; Yang, X.; Zhang, J.; Liu, Y.; Ye, N. Identifying QTLs for Grain Size in a Colossal Grain Rice (*Oryza sativa* L.) Line, and Analysis of Additive Effects of QTLs. *Int. J. Mol. Sci.* **2022**, *23*, 3526. https://doi.org/10.3390/ijms23073526

Academic Editor: Luigi Cattivelli

Received: 4 March 2022
Accepted: 21 March 2022
Published: 24 March 2022

Publisher's Note: MDPI stays neutral with regard to jurisdictional claims in published maps and institutional affiliations.

Abstract: Grain size is an important component of quality and harvest traits in the field of rice breeding. Although numerous quantitative trait loci (QTLs) of grain size in rice have been reported, the molecular mechanisms of these QTLs remain poorly understood, and further research on QTL observation and candidate gene identification is warranted. In our research, we developed a suite of F_2 intercross populations from a cross of 9311 and CG. These primary populations were used to map QTLs conferring grain size, evaluated across three environments, and then subjected to bulked-segregant analysis-seq (BSA-seq). In total, 4, 11, 12 and 14 QTLs for grain length (GL), grain width (GW), 1000-grain weight (TGW), and length/width ratio (LWR), respectively, were detected on the basis of a single-environment analysis. In particular, over 200 splicing-related sites were identified by whole-genome sequencing, including one splicing-site mutation with G>A at the beginning of intron 4 on *Os03g0841800* (*qGL3.3*), producing a smaller open reading frame, without the third and fourth exons. A previous study revealed that the loss-of-function allele caused by this splicing site can negatively regulate rice grain length. Furthermore, *qTGW2.1* and *qGW2.3* were new QTLs for grain width. We used the near-isogenic lines (NILs) of these GW QTLs to study their genetic effects on individuals and pyramiding, and found that they have additive effects on GW. In summary, these discoveries provide a valuable genetic resource, which will facilitate further study of the genetic polymorphism of new rice varieties in rice breeding.

Keywords: rice; grain size; QTL; BSA-seq; splicing; additive effects

1. Introduction

Rice (*Oryza sativa* L.) is one of the crucial and staple cereal crops in the global village and provides over 21 percent of the food needed by humankind [1]. There are many advantages (published genome sequence, smaller genome size, high efficiency of genetic transformation technology, and rich genetic resources) of rice, making it one of the most popular model plants for studying plant biology, especially among monocotyledons [2]. In 2050, the global population is forecast to increase to 9.7 billion, and we will require more food to sustainably feed the world [3]. To address the food crisis, it is particularly important to increase the yield of rice and other food crops.

Effective tiller number per plant, grain number per tiller, and grain weight are the key factors controlling rice yield. Grain weight is predominantly controlled by grain shape [4]. To date, several major quantitative trait loci (QTLs) and genes controlling grain size have been cloned and characterized (http://www.gramene.org/; accesed on 12 August 2021), most of which play a role in affecting the yield and appearance quality in rice [1,2]. These QTLs and genes, which encode different kinds of proteins, regulate grain size by affecting cell expansion and proliferation in the seed coat or the growth of the endosperm [5].

It has been reported that there are a few genes that affect grain size by controlling cell proliferation [5]. GW2 is a RING-type E3 ubiquitin ligase that negatively controls grain width by anchoring substrates to proteasomes, to regulate the proteolytic process [6]. *GW5* is an important gene for controlling grain width and regulating the ubiquitin–proteasome pathway [7]. Moreover, GS3 is a part of the G protein signaling pathway, which is associated with rice grain size [8,9]. *qGL3* regulates grain length antagonistically and contains two Kelch motifs that are necessary and sufficient for the negative regulatory function of OsPPKL1 [10]. *SG1* functions in the elongation of seeds, by decreasing cellular proliferation and reducing responses to brassinolides (BRs) [11]. *SMG1* encodes a mitogen-activated protein kinase kinase 4 (OsMKK4), which regulates cell proliferation and grain size in rice. OsMKK4 may act as a linkage factor between the BR and MAPK pathways in grain development [12]. OsMKKK10, OsMKK4, and OsMAPK6 work in a common signaling pathway, to regulate the grain size of rice by enhancing cell proliferation [13]. *OsUBP15/LG1* affects grain width by regulating cell proliferation, and a ubiquitin-specific protease15 that it encodes has deubiquitination activity. The results of genetic analysis indicated that *GW2* and *LG1* have some relationship in affecting rice grain width [14]. All of these genes affect grain size by controlling cell proliferation. In addition, several genes affect grain size by controlling cell expansion [5]. *OsBRI1* (*d61*) encodes a putative BR receptor kinase that has a regulatory effect on rice growth and development [15]. SRS5 is an alpha-tubulin protein and independently controls cell elongation in the BR signaling pathway [16]. *GL7* and its negative regulator together regulate longitudinal cell elongation [17]. GS2/OsGRF4 is a transcription factor that enhances grain weight and yield, primarily by regulating cell expansion [18]. OsmiR396 could not regulate the expression of mutated OsGRF4, and Os-GRF4 directly interacts with OsGIF1. Thus, there is an OsmiR396–OsGRF4–OsGIF1 module that regulates grain size [19]. As a plant-specific transcription factor, GLW7 affects cell size by controlling cell expansion in the grain hull [20]. *WTG1* encodes a functional deubiquitinating enzyme and affects cell expansion in spikelet hulls, to determine grain size [21].

Alternative splicing (AS) mainly regulates plant development and stress responses at the posttranscriptional level and is one of the molecular mechanisms through which plants regulate their gene interaction network [22,23]. Omics technologies have been diffusely used to explore plant growth, organization development, and molecular characteristics [24–26]; these technologies provide large amounts of information and can be used to look for alternative splicing sites. Genome-wide transcriptome mapping has shown that the range of alternative splicing in plants ranges from 42 to 61%. During different stages of plant development and responses to environmental stress, many genes can produce different alternative splicing products to participate in particular developmental processes and responses to various environments [27]. Previous studies have found that approximately 48% of rice genes show alternative splicing patterns [28]. In rice, AS participates in the circadian clock [29], grain quality [30], grain size [31], and so on. Some important stress-response genes of AS, experienced in response to stress, are conserved in different plant species [32–34].

In our study, we compared the whole-genome sequence of CG (a colossal grain rice) with that of Nipponbare and found many differences. Among them, AS may play an important role in the particularly large phenotypic differences between CG and the other japonica rice. Meanwhile, we constructed an F_2 population using a cross between 9311 and CG, to locate QTLs correlated with grain size. In light of the QTL mapping results, we developed a series of different grain size NILs, by successively backcrossing and selfing CG with 9311 as the recurrent parent. The objective of developing these NILs was to analyze the additive and positive effects on the traits. In the future, the results of the current research will be useful for exploring new genes related to grain size.

2. Results

2.1. Phenotypic Variation and Correlation Analysis

The two parents, 9311 and CG, showed highly significant differences in all grain-related traits (GL, GW, LWR, and TGW). The grain size of CG and 9311 hybrid F_1 plants was large and full, indicating that the large-grain phenotype was superior to the small-grain phenotype (Table 1; Figure 1). CG had an extra-large grain size per 1000-grain weight (over 65 g) in comparison to 9311 (approximately 32 g). For the other grain-related traits, there was also a significant difference between the parents in all environments (Table 1). This result also indicates that CG can show good yield traits of grain in different environments. All grain-related traits of the $F_{2:3}$ populations exhibited a typical and continuous normal distribution in the three investigated environments (Figure 2), indicating a quantitative inheritance suitable for QTL identification. For all grain-related traits, transgressive inheritance was found in all $F_{2:3}$ populations. Nevertheless, the plants produced by transgressive inheritance exhibited more extreme grain-related traits in Jining, demonstrating that these grain-related phenotypes were potentially regulated by the environment.

To explore the correlations between these traits, we analyzed the Pearson correlation coefficient of the F_2 population traits using IBM SPSS Statistics 26. GL and GW displayed significant positive correlations with TGW. However, LWR and TGW showed no significant correlations (Table 2).

Table 1. Phenotypic values of grain size in $F_{2:3}$ populations.

Trait	Env.	Parents		F₂:₃ Families				h^2_b (%)
		CG	9311	Mean ± SD	Range	Skewness	Kurtosis	
GL (mm)	15JN	13.92 ****	9.66	11.33 ± 0.96	9.18–15.24	0.340	0.431	
	18HN	13.42 ****	9.55	11.35 ± 0.96	9.51–14.75	0.445	0.037	97.7
	19HUN	13.98 ****	9.81	10.59 ± 0.92	8.60–13.91	0.470	0.458	
GW (mm)	15JN	4.24 ****	2.78	3.28 ± 0.38	2.43–4.55	0.214	−0.189	
	18HN	4.22 ****	2.85	3.46 ± 0.33	2.76–4.50	0.542	−0.188	98.5
	19HUN	4.31 ****	2.81	3.05 ± 0.31	2.44–3.95	0.478	0.062	
LWR	15JN	3.29 ***	3.48	3.48 ± 0.39	2.30–4.81	0.182	0.003	
	18HN	3.18 ***	3.36	3.31 ± 0.37	2.42–4.33	−0.007	−0.239	93.9
	19HUN	3.25 ****	3.49	3.50 ± 0.41	2.49–4.67	0.256	0.015	
TGW (g)	15JN	69.25 ****	32.15	42.71 ± 7.57	22.00–64.34	−0.048	−0.303	
	18HN	65.00 ****	32.25	45.23 ± 6.22	31.30–70.10	0.385	0.776	90.6
	19HUN	67.09 ****	31.58	44.83 ± 6.35	28.86–67.50	0.470	0.469	

15JN, Jining 2015; 18HN, Hainan 2018; 19HUN, Hunan 2019. GL, grain length (mm); GW, grain width (mm); LWR, length/width ratio; TGW, thousand-grain weight (g); SD, standard deviation; Env. represents environments; h^2_b represents broad-sense heritability. Student's t-test was used to generate p values; *** $p < 0.001$; **** $p < 0.0001$.

2.2. Identification of TGW QTLs Using QTL-Seq

The NGS-based high-throughput sequencing of the two parental genotypes, B-pool and S-pool, resulted in 34,835,652, 31,949,878, 127,452,818, and 110,758,536 high-quality short reads, covering 89.34–96.92% of the Nipponbare reference genome (Table S1). GATK software was used to analyze the SNPs in the results. Finally, we obtained 640,595 high-quality SNPs from polymorphisms in two parents and homozygotes for each parent. The Δ (SNP index) and SNP index were computed using these SNPs. The average SNP index across a 100-kb genomic range was calculated for the S-pool and B-pool using a 1-kb sliding window and mapped by genome location, then Δ (SNP index) was calculated between the two extreme bulk samples (Figure 3). A 95% (blue) confidence level was chosen as the screening threshold, and windows larger than the threshold were selected as our candidate intervals. Three regions were detected on chromosome 2, two regions were detected on chromosome 3, and three regions were detected on chromosome 5. As a result, these regions may contain major QTLs controlling TGW. Nevertheless, these regions contained abundant candidate SNPs, which made it difficult to find candidate genes responsible for TGW.

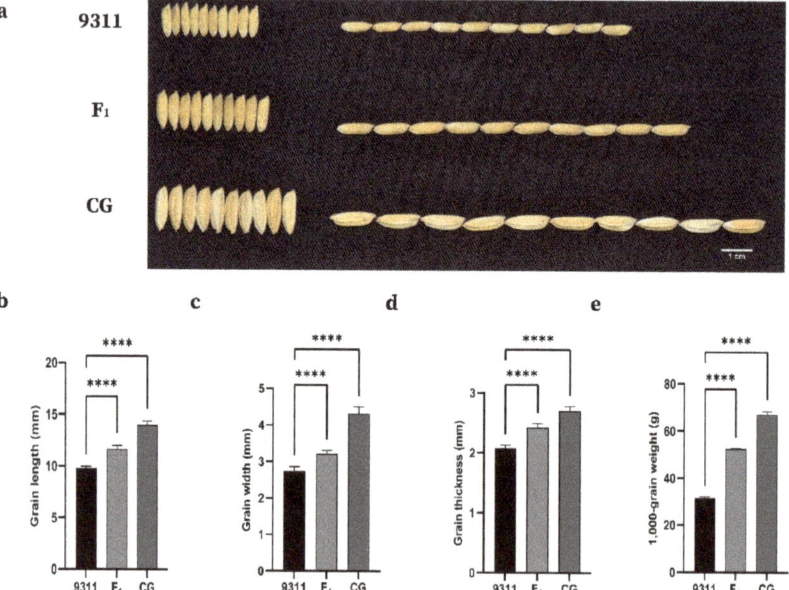

Figure 1. Grain size of two parents and F_1 individuals derived from the cross between CG and 9311. (**a**) Grain phenotypes of 9311, F_1, and CG, Bar, 1 cm. (**b**–**e**) Grain length, width, thickness, and 1000-grain weight of 9311, F_1, and CG. Student's *t*-test was used to generate *p* values; **** *p* < 0.0001.

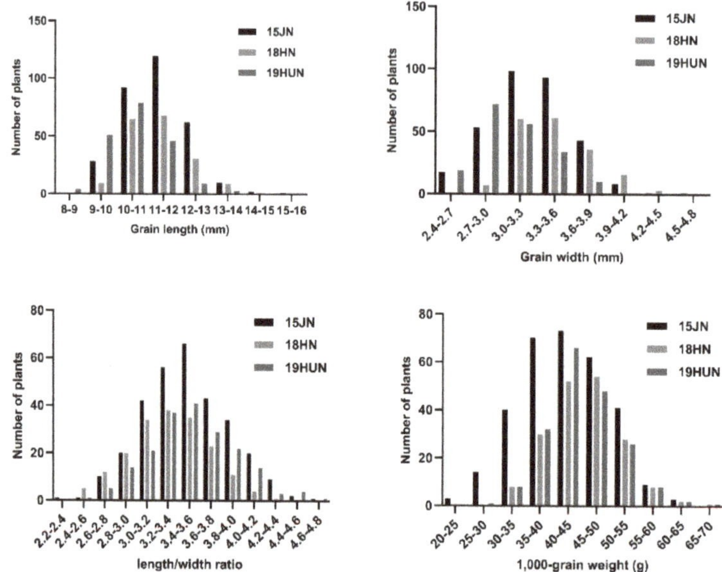

Figure 2. The frequency of the four grain traits in the F_2 population. The frequency of grain length, width, length/width ratio, and 1000–grain weight in the F_2 population.

Table 2. Phenotypic correlation between four grain traits (GL, GW, TGW, and LWR) of the $F_{2:3}$ populations across three environments.

Trait	Env.	GL	GW	TGW	LWR
GL	15JN	1			
	18HN	1			
	19HUN	1			
GW	15JN	0.385 **	1		
	18HN	0.163 *	1		
	19HUN	0.241 **	1		
TGW	15JN	0.740 **	0.631 **	1	
	18HN	0.751 **	0.627 **	1	
	19HUN	0.790 **	0.708 **	1	
LWR	15JN	0.339 **	−0.728 **	−0.114 *	1
	18HN	0.603 **	−0.683 **	0.046	1
	19HUN	0.536 **	−0.682 **	−0.027	1

* $p < 0.05$, ** $p < 0.01$ Env. represents environments; 15JN, 18HN, and 19HUN represent Jining in 2015, Hainan in 2018, and Hunan in 2019, respectively.

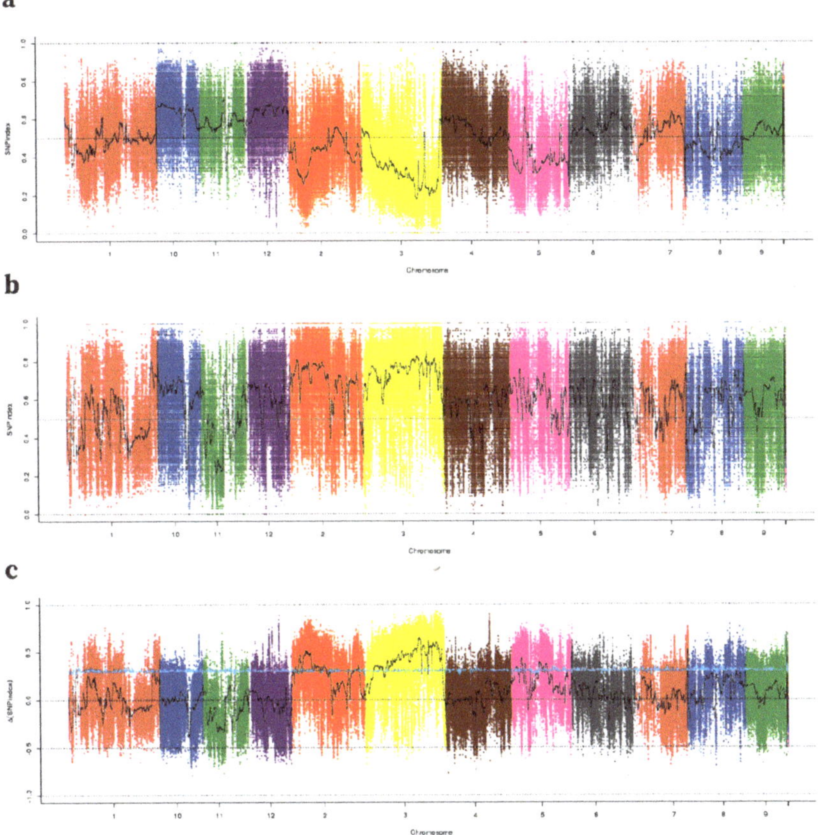

Figure 3. SNP-index Manhattan plot of B-pool, S-pool, and Δ (SNP-index) from the BSA analysis. SNP-index graphs of (**a**) big-grain bulk and (**b**) small-grain bulk. (**c**) Δ(SNP-index) graph. The X-axis shows the position of the 12 chromosomes and the Y-axis shows the SNP-index (**a**,**b**) and Δ (SNP-index). The blue line is the screening threshold at a 95% confidence level. SNPs are set to different colors to help distinguish chromosomal boundaries.

2.3. QTL Mapping of Grain Size in F_2 Populations

In the $F_{2:3}$ population, up to forty-one QTLs for rice grain size were detected separately in each environment and spread on eight chromosomes (Table S2; Figure 4). Each QTL explained 2.35% to 53.61% of the phenotypic variance.

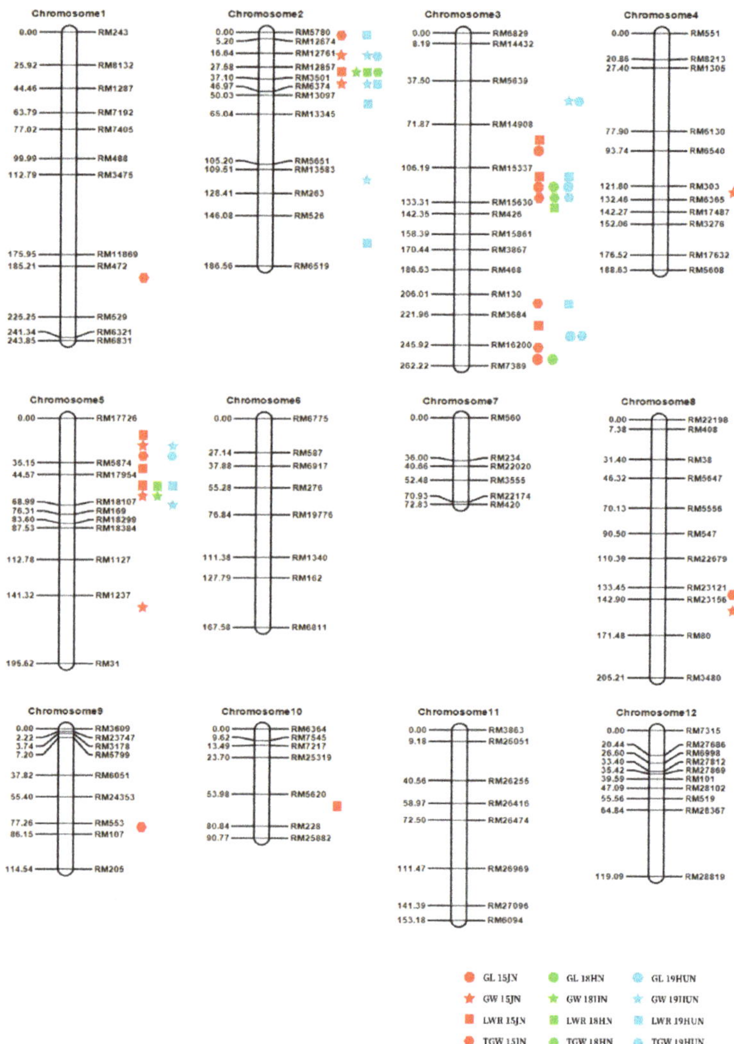

Figure 4. Distribution of identified quantitative trait loci (QTL) for grain size and grain weight on genetic linkage maps in F_2 populations. The marks denote peak positions of QTL; 15JN, 18HN, and 19HUN represent Jining in 2015, Hainan in 2018 and Hunan in 2019, respectively. Numbers on the left side are the genetic distances between two flanking markers with the unit centiMorgan (cM).

For GL, four QTLs were discovered on only chromosome 3 across the three tested environments (Table S2; Figure 4). This finding suggested that chromosome 3 is the key chromosome controlling grain length. The major QTL *qGL3.2* overlapped in three environments, accounting for 22.23–53.61% of the phenotypic variation. Another major QTL, *qGL3.4*, explained 12.11% and 11.17% of the phenotypic variation in the two environments, separately.

For GW, 11 QTLs were mapped on chromosomes 2, 3, 4, 5, and 8 (Table S2; Figure 4). Among them, four QTLs were expressed across two environments, and seven QTLs were specific to one location. The proximity of the two QTL (*qGW2.1* and *qGW2.2*) regions indicated that they might represent one QTL detected in different environments. Three major QTLs on chromosome 2 (*qGW2.1*, *qGW2.2* and *qGW2.3*) were identified and explained 9.82–48.61% of the phenotypic variance. Moreover, three major QTLs on chromosome 5 (*qGW5.1*, *qGW5.2* and *qGW5.3*) were detected and revealed high LOD scores ranging from 10.34 to 23.27.

For the LWR, a total of 14 QTLs were found and distributed on four chromosomes, with PVEs of 2.51–33.68% (Table S2; Figure 4). Among them, three QTLs (*qLWR2.2*, *qLWR3.2* and *qLWR5.3*) were significant in multiple environments, and only one QTL, *qLWR5.3*, was identified across all three environments. In a single environment, the largest number of QTLs was identified in Jining in 2015.

For TGW, 12 QTLs were mapped on chromosomes 1, 2, 3, 5, 8, and 9 (Table S2; Figure 4). *qTGW3.2* and *qTGW5.1* were consistently observed in multiple environments, and others were environment-specific QTLs. A major QTL, *qTGW3.2*, was constantly resolved in all three environments, explaining 24.01–33.47% of the phenotypic variance.

It is worth noting that some QTLs were located in nearly the same region (Table S2). The area between RM12674 and RM3501 on chromosome 2 was connected with *qGW2.1*, *qGW2.2*, *qLWR2.2*, *qTGW2.2* and *qTGW2.3*; and the hereditary effects of these QTLs might all result from the major gene *GW2* and other new minor genes. Combined with the results of whole-genome sequencing, it was found that a 1-bp deletion resulted in a premature stop codon in exon 4 of *GW2*. DNA sequencing also showed this difference in *GW2* between CG and 9311. Based on the sequencing results and EnsemblPlants (http://plants.ensembl.org; accessed on 16 May 2021) data, we predicted the amino acid sequences of GW2 in the coding regions of 9311, CG and Nipponbare (Figure S1). Other major QTLs for GL (*qGL3.3*, *qGL3.4*, *qLWR3.5*, *qTGW3.4* and *qTGW3.5*) were located in the same region as *GL3.3*. We also found base mutation-induced alternative splicing of *OsSK41/OsGSK5* in whole-genome sequencing results. These results are consistent with those previously published [31]. These results also demonstrated the accuracy of the high-throughput sequencing results.

2.4. Construct NIL and Analysis of Additive Effects of Some QTLs

To assess the adjustment effects of several adjacent QTLs on chromosome 2, we developed four NILs in the BC_3F_5 population by marker-assisted selection (Figure 5a). NIL-0 (similar to 9311), NIL-1 (containing *qLWR2.1/qTGW2.1*), NIL-2 (*qGW2.1* and *qGW2.2*), and NIL-3 (*qGW2.1*, *qGW2.2* and *qGW2.3*) had positive additive effects on grain width (Figure 5b,c). Thus, compared to NIL-0, NIL-2 and NIL-3 had significantly increased grain width, by 22.9% and 33.2%, respectively. The 1000-grain weights of the two were almost the same and more than 33% higher than that of NIL-0, which might be due to the slightly longer grain length of NIL-2 compared to NIL-3 (Figure 5f,g). Moreover, we constructed NIL-1 using the tightly-linked molecular markers RM5780 and RM12674. Although the locus in NIL-1 was not detected in grain width alone, it still increased grain width, and we suspected that it might be a minor QTL covered by *GW2*.

2.5. Candidate Gene Analysis of the qTGW2.1 Target Region

Since this locus was a minor QTL, we could only perform fine mapping by combining the constructed NIL-1 with high-throughput sequencing. To avoid missing the possible mutation sites, we expanded the screening range and finally selected 5,350,000–5,780,000 bp on chromosome 2 as the candidate interval for *qTGW2.1*. SNPs and InDels existing in the above interval were screened by whole-genome sequencing results, and the genes at the loci causing frameshift, nonsynonymous, or alternative splicing were preferentially selected as candidate genes; 39 genes were finally screened (Table S3).

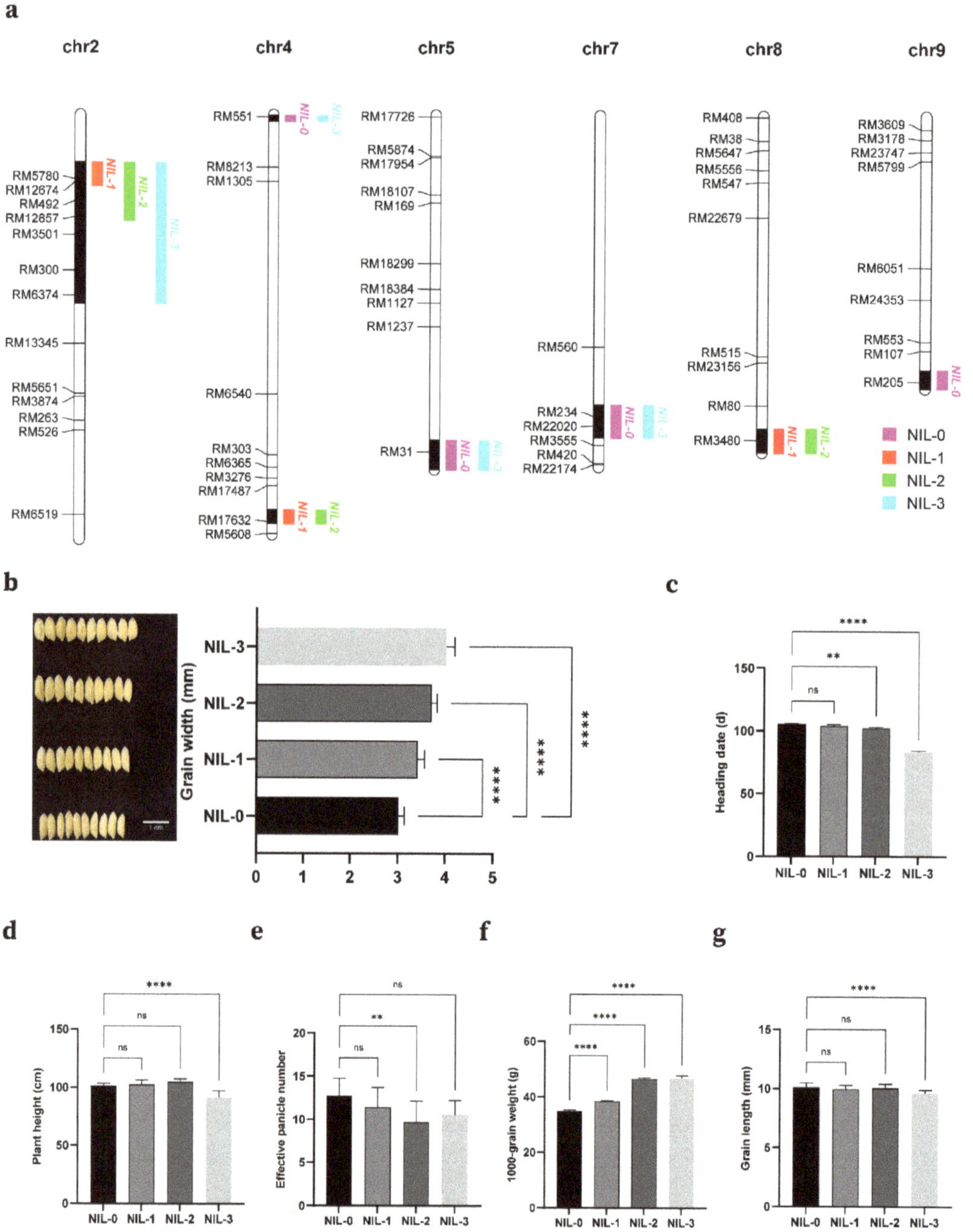

Figure 5. Yield related agronomic traits of the NILs and additive effects of grain width QTLs in NILs. (**a**) The introgression segments of four NILs genotype. (**b**) The additive effect of grain width QTLs, Bar, 1 cm. (**c–g**) Yield-related agronomic traits (heading date, plant height, effective panicle number, 1000-grain weight, and grain length) of the NILs. Student's *t*-test was used to generate *p* values; ns, no significant difference, **** *p* < 0.0001, ** *p* < 0.01.

3. Discussion

3.1. Strategy: An Efficient and Economical Approach for QTL Mapping

Rice yield is closely related to grain size. A few QTLs of rice grain size have been detected and characterized [35]. Previously, traditional BSA was shown to be an elegant method to detect molecular markers that are closely related to QTLs or target genes for an assigned phenotype [36]. It should be noted that the practicality of DNA markers is the major factor constraining their usefulness. Moreover, genotyping of the bulked DNAs expends considerable energy and money.

In recent years, high-throughput sequencing technologies have been gradually applied to map genes. Combining high-throughput sequencing and BSA has been shown to be an effective and economical method for QTL mapping [37]. Some successful studies have been reported in rice [37–40].

In this study, we not only used SNPs obtained by high-throughput sequencing technology for primary mapping of grain weight, but also developed the detected InDels into molecular markers, which contributed to marker-assisted selection and fine mapping of the newly discovered QTLs.

3.2. Novel QTLs Are Identified for Grain Size

Considering the great difference in grain size between CG and 9311, we used the F_2 population to map QTLs in different environments. Finally, up to 41 QTLs related to grain size were detected in these populations. We contrasted corresponding and colocalized QTLs with previously reported QTLs for similar or identical traits, according to the physical positions of markers closely associated with the base linkage map. Most of the QTLs detected in our study were found to be located at similar or identical chromosomal regions as previously reported [6,7,10,31,41]. *qGW2.1* and *qGW2.2* affect grain width and are linked to RM12857 (8.6 Mb) and may be identical to *GW2* (8.1 Mb) [6]. *qGL3.3* and *qGL3.4*, which affect grain length, are linked to RM16200 (35.6 Mb) and may be the same as *GL3.3* (35.4 Mb) [31]. We verified these reported major QTLs using whole-genome sequencing results.

Furthermore, comparing the other QTLs in the three regions, we regarded some of them as having the same loci. For instance, *qGL3.2*, *qLWR3.2*, *qLWR3.3* and *qTGW3.2* were mapped to the same location as *OsPPKL1* [10]. *qGW5.1*, *qLWR5.1*, *qLWR5.2* and *qTGW5.1* were mapped to the same location as *GS5* [41]. *qGW5.2*, *qGW5.3* and *qLWR5.3* were mapped to the same location as *GW5* [7].

In addition, we identified two valuable new loci (*qTGW2.1* and *qGW2.3*) and fine mapped one of them. There are many candidate genes related to rice growth and development in *qTGW2.1*. For example, *Os02g0202000* (*OsWR1*) is an ethylene response factor that regulates wax synthesis and drought tolerance [42]. *Os02g0202400* (*OsBT1*) encodes an ADP-glucose transporter protein that controls the synthesis of starch during grain development [43]. *Os02g0202800* encodes the FAR1 domain-containing protein. In Arabidopsis, FAR1 and its homologue FHY3 have crucial functions in plant growth and development [44].

The above QTLs showed stability in different genetic backgrounds and environments and could be used for MAS. The detected alleles are crucial for NILs based on the 9311 genetic background in biodiversity research and molecular breeding. A number of the major QTLs will be further applied for fine mapping, cloning, and analysis of their genetic mechanisms.

3.3. Alternative Splicing in the Mutant CG

We compared the whole-genome sequencing results with the Nipponbare reference genome and identified 222 splicing sites, including the published gene *OsGSK5* (Figure 6; Table S4). *OsGSK5* can encode a GSK3/SHAGGY-like kinase and negatively affect GL and TGW by regulating cell size and number. Other splicing sites may also have some functions in regulating rice growth and development and responding to stress.

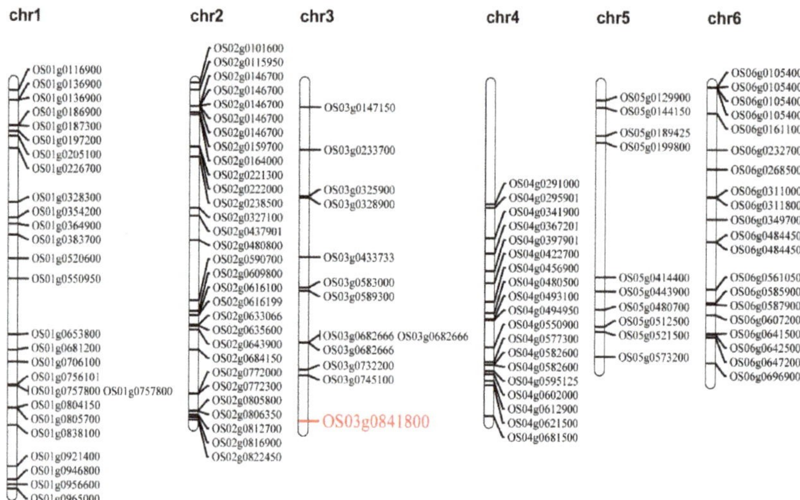

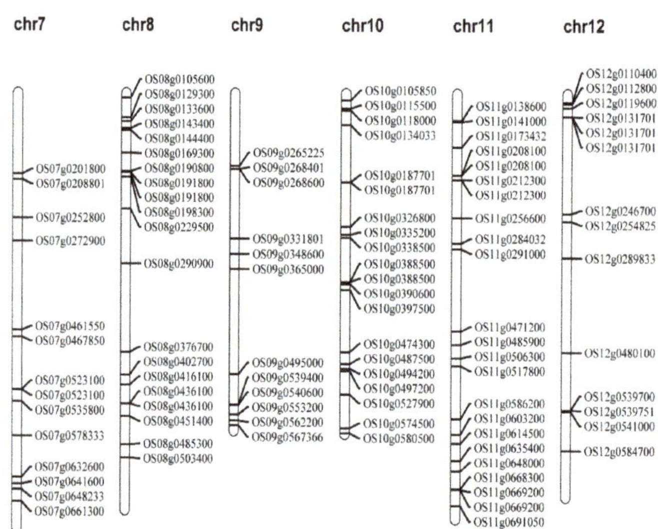

Figure 6. Distribution of identified splicing sites, which compared the whole genome sequencing of CG with the Nipponbare reference genome.

Whole-genome and transcriptome datasets of different rice varieties can serve as a powerful resource for the biological interpretation of trait-related loci, splicing isoform ratios and phenotypic results to help produce high-yielding rice varieties [45]. In future studies, transcriptome datasets of CG can be used to verify these splicing sites. We will use The Rice Annotation Project Database (https://rapdb.dna.affrc.go.jp/; accessed on 22 May 2021) to analyze and predict the function of the genes where these splicing sites reside. Furthermore, the splicing variants between CG, 9311 and Nipponbare were detected by a semiquantitative technique, and their phenotypic effects on different rice varieties were determined.

3.4. NIL-3 Is a Possibly Beneficial Resource for Rice Breeding

The results of planting in Hainan showed that the plant height and the heading stage of NIL-3 were significantly different from those of other NILs, which may be due to the presence of a gene near *qGW2.3* that controls plant height and heading stage (Figure 5d,e). In subsequent studies, we will construct a near-isogenic line of this locus, which will be used to locate genes controlling grain width and heading stage.

Yangdao 6 (9311) is a single cropping indica rice variety with disease resistance, high quality, and high yield. The combinations of Yueyou 938, Guangliangyou 6, and Liangyoupeijiu were bred with Yangdao 6 as the male parent, which also had a good quality, high yield, and disease resistance and presented good prospects for popularization and application. These results indicate that a high level of inbred lines might be beneficial for improving the level of hybrid rice breeding. The NIL-3 constructed in this study, not only greatly shortened the growth period of 9311, but also promoted grain shape and yield, which may provide germplasm resources for superior rice breeding and improvement.

4. Materials and Methods

4.1. Plant Materials

We found a colossal grain rice (CG) of the japonica rice variety Azucena background, which was mutagenized with ethyl methane sulfonate (EMS) to produce a mutant library for screening large grains. After the seed of CG was harvested and planted for multiple generations, its grain phenotype became stable.

For the QTL analysis, an F_2 population was derived from a cross between the japonica cultivar CG and the indica cultivar 9311 (small, slender grains). The seeds of the F_2 population were planted in different times and places, while 305 F_2 plants were harvested in Jining in 2015, 183 F_2 plants were reaped in Hainan in 2018, and 192 F_2 plants were successfully self-pollinated in Hunan in 2019, respectively.

From the BC_3F_5 generation, four corresponding NILs were screened by molecular marker-assisted selection (MAS).

4.2. Phenotypic Measurements and Heritability Estimation

In this study, four grain size traits were measured for all populations: grain length (GL; mm), grain width (GW; mm), length/width ratio (LWR), and 1000-grain weight (TGW; g). Grain length, grain width, length/width ratio, and 1000-grain weight were estimated using 50 grains randomly chosen from grains of individual plants. These traits were calculated by the average of three replicated measurements using Wanshen SC-G automatic seed test analysis and a 1000-grain weight instrument. In addition, descriptive statistics, such as kurtosis and skewness, were procured for F_2 using the statistical functions of Microsoft Excel 2013, and statistical analysis, such as Student's t test, was conducted using SPSS statistics 26.0 for Windows (IBM, Armonk, NY, USA).

The broad-sense heritability ($h^2{}_b$) for each trait was estimated according to a previously described method [46].

4.3. Acquisition and Analysis of BSA Data

We extracted DNA samples from rice leaves and constructed two DNA pools for Pool-seq, according to a previously described method [37,47]. Among them, there was little change in pool size. We constructed two DNA pools (large pool and small pool) using 50 plants from the 2015 Jining experiment. The average sequencing depth of sequencing libraries was over $10\times$ for the parental pools and over $40\times$ for the two descendant pools.

Library construction and analysis of BSA data were performed using Novogene (Novogene, Beijing, China). Detailed sequencing and analysis methods can be found in previous descriptions [37,48–50].

4.4. DNA Extraction, Molecular Marker Development and PCR Protocol

We extracted total genomic DNA samples from individual plants using the cetyl-trimethyl-ammonium bromide (CTAB) method. Subsequently, the DNA samples were dissolved in double-distilled H_2O. The quality was determined by 1% agarose gel mixed with ethidium bromide.

The insertion/deletion primers (InDels) were developed using Primer 5.0 based on polymorphisms between the whole-genome sequences of CG and 9311. Primers for amplifying gene fragments were also developed using Primer 5.0. A total of 482 simple sequence repeat (SSR) molecular markers were downloaded from the Gramene website (http://www.gramene.org/archive; accesed on 17 June 2015) and tested with population and parents for polymorphism. However, only 119 of them could be used for QTL mapping and screening target lines for developing near-isogenic lines (NILs).

Total genomic DNA samples were diluted to 100 ng/μL, and PCR amplification was enforced by Vazyme's protocol with 2× Taq Master Mix (Dye Plus) in a 15-μL reaction volume. Denatured amplified products were then separated on 4% polyacrylamide gels with ethidium bromide.

4.5. Construction of Genetic Linkage Maps and QTL Analysis

In light of the genotype of the F_2 population, QTL IciMapping 4.2 software with the MAP function was used to map QTLs based on 3 years of phenotypic data.

QTL analysis was carried out by the Inclusive Composite Interval Mapping of Additive (ICIM-ADD) module within QTL IciMapping 4.2. The chromosomal location figures of NILs and AS sites were drawn by Mapchart 2.32 [51]. The arguments for detecting putative QTLs and additive QTLs were set according to a previously described method [52].

5. Conclusions

Grain size is an important part of quality and harvest traits in rice breeding. Increasing rice yield and ensuring food security are together considered the direction of our future efforts [3]. The QTLs detected in this study will be helpful to elucidate the molecular mechanism of grain size adjustment. In particular, NIL-3 will provide breeding material for superior rice. Furthermore, SNPs identified at splice sites or splicing-related proteins in the present study may play an important role in particularly large phenotypic differences between CG and the other japonica rice.

Supplementary Materials: The following supporting information can be downloaded at: https://www.mdpi.com/article/10.3390/ijms23073526/s1.

Author Contributions: Conceptualization, M.C., Y.L. and N.Y.; methodology, X.H. (Xuanxuan Hou), Y.C., X.H. (Xin Hou), Z.J., X.Y. and J.Z.; software, X.H. (Xuanxuan Hou) and Y.C.; resources, X.H. (Xuanxuan Hou), J.Z., M.C., Y.L. and N.Y.; data curation, X.H. (Xuanxuan Hou), Y.C., X.H. (Xin Hou) and Z.J.; writing—original draft preparation, X.H. (Xuanxuan Hou), M.C., Y.L. and N.Y.; writing—review and editing, M.C., Y.L. and N.Y. project administration, J.Z., M.C., Y.L. and N.Y.; funding acquisition, N.Y. All authors have read and agreed to the published version of the manuscript.

Funding: This research was funded by the National Natural Science Foundation of China (31971924, U2106230, 32001452 and 32171927), Natural Science Foundation of Jiangsu Province (SBK2020042924), Platform Funding for Guangdong Provincial Enterprise Key Laboratory of Seed and Seedling Health Management Technology (2021B1212050011), Science Technology and Innovation Committee of Shenzhen (GJHZ20190821160401654), Program for Scientific Research Innovation Team of Young Scholar in Colleges and Universities of Shandong Province (2019KJE011), Postgraduate Scientific Research Innovation Project of Hunan Province (CX20190496) and the Hong Kong Re-search Grant Council (AoE/M-05/12, AoE/M-403/16,GRF14160516, 14177617, 12100318).

Institutional Review Board Statement: Not applicable.

Informed Consent Statement: Not applicable.

Data Availability Statement: All of the data generated or analyzed during this study are included in this published article.

Conflicts of Interest: The authors declare no conflict of interest.

References

1. Huang, R.; Jiang, L.; Zheng, J.; Wang, T.; Wang, H.; Huang, Y.; Hong, Z. Genetic bases of rice grain shape: So many genes, so little known. *Trends Plant Sci.* **2013**, *18*, 218–226. [CrossRef] [PubMed]
2. Ashikari, M.; Matsuoka, M. Identification, isolation and pyramiding of quantitative trait loci for rice breeding. *Trends Plant Sci.* **2006**, *11*, 344–350. [CrossRef] [PubMed]
3. Cole, M.B.; Augustin, M.A.; Robertson, M.J.; Manners, J.M. The science of food security. *NPJ Sci. Food* **2018**, *2*, 14. [CrossRef] [PubMed]
4. Xing, Y.; Zhang, Q. Genetic and molecular bases of rice yield. *Annu. Rev. Plant Biol.* **2010**, *61*, 421–442. [CrossRef] [PubMed]
5. Li, N.; Li, Y. Signaling pathways of seed size control in plants. *Curr. Opin. Plant Biol.* **2016**, *33*, 23–32. [CrossRef] [PubMed]
6. Song, X.J.; Huang, W.; Shi, M.; Zhu, M.Z.; Lin, H.X. A QTL for rice grain width and weight encodes a previously unknown RING-type E3 ubiquitin ligase. *Nat. Genet.* **2007**, *39*, 623–630. [CrossRef] [PubMed]
7. Weng, J.; Gu, S.; Wan, X.; Gao, H.; Guo, T.; Su, N.; Lei, C.; Zhang, X.; Cheng, Z.; Guo, X.; et al. Isolation and initial characterization of GW5, a major QTL associated with rice grain width and weight. *Cell Res.* **2008**, *18*, 1199–1209. [CrossRef] [PubMed]
8. Fan, C.; Xing, Y.; Mao, H.; Lu, T.; Han, B.; Xu, C.; Li, X.; Zhang, Q. GS3, a major QTL for grain length and weight and minor QTL for grain width and thickness in rice, encodes a putative transmembrane protein. *Theor. Appl. Genet.* **2006**, *112*, 1164–1171. [CrossRef] [PubMed]
9. Mao, H.; Sun, S.; Yao, J.; Wang, C.; Yu, S.; Xu, C.; Li, X.; Zhang, Q. Linking differential domain functions of the GS3 protein to natural variation of grain size in rice. *Proc. Natl. Acad. Sci. USA* **2010**, *107*, 19579–19584. [CrossRef]
10. Zhang, X.; Wang, J.; Huang, J.; Lan, H.; Wang, C.; Yin, C.; Wu, Y.; Tang, H.; Qian, Q.; Li, J.; et al. Rare allele of OsPPKL1 associated with grain length causes extra-large grain and a significant yield increase in rice. *Proc. Natl. Acad. Sci. USA* **2012**, *109*, 21534–21539. [CrossRef] [PubMed]
11. Nakagawa, H.; Tanaka, A.; Tanabata, T.; Ohtake, M.; Fujioka, S.; Nakamura, H.; Ichikawa, H.; Mori, M. Short grain1 decreases organ elongation and brassinosteroid response in rice. *Plant Physiol.* **2012**, *158*, 1208–1219. [CrossRef] [PubMed]
12. Duan, P.; Rao, Y.; Zeng, D.; Yang, Y.; Xu, R.; Zhang, B.; Dong, G.; Qian, Q.; Li, Y. Small grain 1, which encodes a mitogen-activated protein kinase kinase 4, influences grain size in rice. *Plant J.* **2014**, *77*, 547–557. [CrossRef] [PubMed]
13. Xu, R.; Duan, P.; Yu, H.; Zhou, Z.; Zhang, B.; Wang, R.; Li, J.; Zhang, G.; Zhuang, S.; Lyu, J.; et al. Control of grain size and weight by the OsMKKK10-OsMKK4-OsMAPK6 signaling pathway in rice. *Mol. Plant* **2018**, *11*, 860–873. [CrossRef]
14. Shi, C.; Ren, Y.; Liu, L.; Wang, F.; Zhang, H.; Tian, P.; Pan, T.; Wang, Y.; Jing, R.; Liu, T.; et al. Ubiquitin specific protease 15 has an important role in regulating grain width and size in rice. *Plant Physiol.* **2019**, *180*, 381–391. [CrossRef] [PubMed]
15. Yamamuro, C.; Ihara, Y.; Xiong, W.; Noguchi, T.; Fujioka, S.; Takatsuto, S.; Ashikari, M.; Matsuoka, K.M. Loss of function of a rice brassinosteroid insensitive1 homolog prevents internode elongation and bending of the lamina joint. *Plant Cell* **2000**, *12*, 1591–1605. [CrossRef]
16. Segami, S.; Kono, I.; Ando, T.; Yano, M.; Kitano, H.; Miura, K.; Iwasaki, Y. Small and round seed 5 gene encodes alpha-tubulin regulating seed cell elongation in rice. *Rice* **2012**, *5*, 4. [CrossRef]
17. Wang, Y.; Xiong, G.; Hu, J.; Jiang, L.; Yu, H.; Xu, J.; Fang, Y.; Zeng, L.; Xu, E.; Xu, J.; et al. Copy number variation at the GL7 locus contributes to grain size diversity in rice. *Nat. Genet.* **2015**, *47*, 944–948. [CrossRef] [PubMed]
18. Hu, J.; Wang, Y.; Fang, Y.; Zeng, L.; Xu, J.; Yu, H.; Shi, Z.; Pan, J.; Zhang, D.; Kang, S.; et al. A rare allele of GS2 enhances grain size and grain yield in rice. *Mol. Plant* **2015**, *8*, 1455–1465. [CrossRef]
19. Li, S.; Gao, F.; Xie, K.; Zeng, X.; Cao, Y.; Zeng, J.; He, Z.; Ren, Y.; Li, W.; Deng, Q.; et al. The OsmiR396c-OsGRF4-OsGIF1 regulatory module determines grain size and yield in rice. *Plant Biotechnol. J.* **2016**, *14*, 2134–2146. [CrossRef] [PubMed]
20. Si, L.; Chen, J.; Huang, X.; Gong, H.; Luo, J.; Hou, Q.; Zhou, T.; Lu, T.; Zhu, J.; Shangguan, Y.; et al. OsSPL13 controls grain size in cultivated rice. *Nat. Genet.* **2016**, *48*, 447–456. [CrossRef] [PubMed]
21. Huang, K.; Wang, D.; Duan, P.; Zhang, B.; Xu, R.; Li, N.; Li, Y. Wide and thick grain 1, which encodes an otubain-like protease with deubiquitination activity, influences grain size and shape in rice. *Plant J.* **2017**, *91*, 849–860. [CrossRef]
22. Chen, M.X.; Mei, L.C.; Wang, F.; Boyagane Dewayalage, I.K.W.; Yang, J.F.; Dai, L.; Yang, G.F.; Gao, B.; Cheng, C.L.; Liu, Y.G.; et al. PlantSPEAD: A web resource towards comparatively analysing stress-responsive expression of splicing-related proteins in plant. *Plant Biotechnol. J.* **2021**, *19*, 227–229. [CrossRef]
23. Chen, M.X.; Zhang, K.L.; Zhang, M.; Das, D.; Fang, Y.M.; Dai, L.; Zhang, J.; Zhu, F.Y. Alternative splicing and its regulatory role in woody plants. *Tree Physiol.* **2020**, *40*, 1475–1486. [CrossRef] [PubMed]
24. Chen, M.X.; Zhang, Y.; Fernie, A.R.; Liu, Y.G.; Zhu, F.Y. SWATH-MS-based proteomics: Strategies and applications in plants. *Trends Biotechnol.* **2021**, *39*, 433–437. [CrossRef]
25. Song, T.; Das, D.; Ye, N.H.; Wang, G.Q.; Zhu, F.Y.; Chen, M.X.; Yang, F.; Zhang, J.H. Comparative transcriptome analysis of coleorhiza development in japonica and Indica rice. *BMC Plant Biol.* **2021**, *21*, 514. [CrossRef]
26. Shen, C.C.; Chen, M.X.; Xiao, T.; Zhang, C.; Shang, J.; Zhang, K.L.; Zhu, F.Y. Global proteome response to Pb(II) toxicity in poplar using SWATH-MS-based quantitative proteomics investigation. *Ecotoxicol. Environ. Saf.* **2021**, *220*, 112410. [CrossRef]

27. Shang, X.; Cao, Y.; Ma, L. Alternative splicing in plant genes: A means of regulating the environmental fitness of plants. *Int. J. Mol. Sci.* **2017**, *18*, 432. [CrossRef] [PubMed]
28. Lu, T.; Lu, G.; Fan, D.; Zhu, C.; Li, W.; Zhao, Q.; Feng, Q.; Zhao, Y.; Guo, Y.; Li, W.; et al. Function annotation of the rice transcriptome at single-nucleotide resolution by RNA-seq. *Genome Res.* **2010**, *20*, 1238–1249. [CrossRef]
29. Filichkin, S.A.; Mockler, T.C. Unproductive alternative splicing and nonsense mRNAs: A widespread phenomenon among plant circadian clock genes. *Biol. Direct* **2012**, *7*, 20. [CrossRef] [PubMed]
30. Prathepha, P. Identification of variant transcripts of waxy gene in non-glutinous rice (*O. sativa* L.) with different amylose content. *Pak. J. Biol.* **2007**, *10*, 2500–2504. [CrossRef] [PubMed]
31. Xia, D.; Zhou, H.; Liu, R.; Dan, W.; Li, P.; Wu, B.; Chen, J.; Wang, L.; Gao, G.; Zhang, Q.; et al. GL3.3, a novel QTL encoding a GSK3/SHAGGY-like kinase, epistatically interacts with GS3 to produce extra-long grains in rice. *Mol. Plant* **2018**, *11*, 754–756. [CrossRef] [PubMed]
32. Matsukura, S.; Mizoi, J.; Yoshida, T.; Todaka, D.; Ito, Y.; Maruyama, K.; Shinozaki, K.; Yamaguchi-Shinozaki, K. Comprehensive analysis of rice DREB2-type genes that encode transcription factors involved in the expression of abiotic stress-responsive genes. *Mol. Genet. Genom.* **2010**, *283*, 185–196. [CrossRef]
33. Cheng, Q.; Zhou, Y.; Liu, Z.; Zhang, L.; Song, G.; Guo, Z.; Wang, W.; Qu, X.; Zhu, Y.; Yang, D.; et al. An alternatively spliced heat shock transcription factor, OsHSFA2dI, functions in the heat stress-induced unfolded protein response in rice. *Plant Biol.* **2015**, *17*, 419–429. [CrossRef] [PubMed]
34. Chen, M.X.; Zhang, K.L.; Gao, B.; Yang, J.F.; Tian, Y.; Das, D.; Fan, T.; Dai, L.; Hao, G.F.; Yang, G.F.; et al. Phylogenetic comparison of 5′ splice site determination in central spliceosomal proteins of the U1-70K gene family, in response to developmental cues and stress conditions. *Plant J.* **2020**, *103*, 357–378. [CrossRef] [PubMed]
35. Zuo, J.; Li, J. Molecular genetic dissection of quantitative trait loci regulating rice grain size. *Annu. Rev. Genet.* **2014**, *48*, 99–118. [CrossRef]
36. Michelmore, R.W.; Kesseli, I.P.V. Identification of markers linked to disease-resistance genes by bulked segregant analysis: A rapid method to detect markers in specific genomic regions by using segregating populations. *Proc. Natl. Acad. Sci. USA* **1991**, *88*, 9828–9832. [CrossRef] [PubMed]
37. Qi, L.; Ding, Y.; Zheng, X.; Xu, R.; Zhang, L.; Wang, Y.; Wang, X.; Zhang, L.; Cheng, Y.; Qiao, W.; et al. Fine mapping and identification of a novel locus qGL12.2 control grain length in wild rice (*Oryza rufipogon* Griff.). *Theor. Appl. Genet.* **2018**, *131*, 1497–1508. [CrossRef] [PubMed]
38. Abe, A.; Kosugi, S.; Yoshida, K.; Natsume, S.; Takagi, H.; Kanzaki, H.; Matsumura, H.; Yoshida, K.; Mitsuoka, C.; Tamiru, M.; et al. Genome sequencing reveals agronomically important loci in rice using MutMap. *Nat. Biotechnol.* **2012**, *30*, 174–178. [CrossRef] [PubMed]
39. Yang, Z.; Huang, D.; Tang, W.; Zheng, Y.; Liang, K.; Cutler, A.J.; Wu, W. Mapping of quantitative trait loci underlying cold tolerance in rice seedlings via high-throughput sequencing of pooled extremes. *PLoS ONE* **2013**, *8*, e68433. [CrossRef]
40. Takagi, H.; Tamiru, M.; Abe, A.; Yoshida, K.; Uemura, A.; Yaegashi, H.; Obara, T.; Oikawa, K.; Utsushi, H.; Kanzaki, E.; et al. MutMap accelerates breeding of a salt-tolerant rice cultivar. *Nat. Biotechnol.* **2015**, *33*, 445–449. [CrossRef]
41. Li, Y.; Fan, C.; Xing, Y.; Jiang, Y.; Luo, L.; Sun, L.; Shao, D.; Xu, C.; Li, X.; Xiao, J.; et al. Natural variation in GS5 plays an important role in regulating grain size and yield in rice. *Nat. Genet.* **2011**, *43*, 1266–1269. [CrossRef]
42. Wang, Y.; Wan, L.; Zhang, L.; Zhang, Z.; Zhang, H.; Quan, R.; Zhou, S.; Huang, R. An ethylene response factor OsWR1 responsive to drought stress transcriptionally activates wax synthesis related genes and increases wax production in rice. *Plant Mol. Biol.* **2012**, *78*, 275–288. [CrossRef]
43. Cakir, B.; Shiraishi, S.; Tuncel, A.; Matsusaka, H.; Satoh, R.; Singh, S.; Crofts, N.; Hosaka, Y.; Fujita, N.; Hwang, S.K.; et al. Analysis of the rice ADP-glucose transporter (OsBT1) indicates the presence of regulatory processes in the amyloplast stroma that control ADP-glucose flux into starch. *Plant Physiol.* **2016**, *170*, 1271–1283.
44. Ma, L.; Li, G. FAR1-related sequence (FRS) and FRS-related factor (FRF) family proteins in arabidopsis growth and development. *Front. Plant Sci.* **2018**, *9*, 692. [CrossRef]
45. Khokhar, W.; Hassan, M.A.; Reddy, A.S.N.; Chaudhary, S.; Jabre, I.; Byrne, L.J.; Syed, N.H. Genome-wide identification of splicing quantitative trait loci (sQTLs) in diverse ecotypes of *Arabidopsis thaliana*. *Front. Plant Sci.* **2019**, *10*, 1160. [CrossRef]
46. Raihan, M.S.; Liu, J.; Huang, J.; Guo, H.; Pan, Q.; Yan, J. Multi-environment QTL analysis of grain morphology traits and fine mapping of a kernel-width QTL in Zheng58 x SK maize population. *Theor. Appl. Genet.* **2016**, *129*, 1465–1477. [CrossRef]
47. Murray, M.G.; Thompson, W.F. Rapid isolation of high molecular weight plant DNA. *Nucl Acids Res.* **1980**, *8*, 4321–4326. [CrossRef]
48. Li, H.; Durbin, R. Fast and accurate short read alignment with burrows-wheeler transform. *Bioinformatics* **2009**, *25*, 1754–1760. [CrossRef]
49. McKenna, A.; Hanna, M.; Banks, E.; Sivachenko, A.; Cibulskis, K.; Kernytsky, A.; Garimella, K.; Altshuler, D.; Gabriel, S.; Daly, M. The genome analysis toolkit: A MapReduce framework for analyzing next-generation DNA sequencing data. *Genome Res.* **2010**, *20*, 1297–1303. [CrossRef]
50. Takagi, H.; Abe, A.; Yoshida, K.; Kosugi, S.; Natsume, S.; Mitsuoka, C.; Uemura, A.; Utsushi, H.; Tamiru, M.; Takuno, S.; et al. QTL-seq: Rapid mapping of quantitative trait loci in rice by whole genome resequencing of DNA from two bulked populations. *Plant J.* **2013**, *74*, 174–183. [CrossRef]
51. Voorrips, R.E. MapChart: Software for the graphical presentation of linkage maps and QTLs. *J. Hered.* **2002**, *93*, 77–78. [CrossRef]
52. Meng, L.; Li, H.H.; Zhang, L.Y.; Wang, J.K. QTL IciMapping: Integrated software for genetic linkage map construction and quantitative trait locus mapping in biparental populations. *Crop. J.* **2015**, *3*, 269–283. [CrossRef]

International Journal of
Molecular Sciences

Article

Moderate Soil Drying-Induced Alternative Splicing Provides a Potential Novel Approach for the Regulation of Grain Filling in Rice Inferior Spikelets

Zhenning Teng [1,2,†], Qin Zheng [1,†], Bohan Liu [1,3], Shuan Meng [1,3], Jianhua Zhang [4,*] and Nenghui Ye [1,3,*]

1 College of Agronomy, Hunan Agricultural University, Changsha 410128, China; sailingtzn@163.com (Z.T.); zhengqin@stu.hunau.edu.cn (Q.Z.); liubohan@hunau.edu.cn (B.L.); mengshuanz@outlook.com (S.M.)
2 Shenzhen Research Institute, The Chinese University of Hong Kong, Shenzhen 518057, China
3 Key Laboratory of Crop Physiology and Molecular Biology, Ministry of Education, Hunan Agricultural University, Changsha 410128, China
4 Department of Biology, Hong Kong Baptist University, Kowloon, Hong Kong 999077, China
* Correspondence: jzhang@hkbu.edu.hk (J.Z.); nye@hunau.edu.cn (N.Y.)
† These authors contributed equally to this work.

Abstract: Poor grain filling of inferior spikelets, especially in some large-panicle rice varieties, is becoming a major limitation in breaking the ceiling of rice production. In our previous studies, we proved that post-anthesis moderate soil drying (MD) was an effective way to promote starch synthesis and inferior grain filling. As one of the most important regulatory processes in response to environmental cues and at different developmental stages, the function of alternative splicing (AS) has not yet been revealed in regulating grain filling under MD conditions. In this study, AS events at the most active grain-filling stage were identified in inferior spikelets under well-watered control (CK) and MD treatments. Of 16,089 AS events, 1840 AS events involving 1392 genes occurred differentially between the CK and MD treatments, many of which function on spliceosome, ncRNA metabolic process, starch, and sucrose metabolism, and other functions. Some of the splicing factors and starch synthesis-related genes, such as SR protein, hnRNP protein, *OsAGPL2*, *OsAPS2*, *OsSSIVa*, *OsSSIVb*, *OsGBSSII*, and *OsISA1* showed differential AS changes under MD treatment. The expression of miR439f and miR444b was reduced due to an AS event which occurred in the intron where miRNAs were located in the MD-treated inferior spikelets. On the contrary, *OsAGPL2*, an AGPase encoding gene, was alternatively spliced, resulting in different transcripts with or without the miR393b binding site, suggesting a potential mechanism for miRNA-mediated gene regulation on grain filling of inferior spikelets in response to MD treatment. This study provides some new insights into the function of AS on the MD-promoted grain filling of inferior spikelets, and potential application in agriculture to increase rice yields by genetic approaches.

Keywords: alternative splicing; microRNA; moderate soil drying; inferior spikelets; rice

Citation: Teng, Z.; Zheng, Q.; Liu, B.; Meng, S.; Zhang, J.; Ye, N. Moderate Soil Drying-Induced Alternative Splicing Provides a Potential Novel Approach for the Regulation of Grain Filling in Rice Inferior Spikelets. *Int. J. Mol. Sci.* **2022**, 23, 7770. https://doi.org/10.3390/ijms23147770

Academic Editor: Zhijun Cheng

Received: 13 June 2022
Accepted: 12 July 2022
Published: 14 July 2022

Publisher's Note: MDPI stays neutral with regard to jurisdictional claims in published maps and institutional affiliations.

1. Introduction

Rice (*Oryza sativa* L.) is a staple food for more than half of the world's population. However, its current levels of production cannot meet the demand driven by rapid population growth and economic development [1]. Hence, there is an urgent need to improve global rice production. China officially launched the Super Rice Breeding Program in 1996, aiming to cultivate new rice varieties with a high yield [2]. However, most of the super rice cultivars failed to achieve the high yield, mainly due to the poor grain filling of the later-flowering inferior spikelets [3,4].

The inferior spikelets flower later, exhibit a slower rate of increase in dry weight during grain development, and have a lower grain weight [4]. Many measures, such as moderate soil drying (MD) irrigation [5–7], rising CO_2 concentration, and applications of plant hormones [8–10] have been used to improve the grain filling of inferior grains. The main component of rice endosperm is starch, and defects in starch synthesis will usually lead to incomplete grain filling. Four classes of enzymes, including ADP-glucose pyrophosphorylase (AGP), starch synthase (SS), starch branching enzyme (SBE), and starch debranching enzyme (DBE) are the most important enzymes in rice grain filling, which determine the grain weight and starch quality [11,12]. Numerous sources have shown that the manipulation of the starch biosynthesis pathway by modern molecular genetic techniques will alter the grain filling. It is also well demonstrated that grain filling is sensitive to different environmental cues [13,14], of which MD treatment has been proven to be a highly effective means of increasing inferior grain filling and grain yield by improving the key enzymes in sucrose-to-starch conversion [5,6,15]. The enzyme activities, especially sucrose synthase (SUS) and AGP in the inferior spikelets, are significantly enhanced by MD treatment, resulting in am improved inferior grain filling rate, grain weight, and yield [7,16]. Antagonism or synergistic interaction between ABA, ethylene, GA, and IAA play a critical role during grain filling under MD treatment [5,7,9]. Although it has been demonstrated that MD treatment regulated the hormonal balance, which thus played facilitatory roles during the grain filling in the inferior spikelets, the critical mechanism underlying the regulation of starch biosynthesis has yet to be well established.

Alternative splicing (AS), a process that generates multiple distinct transcripts from a single multi-exon gene, is prevalent in plants and responds to environmental changes and stress treatments [17,18], including ultraviolet (UV) irradiation [19], temperature stress (cold and heat) [20,21], and cadmium stress [22]. As an important factor in gene regulation, AS is an emerging research area related to post-transcriptional regulation [23], and significantly affects crop growth and development. For instance, the pre-mRNA splicing of the *OsFCA* gene controls the developmental switch from the vegetative to the reproductive phase in Arabidopsis [24]. In addition, the *Waxy* gene encodes a granule-bound starch synthase that is necessary for the synthesis of amylose in endosperm. Alternative splicing, caused by a single mutation in the 5' splice site of *Waxy*, results in a reduced level of amylose [25,26]. Most recently, AS of *OsGS1;1* affected grain filling by regulating the amylose content and sugar metabolism [27]. These studies indicate that AS affects plant development and is also involved in regulating grain filling.

MicroRNAs (miRNAs), another post-transcriptional regulation mechanism, can be classified as either intergenic miRNAs or intronic miRNAs [24]. In animals, pre-mRNA splicing has been shown to participate in both intergenic and intronic miRNA processing [28,29]. This mechanism has also been conducted in plant miRNA primary transcripts [30,31]. MiRNAs play roles usually through regulating their target genes [32], which would be interrupted by the AS-induced disruption of miRNA binding sites [33]. It was shown that the additional regulation conferred by alternative splicing may link spliceosome activity to the regulation of certain miRNA–target interactions [33]. Recent works suggest that miRNAs, including miR1861, miR1432, and miR397, contribute to grain filling by regulating the starch synthesis and phytohormone biosynthesis in response to MD treatment [34–37]. However, the function of AS and its interaction with the miRNAs induced by MD treatment in regulating grain filling has not been reported. The aim of the present investigation was to identify AS events and the regulatory mechanism of grain filling in rice inferior spikelets under MD treatment at the most active stage of grain filling. We expect to propose a mechanism of MD-induced AS and its interaction with miRNAs in regulating the grain filling of inferior spikelets.

2. Results

2.1. AS Events of Rice Inferior Spikelets in Response to Moderate Soil Drying during Grain Filling

RNA-seq and small-RNA analysis at the most active grain filling stage (9 days after anthesis (DAA)) revealed that both starch synthesis and phytohormone biosynthesis are both regulated directly by MD treatment and indirectly regulated through differentially expressed miRNAs in inferior spikelets in response to MD treatment [7,34]. Another significant advantage of RNA-seq analysis is that differences in splicing can be detected from the sequences of the various transcripts. In this study, rMATS was utilized to identify the frequency of the different classes of differential splicing in rice inferior spikelets in response to MD treatment during grain filling.

There are several AS mechanisms, including skipped exon (SE), alternative 5′ splice site (A5SS), alternative 3′ splice site (A3SS), mutually exclusive exon (MXE), and retained intron (RI) (Figure S1, Supplementary Materials). Each of these AS events can result in distinct transcripts, and hence diverse biological functions. The AS events of the rice inferior spikelets were analyzed using the rMATS software (http://rnaseq-mats.sourceforge.net/index.html; Version 4.1.0, accessed on 11 July 2022). As shown in Figure 1A and Table S1, Supplementary Materials, large numbers of AS events (CK, 14,264; MD, 15,788) were detected in the rice inferior spikelets during grain filling. Intriguingly, all of the AS types, including SE, A5SS, A3SS, MXE and RI, were significantly increased under MD treatment. Moreover, SE represented the largest proportion of AS events, at 45.57%. The percentages of MXE, A5SS, RI, and A3SS over the total AS event types were 2.42%, 12.38%, 17.44%, and 22.18%, respectively (Figure 1B).

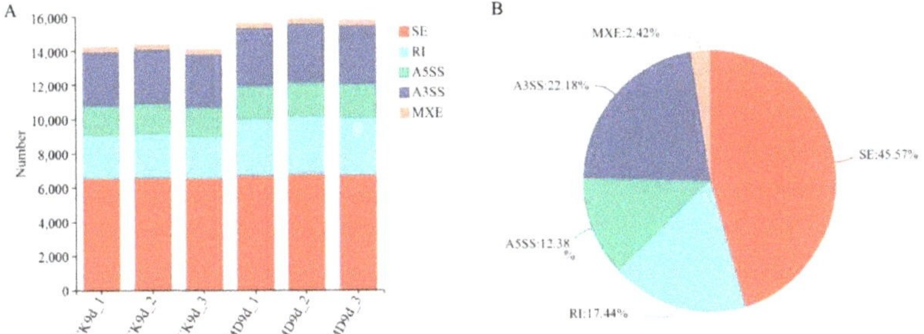

Figure 1. Summary of AS type of rice inferior spikelets under CK and MD conditions. (**A**) Number of the identified AS events. SE, skipped exon; A5SS, alternative 5′ splice site; A3SS, alternative 3′ splice site; MXE, mutually exclusive exon; RI, retained intron; (**B**) Summary of AS type, represented as percentages.

2.2. MD-Induced AS Might Be Involved in Regulating Grain Filling of Rice Inferior Spikelets

On detailed analysis of the AS events, we found that the known AS events were the major event of rice inferior spikelets, accounting for 85.89% of the total AS events (Figure 2A). A total of 1840 differentially alternative splicing (DAS) events involving 1392 genes between the CK and MD treatments in inferior grains were obtained (Figure 2B). Compared with all of the AS types (Figure 1B), the percentages of A3SS and A5SS of DAS types increased by 8.25–9.14%, while MXE, RI, and SE had a 1.31–14.40% reduction (Figure 2C), suggesting that the AS events in inferior spikelets showed various susceptibilities to MD treatment.

Gene ontology (GO) and the Kyoto Encyclopedia of Genes and Genomes (KEGG) enrichment analysis (Figure 2D,E) were performed to identify the potential functions of these DAS genes. Of these, based on the GO database, the GO terms in the DAS genes were enriched in functions of biological process, including "Branched-chain amino acid metabolic process", "Hexose metabolic process", "Starch metabolic process", "ncRNA metabolic process", "Monosaccharide metabolic process", "Starch biosynthetic process" and "RNA processing" (Figure 2D). The KEGG analysis of the DAS genes revealed that most of the genes were enriched in functions of "metabolism" and "genetic information processing". "Other glycan degradation", "RNA transport", "Spliceosome" and "Starch and sucrose metabolism" were the most enriched KEGG pathways (Figure 2E). We then investigated the DAS genes related to the spliceosome pathway (Figure S2, Supplementary Materials), which showed that the parts of the splicing factor encoding genes and the RNA recognition motif containing proteins were affected in the MD-treated grains

(Figure S2, Supplementary Materials), such as serine/arginine-rich (SR) protein subfamily gene (LOC_Os07g47630), RS domain with zinc knuckle protein (RSZ) subfamily gene (LOC_Os02g54770), plant-specific SC35-like splicing factor (SCL) (LOC_Os12g38430), heterogeneous nuclear ribonucleoprotein particle (hnRNP) protein gene (LOC_Os02g12850), and pre-mRNA-splicing factor SF2 (LOC_Os01g21420) (Figure 3). All of those genes may greatly alter the protein isoforms in comparison with the control group, and function as essential factors for constitutive and alternative splicing [38–40], which might in turn explain the increase in the AS events under MD treatment (Figure 1).

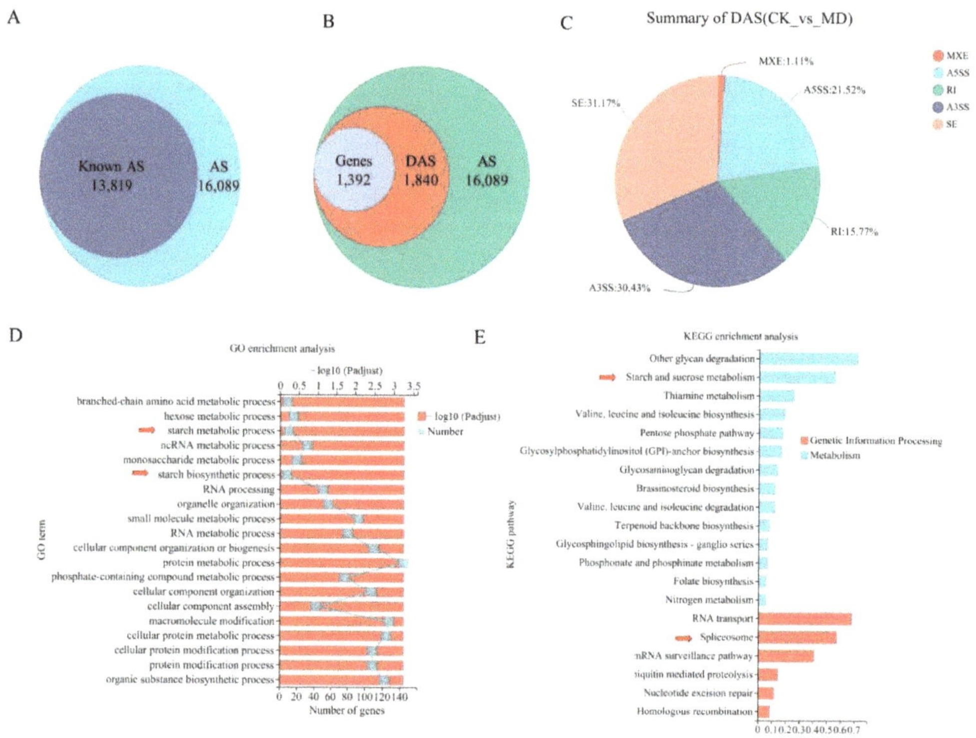

Figure 2. Identification of DAS events and most enriched GO and KEGG pathways of DAS genes in inferior kernels in the comparison of CK and MD at 9 DAA. (**A**) Proportion of known AS in all AS events; (**B**) Proportion of DAS in all AS events; (**C**) Summary of DAS type, represented as percentages; (**D**) The top 20 most enriched GO pathways of DAS genes; (**E**) The top 20 most enriched KEGG pathways of DAS genes. Several prominent signaling pathways were annotated with red arrow.

Given the GO and KEGG analysis results and the essential role of starch metabolism in grain filling, we focused on the DAS events of the starch synthesis-related genes and found that several DAS genes involved in starch synthesis had several AS forms under CK and MD treatments (Figure 4; File S1). ADP-glucose pyrophosphorylase genes (*OsAGPL2*, LOC_Os01g44220; *OsAPS2*, LOC_Os08g25734), key genes in regulating starch synthesis and grain filling [41] were reported that allosteric regulation on translated proteins of those genes has altered the catalytic activity of the cytoplasmic AGPase and starch biosynthesis [42]. In the present study, the several AS forms of the AGPase genes were identified under CK and MD treatments, including the known and novel AS events (Figure 4; File S1). Furthermore, several variants of other starch synthase-related gene transcripts, including *OsAGPL2*, *OsAPS2*, starch branching enzyme gene (*OsSBEI*, LOC_Os06g51084), granule-bound starch synthase gene (*OsGBSSII*, LOC_Os07g22930), soluble starch synthase genes

(*OsSSIVa*, LOC_Os01g52250; *OsSSIVb*, LOC_Os05g45720), phosphoglucose isomerase gene (*OsPgi*, LOC_Os08g37380), isoamylase gene (*OsISA1*, LOC_Os08g40930; *OsISA3*, LOC_Os09g29404) and glucan phosphatase gene (*OsSEX4*, LOC_Os03g01750). These results indicated that the promotion of starch biosynthesis by MD treatment in inferior spikelets was partially mediated by the MD-induced AS during grain filling.

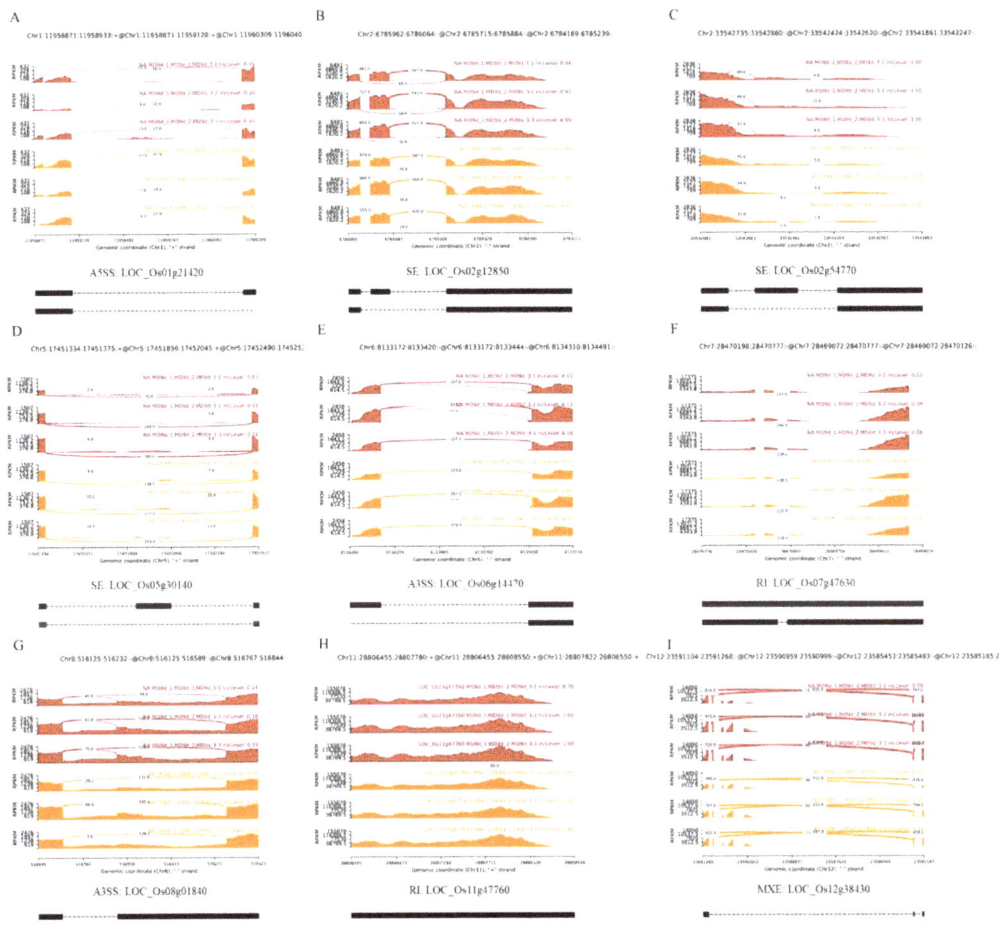

Figure 3. Quantitative visualization (Sashimi plot) of some DAS events of spliceosome pathway genes of rice inferior spikelets in the comparison of CK and MD at 9 DAA. (**A**) LOC_01g21420; (**B**) LOC_02g12850; (**C**) LOC_02g54770; (**D**) LOC_05g30140; (**E**) LOC_06g14470; (**F**) LOC_07g47630; (**G**) LOC_08g01840; (**H**) LOC_11g47760; (**I**) LOC_12g38430. Each track visualizes the splicing event within the biological replicates for the CK (orange) and MD (red) treatment samples. Count values on curved lines describe the coverage within the splice junction. The left scale presents the coverage depth in the range of the AS region. IncLevel are presented on the right side for each track. AS models are indicated below the figure.

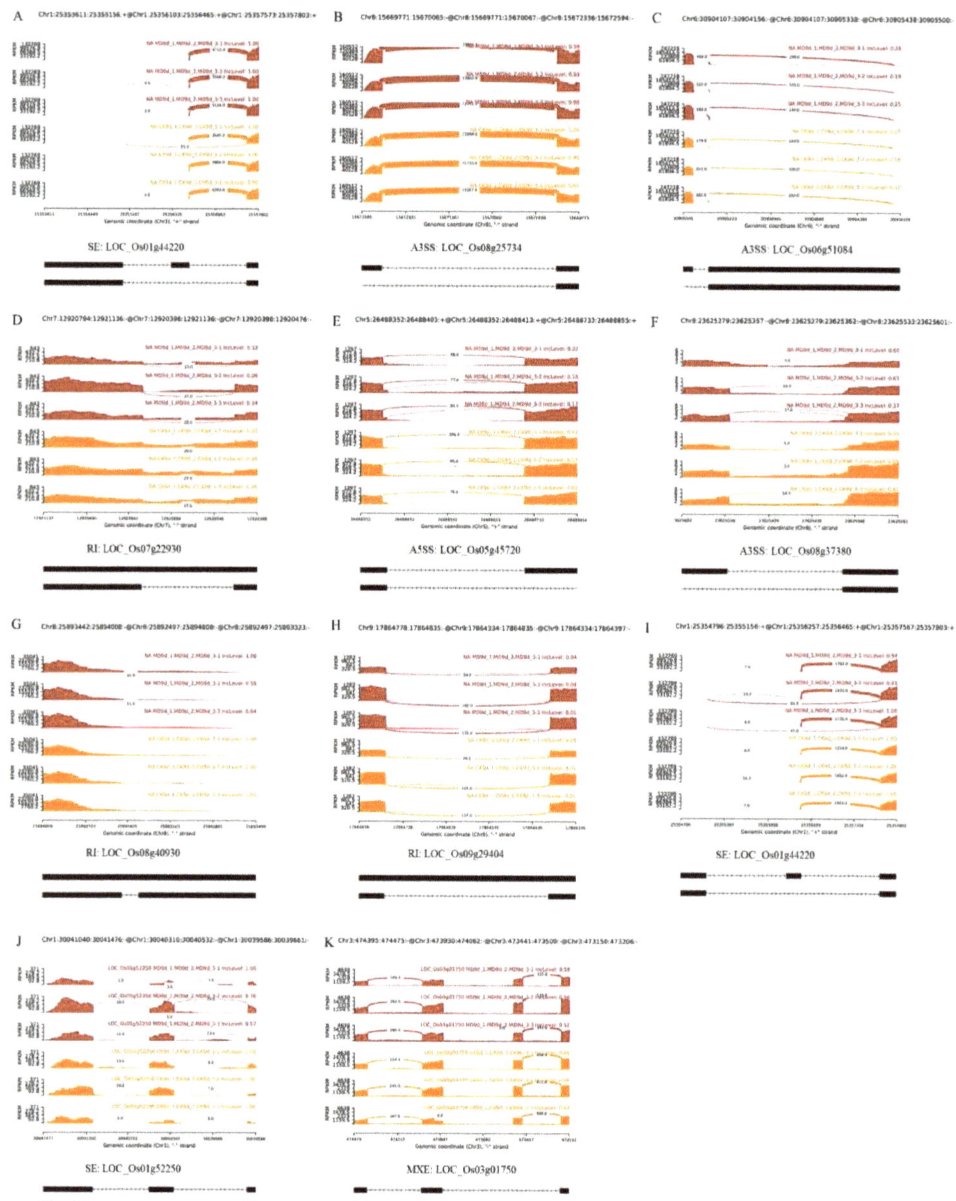

Figure 4. Quantitative visualization (Sashimi plot) of some DAS events of starch synthesis-related genes of rice inferior spikelets in the comparison of CK and MD at 9 DAA. Known AS events (**A–H**) and novel AS events (**I–K**) of starch synthesis-related genes in inferior kernels under CK and MD treatments. (**A**) LOC_01g44220; (**B**) LOC_08g25734; (**C**) LOC_06g51084; (**D**) LOC_07g22930; (**E**) LOC_05g45720; (**F**) LOC_08g37380; (**G**) LOC_08g40930; (**H**) LOC_09g29404; (**I**) LOC_01g44220; (**J**) LOC_01g52250; (**K**) LOC_03g01750.

2.3. MD-Induced AS Event Influenced Primary miRNA Expression

The biogenesis of miRNAs relies on the coupled interaction of Pol-II-mediated pre-mRNA transcription and intron excision [43], during which the accurate splicing of the intron is critical for the efficient processing of the mRNA [21]. For this reason, we identified all of the RI-type AS events and the related intronic miRNAs in the inferior spikelets of rice under MD treatments. Hundreds of the RI-type AS events are organized in each chromosome, four of which identified a significant association between the AS events and intronic miRNAs (Figure 5). MiR439f, miR1847, miR444b, and miR1867 are located within the intronic regions of LOC_Os01g35930, LOC_Os01g36640, LOC_Os02g36924, and LOC_Os03g53190, respectively (Figure 5). The expression of miR439f and miR444b were reduced in MD-treated inferior spikelets, detected by small RNA-seq in our previous study (Figure S4, Supplementary Materials) [34]. Those intronic miRNAs putative targets were predicted in previous studies [34,44], including MADS-box family gene (MADS) genes, AP2-like ethylene-responsive transcription factor genes, NAC domain-containing protein genes, and other transcription factors or genes. A granule-bound starch synthase gene (*OsGBSSII*) was found to be regulated by miR1867, as revealed by the 5′-RACE and degradome analysis [45]. These results indicate that AS acts as a regulatory mechanism for the miRNA processing in response to MD treatment, and might regulate grain filling via their target genes.

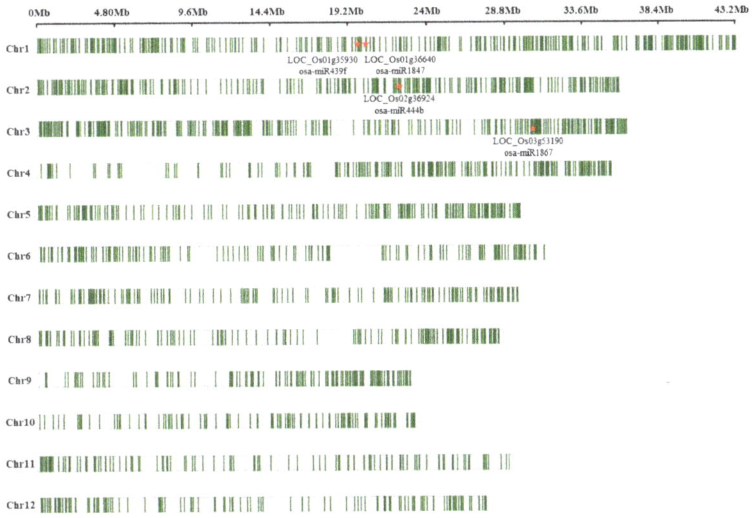

Figure 5. Distribution of RI-type AS events and the related intronic miRNAs in inferior spikelets of rice under CK and MD treatments. The location of the RI-type AS events with intronic miRNAs are indicated with red stars.

2.4. Identification of miRNA Binding Sites Disturbed by AS of Rice Inferior Spikelets under Moderate Soil Drying Post-Anthesis

Plant miRNAs recognize their target mRNAs through perfect or near perfect base pairing, which can be blocked by disrupting the miRNA binding sites (MBS) by AS [33,46]. *OsAGPL2*, a key gene in regulating starch synthesis, has seven different transcripts with a varying length of 5′ UTR (ranging from 218 bp to 600 bp) by AS (Figure 6). Based on the transcriptome results and AsmiR tools (http://forestry.fafu.edu.cn/bioinfor/db/ASmiR/; accessed on 1 April 2022), there was an AS region in the interval of 25,354,183–25,355,073 bp on Chr1 at the 5′ UTR of *OsAGPL2* that contains a functional binding site for miR393b. Several alternatively spliced transcripts of rice inferior spikelets do not contain the MBSs. Therefore, the miR393b-*OsAGPL2* interaction network in rice spikelets might be disrupted by AS on the binding sites under CK and MD treatments, which could be a potential novel mechanism of MD-promoted grain filling of inferior spikelets.

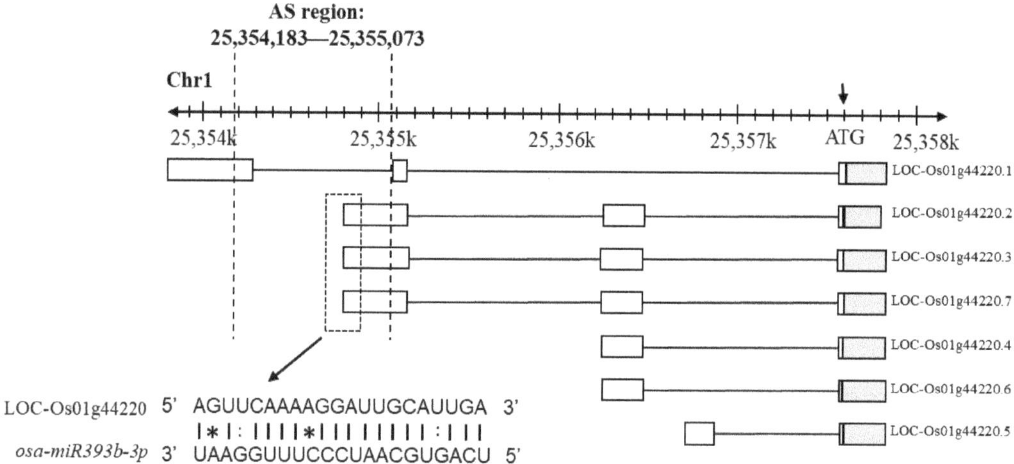

Figure 6. Alternative splicing of the miR393b binding site of *OsAGPL2* 5′ UTR of rice inferior spikelets under moderate soil drying post-anthesis. White boxes represent noncoding exons and shaded boxes represent coding exons. Dashes indicate Watson–Crick base pairing, colons indicate G-U base pairing, and asterisks indicate other non-canonical pairs.

3. Discussion

3.1. DAS Events in Response to MD Treatment Participate in Starch Biosynthesis and Contribute to Increased Grain Filling in Inferior Spikelets

The AS of the pre-mRNAs from multiexon genes allows organisms to increase their coding potential and regulates the gene expression through multiple mechanisms. AS is involved in most of the plant processes, including plant growth, development, and responses to external cues [47]. The function of AS on grain filling and yield has been a research hotspot for many years [26,27,48–52]. A single mutation at the 5′splice site of *Waxy* affects the alternative splicing of its pre-mRNA, resulting in the reduced levels of amylose [25]. The AS of *OsbZIP58* may contribute to heat tolerance, and have an effect on some of the starch-hydrolyzing α-amylase genes during grain filling [52]. Alternative splicing of *OsLG3b* has also been reported to control grain length and yield in rice [50]. When *TaGS3* undergoes AS, it produces five splicing variants, resulting in opposite effects on grain weight and grain size [51]. Recently, a study also demonstrated that the AS of *GS1;1* affects grain amylose content and sugar metabolism in rice [27]. In the present study, a total of 1840 DAS events, including *OsbZIP58*, *OsLG3b*, and *OsAGPL2*, were identified under CK and MD treatments in inferior spikelets (File S1), suggesting that the AS events in response to MD treatment also participate in starch biosynthesis and contribute to increased grain yield.

Furthermore, by analyzing the GO and KEGG signaling pathways of the DAS genes, starch biosynthetic and sucrose metabolism were also represented in the most enriched KEGG pathways (Figure 2D,E), including *OsAGPL2*, *OsAPS2*, *OsSSIVa*, *OsSSIVb*, *OsGBSSII*, *OsISA1*, etc. In rice seed endosperm, the cytosolic AGP isoform, the OsAGPS2b/OsAGPL2 complex, catalyzes the limiting step and plays a key role in starch synthesis [41,53]. SSIV is considered important for the initiation of starch granules [54]. *OsISA1* is one of the most important genes determining the starch structures in rice grains, which is directly involved in the synthesis of amylopectin [55]. These results indicate that the MD promotion of starch synthase and biosynthesis was mediated partially by the MD-induced AS during grain filling.

3.2. AS of Pre-mRNAs of Splicing Factors Increases the Complexity of Inferior Grain Filling Response to MD Treatment

SR proteins and hnRNP proteins are the main families of splicing factors, which guide spliceosomal components and thereby the spliceosome to the respective splice sites [47,56,57]. They are essential splicing factors required for both constitutive and alternative splicing [17]. Several reports indicate that various biotic and abiotic stresses influence the AS of pre-mRNAs of many spliceosomal proteins. The SR proteins, in particular, undergo extensive alternate splicing [17,18,20]. Accumulating evidence suggests that manipulating SR protein expression subsequently alters the splicing of other pre-mRNAs, including SR pre-mRNAs [17,57]. In this study, the spliceosome pathway was also one of the most enriched pathways in the inferior spikelet under MD conditions (Figure 2E), leading to an increase in the AS events. The spliceosome-related genes undergo AS to produce multiple transcripts under MD treatments, including SR protein subfamily genes, RSZ subfamily genes, plant-specific SCL subfamily genes, and other splicing factors genes (Figure 3). This result may also in turn explain the increase in the AS events under MD treatment (Figure 1). Thus, the MD-induced AS of pre-mRNAs of splicing regulators increases the complexity of gene regulation, which may also contribute to the increased grain filling in the inferior spikelets.

3.3. MD-Induced AS Provides a Mechanism for the Regulation of miRNA Processing, Leading to Increased Grain Filling in Inferior Spikelets

The AS of pre-mRNAs are widespread in eukaryotes, and generate different mature RNA isoforms from the same primary transcript, ensuring the proper expression of the genome and the higher proteome diversity [33]. Several intronic miRNAs have been discovered in plants [58,59], some of which have potential AS isoforms that may be affected by the AS events triggered under specific conditions [58,60]. The miR400 case nicely illustrates this issue, providing direct evidence that AS acts as a regulatory mechanism for miRNA processing [21]. In this study, we identified all of the RI-type AS events containing miRNAs. Four relating events particularly attracted our attention (Figure 5). MiR439f, miR1847, miR444b, and miR1867 are located within the intronic regions of its host, whose expressions were reduced in MD-treated inferior spikelets (Figure S4, Supplementary Materials), indicating that intronic splicing may regulate the miRNA expression in response to MD treatment. Recently, growing evidence has demonstrated that the miRNAs also play crucial roles in controlling grain filling [34,35,44,61]. As reported previously, miR1867 was highly expressed during grain filling, and can regulate the genes implicated in the starch synthesis pathways in rice [45]. Therefore, we conclude that the MD-induced AS provides a possible mechanism for the regulation of microRNA processing, which may also contribute to the increased grain filling in inferior spikelets.

3.4. AS of Target Genes Increases the Complexity of miRNAs Regulation of Starch Biosynthesis

As key post-transcriptional regulators, miRNAs regulate their target genes by binding to the complementary MBS in the target mRNAs. A higher frequency of AS at MBSs and alternatively spliced MBSs enhances the regulatory complexity of the miRNA-mediated gene networks [33]. For example, AS produces multiple *SPL4* mRNA isoforms, with or without the binding site for miR156, resulting in an accelerated rate of flowering induction and significantly fewer adult leaves [33]. Os01g31870.8, one of the shortest transcript variants of *OsNramp6*, were downregulated by miR7695, which was the only transcript containing MBSs complementary to miR7695 [62]. *OsAGPL2*, a AGPase gene, plays a pivotal role in starch biosynthesis in higher plants, whose activity is directly determined by the population of the allosteric regulation [42]. The genomic data indicated that seven transcript variants of *OsAGPL2* were produced by alternative splicing (Figure 6). Among the various *OsAGPL2* splice variants, only three transcript variants contained complementary sites for miR393b-derived small RNAs, which were located at the 5′ UTR region of those transcript variants. All of those results indicate that the AS of *OsAGPL2* may attenuate miR393b-mediated gene regulation and ultimately lead to altered levels of enzyme activity. Furthermore, the miR393-overexpressed transgenic lines have smaller seed size compared with the wild type [63]. The expression of miR393b-3p was also reduced by MD treatment in the inferior spikelets [34]. Thus, the MD-induced AS of MBS increases the complexity of the miR393b-*OsAGPL2* gene regulation and function on starch biosynthesis regulation.

4. Materials and Methods

4.1. Plant Material and Experimental Design

The plant materials and growth conditions were described in detail in our previous article [7]. Briefly, the experiment was conducted in a greenhouse in the normal rice growing season in Changsha, China. Nipponbare (Nip) was used in this study. The seedings were raised at a hill spacing of 0.2 m × 0.2 m and two seedings per hill. Fertilizer and pesticide treatments were applied, according to normal agricultural practices, as described previously by Zhang et al. [16]. All of the plants were maintained well-watered up to 6 DAA. The soil water potential was monitored in the MD treatment with two tensiometers (Institute of Soil Science, Chinese Academy of Sciences, Nanjing, China) installed at a depth of 30 cm. The pots were not irrigated until the soil water potential reached −25 kPa. The CK plants were used as a blank control. The inferior kernels from each treatment were sampled at 9 DAA with three biological replicates, and preserved in a refrigerator at −80 °C for RNA sequencing.

4.2. RNA-seq Analysis

The RNA sequencing and analysis were performed by Majorbio Bio-pharm Technology Co., Ltd. (Shanghai, China). The data were analyzed on the online platform Majorbio Cloud Platform (www.majorbio.com; accessed on 1 November 2020). Briefly, the total RNA was extracted using Plant RNA Purification Reagent, according to the manufacturer's instructions (Invitrogen, Carlsbad, CA, USA). The construction of the sequencing libraries, analysis, qualification, and paired-end RNA-seq sequencing were described by Teng et al. [7]. The transcriptomic data were submitted to the NCBI, and the relevant accession number is PRJNA728244.

4.3. Alternative Splicing Analysis

The rMATS software with default settings was used to detect the differential alternative splicing events from RNA-seq data [64]. The online platform PlantSPEAD (http://chemyang.ccnu.edu.cn/ccb/database/PlantSPEAD; accessed on 1 November 2020) was used to perform the splicing factors' analyses [65]. The mRNA-Seq alignment files generated by HISAT2 (http://ccb.jhu.edu/software/hisat2/index.shtml; accessed on 1 November 2020) were used as an input for the rMATS analysis. The rice genome annotation project (http://rice.plantbiology.msu.edu/; accessed on 1 November 2020) was used as a reference, with default parameter settings. Finally, the rMATS was used to calculate the *p*-value for the AS events among the different treatments. The DAS events were extracted with a *p*-value ≤ 0.05. Splicing models of some of the DAS events were exported from the rMATS software and processed. To visualize the rMATS results and splicing model of DAS events, Majorbio Cloud Platform was used.

4.4. GO and KEGG Pathway Analyses

In addition, functional enrichment analysis, including GO and KEGG was performed to identify which DAS genes were the top 20 most enriched in GO (http://www.geneontology.org/; accessed on 1 April 2022) and KEGG (http://www.genome.jp/kegg/; accessed on 1 April 2022) database, compared with the whole-transcriptome background.

5. Conclusions

Taken together, the RNA-sequencing and small RNA-sequencing were used to reveal the interaction between AS and miRNAs in inferior spikelets under moderate soil drying.

The results show that post-anthesis moderate soil drying differentially affected the AS of genes, many of which function on spliceosome and starch and sucrose metabolism. The MD-induced AS could be a potential mechanism modulating the expression of intronic miRNAs and their function on grain filling. Alternatively, the AS of MBSs is also a plausible mechanism for miRNA-mediated gene regulation on inferior grain filling under MD treatment. To summarize, this study expands our understanding of the AS function on the MD-promoted grain filling of inferior spikelets.

Supplementary Materials: The following supporting information can be downloaded at: https://www.mdpi.com/article/10.3390/ijms23147770/s1.

Author Contributions: Conceptualization, J.Z. and N.Y.; methodology, Z.T. and Q.Z.; software, Z.T., B.L. and S.M.; resources, Z.T., J.Z. and N.Y.; data curation, Z.T.; writing—original draft preparation, Z.T., Q.Z. and N.Y.; writing—review and editing, J.Z. and N.Y.; project administration, J.Z. and N.Y.; funding acquisition, N.Y. All authors have read and agreed to the published version of the manuscript.

Funding: This research was funded by the National Natural Science Foundation of China (31971924 and 32171927), Natural Science Foundation of Hunan Province (2021JJ30349), Science and Technology Plan of Changsha City (kq2004034), and the Hong Kong Research Grant Council (AoE/M-05/12, AoE/M-403/16, GRF12100318, 12103219, 12103220).

Institutional Review Board Statement: Not applicable.

Informed Consent Statement: Not applicable.

Data Availability Statement: All of the data generated or analyzed during this study are included in this published article.

Conflicts of Interest: The authors declare no conflict of interest.

References

1. Zhang, Q. Strategies for developing green super rice. *Proc. Natl. Acad. Sci. USA* **2007**, *104*, 16402–16409. [CrossRef]
2. Yuan, L.; Denning, G.; Mew, T. Hybrid rice breeding for super high yield. *Denning GL Mew TW Ed.* **1998**, *3*, 10–12.
3. Yang, J.; Peng, S.; Zhang, Z.; Wang, Z.; Visperas, R.M.; Zhu, Q. Grain and dry matter yields and partitioning of assimilates in japonica/indica hybrid rice. *Crop Sci.* **2002**, *42*, 766–772. [CrossRef]
4. Yang, J.; Zhang, J. Grain-filling problem in 'super' rice. *J. Exp. Bot.* **2010**, *61*, 1–5. [CrossRef]
5. Yang, J.; Zhang, J. Grain filling of cereals under soil drying. *New Phytol.* **2006**, *169*, 223–236. [CrossRef]
6. Wang, G.-Q.; Li, H.-X.; Feng, L.; Chen, M.-X.; Meng, S.; Ye, N.-H.; Zhang, J. Transcriptomic analysis of grain filling in rice inferior grains under moderate soil drying. *J. Exp. Bot.* **2019**, *70*, 1597–1611. [CrossRef]
7. Teng, Z.; Yu, H.; Wang, G.; Meng, S.; Liu, B.; Yi, Y.; Chen, Y.; Zheng, Q.; Liu, L.; Yang, J. Synergistic interaction between ABA and IAA due to moderate soil drying promotes grain filling of inferior spikelets in rice. *Plant J.* **2021**, *109*, 1457–1472. [CrossRef]
8. Mohapatra, P.K.; Naik, P.K.; Patel, R. Ethylene inhibitors improve dry matter partitioning and development of late flowering spikelets on rice panicles. *Funct. Plant Biol.* **2000**, *27*, 311–323. [CrossRef]
9. Zhang, H.; Tan, G.; Yang, L.; Yang, J.; Zhang, J.; Zhao, B. Hormones in the grains and roots in relation to post-anthesis development of inferior and superior spikelets in japonica/indica hybrid rice. *Plant Physiol. Biochem.* **2009**, *47*, 195–204. [CrossRef]
10. Chen, Y.; Teng, Z.; Yuan, Y.; Yi, Z.; Zheng, Q.; Yu, H.; Lv, J.; Wang, Y.; Duan, M.; Zhang, J. Excessive nitrogen in field-grown rice suppresses grain filling of inferior spikelets by reducing the accumulation of cytokinin and auxin. *Field Crops Res.* **2022**, *283*, 108542. [CrossRef]
11. Nakamura, Y. Towards a better understanding of the metabolic system for amylopectin biosynthesis in plants: Rice endosperm as a model tissue. *Plant Cell Physiol.* **2002**, *43*, 718–725. [CrossRef]
12. Hannah, L.C.; James, M. The complexities of starch biosynthesis in cereal endosperms. *Curr. Opin. Biotechnol.* **2008**, *19*, 160–165. [CrossRef]
13. Peng, Y.; Chen, Y.; Yuan, Y.; Liu, B.; Yu, P.; Song, S.; Yi, Y.; Teng, Z.; Yi, Z.; Zhang, J. Post-anthesis saline-alkali stress inhibits grain filling by promoting ethylene production and signal transduction. *Food Energy Secur.* **2022**, e384. [CrossRef]
14. Pang, Y.; Hu, Y.; Bao, J. Comparative phosphoproteomic analysis reveals the response of starch metabolism to high-temperature stress in rice endosperm. *Int. J. Mol. Sci.* **2021**, *22*, 10546. [CrossRef]
15. Wang, Z.; Xu, Y.; Chen, T.; Zhang, H.; Yang, J.; Zhang, J. Abscisic acid and the key enzymes and genes in sucrose-to-starch conversion in rice spikelets in response to soil drying during grain filling. *Planta* **2015**, *241*, 1091–1107. [CrossRef]
16. Zhang, H.; Li, H.; Yuan, L.; Wang, Z.; Yang, J.; Zhang, J. Post-anthesis alternate wetting and moderate soil drying enhances activities of key enzymes in sucrose-to-starch conversion in inferior spikelets of rice. *J. Exp. Bot.* **2012**, *63*, 215–227. [CrossRef]

17. Reddy, A.S. Alternative splicing of pre-messenger RNAs in plants in the genomic era. *Annu. Rev. Plant Biol.* **2007**, *58*, 267–294. [CrossRef]

18. Chen, M.-X.; Zhu, F.-Y.; Wang, F.-Z.; Ye, N.-H.; Gao, B.; Chen, X.; Zhao, S.-S.; Fan, T.; Cao, Y.-Y.; Liu, T.-Y. Alternative splicing and translation play important roles in hypoxic germination in rice. *J. Exp. Bot.* **2019**, *70*, 817–833. [CrossRef]

19. Muñoz, M.J.; Santangelo, M.S.P.; Paronetto, M.P.; de la Mata, M.; Pelisch, F.; Boireau, S.; Glover-Cutter, K.; Ben-Dov, C.; Blaustein, M.; Lozano, J.J. DNA damage regulates alternative splicing through inhibition of RNA polymerase II elongation. *Cell* **2009**, *137*, 708–720. [CrossRef]

20. Palusa, S.G.; Ali, G.S.; Reddy, A.S. Alternative splicing of pre-mRNAs of *Arabidopsis serine*/arginine-rich proteins: Regulation by hormones and stresses. *Plant J.* **2007**, *49*, 1091–1107. [CrossRef]

21. Yan, K.; Liu, P.; Wu, C.-A.; Yang, G.-D.; Xu, R.; Guo, Q.-H.; Huang, J.-G.; Zheng, C.-C. Stress-induced alternative splicing provides a mechanism for the regulation of microRNA processing in *Arabidopsis thaliana*. *Mol. Cell* **2012**, *48*, 521–531. [CrossRef]

22. Marrs, K.A.; Walbot, V. Expression and RNA splicing of the maize glutathione S-transferase Bronze2 gene is regulated by cadmium and other stresses. *Plant Physiol.* **1997**, *113*, 93–102. [CrossRef]

23. Chen, M.-X.; Zhang, Y.; Fernie, A.R.; Liu, Y.-G.; Zhu, F.-Y. SWATH-MS-based proteomics: Strategies and applications in plants. *Trends Biotechnol.* **2021**, *39*, 433–437. [CrossRef]

24. Quesada, V.; Macknight, R.; Dean, C.; Simpson, G.G. Autoregulation of FCA pre-mRNA processing controls Arabidopsis flowering time. *EMBO J.* **2003**, *22*, 3142–3152. [CrossRef]

25. Isshiki, M.; Morino, K.; Nakajima, M.; Okagaki, R.J.; Wessler, S.R.; Izawa, T.; Shimamoto, K. A naturally occurring functional allele of the rice waxy locus has a GT to TT mutation at the 5′ splice site of the first intron. *Plant J.* **1998**, *15*, 133–138. [CrossRef]

26. Isshiki, M.; Nakajima, M.; Satoh, H.; Shimamoto, K. dull: Rice mutants with tissue-specific effects on the splicing of the waxy pre-mRNA. *Plant J.* **2000**, *23*, 451–460. [CrossRef]

27. Liu, X.; Tian, Y.; Chi, W.; Zhang, H.; Yu, J.; Chen, G.; Wu, W.; Jiang, X.; Wang, S.; Lin, Z. Alternative splicing of OsGS1;1 affects nitrogen-use efficiency, grain development, and amylose content in rice. *Plant J.* **2022**, *110*, 1751–1762. [CrossRef]

28. Berezikov, E.; Chung, W.-J.; Willis, J.; Cuppen, E.; Lai, E.C. Mammalian mirtron genes. *Molecular cell* **2007**, *28*, 328–336. [CrossRef]

29. Morlando, M.; Ballarino, M.; Gromak, N.; Pagano, F.; Bozzoni, I.; Proudfoot, N.J. Primary microRNA transcripts are processed co-transcriptionally. *Nat. Struct. Mol. Biol.* **2008**, *15*, 902–909. [CrossRef]

30. Hirsch, J.; Lefort, V.; Vankersschaver, M.; Boualem, A.; Lucas, A.; Thermes, C.; d'Aubenton-Carafa, Y.; Crespi, M. Characterization of 43 non-protein-coding mRNA genes in *Arabidopsis*, including the MIR162a-derived transcripts. *Plant Physiol.* **2006**, *140*, 1192–1204. [CrossRef]

31. Kurihara, Y.; Watanabe, Y. *Arabidopsis* micro-RNA biogenesis through Dicer-like 1 protein functions. *Proc. Natl. Acad. Sci. USA* **2004**, *101*, 12753–12758. [CrossRef]

32. Li, C.; Zhang, B. MicroRNAs in control of plant development. *J. Cell. Physiol.* **2016**, *231*, 303–313. [CrossRef]

33. Yang, X.; Zhang, H.; Li, L. Alternative mRNA processing increases the complexity of microRNA-based gene regulation in Arabidopsis. *Plant J.* **2012**, *70*, 421–431. [CrossRef]

34. Teng, Z.; Chen, Y.; Yuan, Y.; Peng, Y.; Yi, Y.; Yu, H.; Yi, Z.; Yang, J.; Peng, Y.; Duan, M.; et al. Identification of microRNAs regulating grain filling of rice inferior spikelets in response to moderate soil drying post-anthesis. *Crop J.* **2021**. [CrossRef]

35. Zhao, Y.F.; Peng, T.; Sun, H.Z.; Teotia, S.; Wen, H.L.; Du, Y.X.; Zhang, J.; Li, J.Z.; Tang, G.L.; Xue, H.W. miR1432-Os ACOT (Acyl-CoA thioesterase) module determines grain yield via enhancing grain filling rate in rice. *Plant Biotechnol. J.* **2019**, *17*, 712–723. [CrossRef]

36. Zhang, Y.-C.; Yu, Y.; Wang, C.-Y.; Li, Z.-Y.; Liu, Q.; Xu, J.; Liao, J.-Y.; Wang, X.-J.; Qu, L.-H.; Chen, F. Overexpression of microRNA OsmiR397 improves rice yield by increasing grain size and promoting panicle branching. *Nat. Biotechnol.* **2013**, *31*, 848–852. [CrossRef]

37. Peng, T.; Lv, Q.; Zhang, J.; Li, J.; Du, Y.; Zhao, Q. Differential expression of the microRNAs in superior and inferior spikelets in rice (*Oryza sativa*). *J. Exp. Bot.* **2011**, *62*, 4943–4954. [CrossRef]

38. Isshiki, M.; Tsumoto, A.; Shimamoto, K. The serine/arginine-rich protein family in rice plays important roles in constitutive and alternative splicing of pre-mRNA. *Plant Cell* **2006**, *18*, 146–158. [CrossRef]

39. Barta, A.; Kalyna, M.; Reddy, A.S. Implementing a rational and consistent nomenclature for serine/arginine-rich protein splicing factors (SR proteins) in plants. *Plant Cell* **2010**, *22*, 2926–2929. [CrossRef]

40. Chen, M.-X.; Zhang, K.-L.; Zhang, M.; Das, D.; Fang, Y.-M.; Dai, L.; Zhang, J.; Zhu, F.-Y. Alternative splicing and its regulatory role in woody plants. *Tree Physiol.* **2020**, *40*, 1475–1486. [CrossRef]

41. Ohdan, T.; Francisco Jr, P.B.; Sawada, T.; Hirose, T.; Terao, T.; Satoh, H.; Nakamura, Y. Expression profiling of genes involved in starch synthesis in sink and source organs of rice. *J. Exp. Bot.* **2005**, *56*, 3229–3244. [CrossRef]

42. Tuncel, A.; Kawaguchi, J.; Ihara, Y.; Matsusaka, H.; Nishi, A.; Nakamura, T.; Kuhara, S.; Hirakawa, H.; Nakamura, Y.; Cakir, B. The rice endosperm ADP-glucose pyrophosphorylase large subunit is essential for optimal catalysis and allosteric regulation of the heterotetrameric enzyme. *Plant Cell Physiol.* **2014**, *55*, 1169–1183. [CrossRef]

43. Ying, S.-Y.; Lin, S.-L. Intronic microRNAs. *Biochem. Biophys. Res. Commun.* **2005**, *326*, 515–520. [CrossRef]

44. Peng, T.; Sun, H.; Qiao, M.; Zhao, Y.; Du, Y.; Zhang, J.; Li, J.; Tang, G.; Zhao, Q. Differentially expressed microRNA cohorts in seed development may contribute to poor grain filling of inferior spikelets in rice. *BMC Plant Biol.* **2014**, *14*, 1–17. [CrossRef]

45. Zheng, Y.; Li, Y.-F.; Sunkar, R.; Zhang, W. SeqTar: An effective method for identifying microRNA guided cleavage sites from degradome of polyadenylated transcripts in plants. *Nucleic Acids Res.* **2012**, *40*, e28. [CrossRef]
46. Chen, X. Small RNAs and their roles in plant development. *Annu. Rev. Cell Dev.* **2009**, *25*, 21–44. [CrossRef]
47. Staiger, D.; Brown, J.W. Alternative splicing at the intersection of biological timing, development, and stress responses. *Plant Cell* **2013**, *25*, 3640–3656. [CrossRef]
48. Castiglioni, P.; Warner, D.; Bensen, R.J.; Anstrom, D.C.; Harrison, J.; Stoecker, M.; Abad, M.; Kumar, G.; Salvador, S.; D'Ordine, R. Bacterial RNA chaperones confer abiotic stress tolerance in plants and improved grain yield in maize under water-limited conditions. *Plant Physiol.* **2008**, *147*, 446–455. [CrossRef]
49. Gao, P.; Quilichini, T.D.; Zhai, C.; Qin, L.; Nilsen, K.T.; Li, Q.; Sharpe, A.G.; Kochian, L.V.; Zou, J.; Reddy, A.S. Alternative splicing dynamics and evolutionary divergence during embryogenesis in wheat species. *Plant Biotechnol. J.* **2021**, *19*, 1624. [CrossRef]
50. Yu, J.; Miao, J.; Zhang, Z.; Xiong, H.; Zhu, X.; Sun, X.; Pan, Y.; Liang, Y.; Zhang, Q.; Abdul Rehman, R.M. Alternative splicing of Os LG 3b controls grain length and yield in japonica rice. *Plant Biotechnol. J.* **2018**, *16*, 1667–1678. [CrossRef]
51. Ren, X.; Zhi, L.; Liu, L.; Meng, D.; Su, Q.; Batool, A.; Ji, J.; Song, L.; Zhang, N.; Guo, L. Alternative Splicing of TaGS3 Differentially Regulates Grain Weight and Size in Bread Wheat. *Int. J. Mol. Sci.* **2021**, *22*, 11692. [CrossRef]
52. Xu, H.; Li, X.; Zhang, H.; Wang, L.; Zhu, Z.; Gao, J.; Li, C.; Zhu, Y. High temperature inhibits the accumulation of storage materials by inducing alternative splicing of OsbZIP58 during filling stage in rice. *Plant Cell Environ.* **2020**, *43*, 1879–1896. [CrossRef]
53. Lee, S.-K.; Hwang, S.-K.; Han, M.; Eom, J.-S.; Kang, H.-G.; Han, Y.; Choi, S.-B.; Cho, M.-H.; Bhoo, S.H.; An, G. Identification of the ADP-glucose pyrophosphorylase isoforms essential for starch synthesis in the leaf and seed endosperm of rice (*Oryza sativa* L.). *Plant Mol. Biol.* **2007**, *65*, 531–546. [CrossRef]
54. Szydlowski, N.; Ragel, P.; Raynaud, S.; Lucas, M.M.; Roldán, I.; Montero, M.; Muñoz, F.J.; Ovecka, M.; Bahaji, A.; Planchot, V. Starch granule initiation in Arabidopsis requires the presence of either class IV or class III starch synthases. *Plant Cell* **2009**, *21*, 2443–2457. [CrossRef]
55. Utsumi, Y.; Utsumi, C.; Sawada, T.; Fujita, N.; Nakamura, Y. Functional diversity of isoamylase oligomers: The ISA1 homo-oligomer is essential for amylopectin biosynthesis in rice endosperm. *Plant Physiol.* **2011**, *156*, 61–77. [CrossRef]
56. Chen, M.X.; Zhang, K.L.; Gao, B.; Yang, J.F.; Tian, Y.; Das, D.; Fan, T.; Dai, L.; Hao, G.F.; Yang, G.F. Phylogenetic comparison of 5′ splice site determination in central spliceosomal proteins of the U1-70K gene family, in response to developmental cues and stress conditions. *Plant J.* **2020**, *103*, 357–378. [CrossRef]
57. Zhang, D.; Chen, M.-X.; Zhu, F.-Y.; Zhang, J.; Liu, Y.-G. Emerging functions of plant Serine/Arginine-Rich (SR) proteins: Lessons from animals. *Crit. Rev. Plant Sci.* **2020**, *39*, 173–194. [CrossRef]
58. Brown, J.W.; Marshall, D.F.; Echeverria, M. Intronic noncoding RNAs and splicing. *Trends Plant Sci.* **2008**, *13*, 335–342. [CrossRef]
59. Yang, G.; Yan, K.; Wu, B.; Wang, Y.; Gao, Y.; Zheng, C. Genomewide analysis of intronic microRNAs in rice and *Arabidopsis*. *J. Genet.* **2012**, *91*, 313–324. [CrossRef]
60. Marquez, Y.; Brown, J.W.; Simpson, C.; Barta, A.; Kalyna, M. Transcriptome survey reveals increased complexity of the alternative splicing landscape in *Arabidopsis*. *Genome Res.* **2012**, *22*, 1184–1195. [CrossRef]
61. Chandra, T.; Mishra, S.; Panda, B.B.; Sahu, G.; Dash, S.K.; Shaw, B.P. Study of expressions of miRNAs in the spikelets based on their spatial location on panicle in rice cultivars provided insight into their influence on grain development. *Plant Physiol. Biochem.* **2021**, *159*, 244–256. [CrossRef] [PubMed]
62. Campo, S.; Peris-Peris, C.; Siré, C.; Moreno, A.B.; Donaire, L.; Zytnicki, M.; Notredame, C.; Llave, C.; San Segundo, B. Identification of a novel micro RNA (mi RNA) from rice that targets an alternatively spliced transcript of the N ramp6 (N atural resistance-associated macrophage protein 6) gene involved in pathogen resistance. *New Phytol.* **2013**, *199*, 212–227. [CrossRef]
63. Bian, H.; Xie, Y.; Guo, F.; Han, N.; Ma, S.; Zeng, Z.; Wang, J.; Yang, Y.; Zhu, M. Distinctive expression patterns and roles of the miRNA393/TIR1 homolog module in regulating flag leaf inclination and primary and crown root growth in rice (*Oryza sativa*). *New Phytol.* **2012**, *196*, 149–161. [CrossRef]
64. Shen, S.; Park, J.W.; Lu, Z.-x.; Lin, L.; Henry, M.D.; Wu, Y.N.; Zhou, Q.; Xing, Y. rMATS: Robust and flexible detection of differential alternative splicing from replicate RNA-Seq data. *Proc. Natl. Acad. Sci. USA* **2014**, *111*, E5593–E5601. [CrossRef] [PubMed]
65. Chen, M.X.; Mei, L.C.; Wang, F.; Dewayalage, I.K.W.B.; Yang, J.F.; Dai, L.; Yang, G.F.; Gao, B.; Cheng, C.L.; Liu, Y.G. PlantSPEAD: A web resource towards comparatively analysing stress-responsive expression of splicing-related proteins in plant. *Plant Biotechnol. J.* **2021**, *19*, 227. [CrossRef] [PubMed]

International Journal of
Molecular Sciences

Article

SWATH-MS-Based Proteomics Reveals the Regulatory Metabolism of Amaryllidaceae Alkaloids in Three *Lycoris* Species

Meng Tang [1,2,†], Chaohan Li [1,†], Cheng Zhang [2], Youming Cai [1], Yongchun Zhang [1], Liuyan Yang [1], Moxian Chen [2,3], Fuyuan Zhu [2] , Qingzhu Li [1,*] and Kehu Li [3,*]

[1] Forestry and Pomology Research Institute, Protected Horticultural Research Institute, Shanghai Key Laboratory of Protected Horticultural Technology, Shanghai Academy of Agricultural Sciences, Shanghai 201403, China
[2] Co-Innovation Center for Sustainable Forestry in Southern China, College of Biology and the Environment, Nanjing Forestry University, Nanjing 210037, China
[3] Key Laboratory of Plant Resource Conservation and Germplasm Innovation in Mountainous Region (Ministry of Education), Collaborative Innovation Center for Mountain Ecology & Agro-Bioengineering (CICMEAB), Institute of Agro-Bioengineering, College of Life Sciences, Guizhou University, Guiyang 550025, China
* Correspondence: liqingzhu@saas.sh.cn (Q.L.); khli@gzu.edu.cn (K.L.)
† These authors contributed equally to this work.

Abstract: Alkaloids are a class of nitrogen-containing alkaline organic compounds found in nature, with significant biological activity, and are also important active ingredients in Chinese herbal medicine. Amaryllidaceae plants are rich in alkaloids, among which galanthamine, lycorine, and lycoramine are representative. Since the difficulty and high cost of synthesizing alkaloids have been the major obstacles in industrial production, particularly the molecular mechanism underlying alkaloid biosynthesis is largely unknown. Here, we determined the alkaloid content in *Lycoris longituba*, *Lycoris incarnata*, and *Lycoris sprengeri*, and performed a SWATH-MS (sequential window acquisition of all theoretical mass spectra)-based quantitative approach to detect proteome changes in the three *Lycoris*. A total of 2193 proteins were quantified, of which 720 proteins showed a difference in abundance between *Ll* and *Ls*, and 463 proteins showed a difference in abundance between *Li* and *Ls*. KEGG enrichment analysis revealed that differentially expressed proteins are distributed in specific biological processes including amino acid metabolism, starch, and sucrose metabolism, implicating a supportive role for Amaryllidaceae alkaloids metabolism in *Lycoris*. Furthermore, several key genes collectively known as OMT and NMT were identified, which are probably responsible for galanthamine biosynthesis. Interestingly, RNA processing-related proteins were also abundantly detected in alkaloid-rich *Ll*, suggesting that posttranscriptional regulation such as alternative splicing may contribute to the biosynthesis of Amaryllidaceae alkaloids. Taken together, our SWATH-MS-based proteomic investigation may reveal the differences in alkaloid contents at the protein levels, providing a comprehensive proteome reference for the regulatory metabolism of Amaryllidaceae alkaloids.

Keywords: alkaloids; galanthamine; *Lycoris*; metabolism; SWATH-MS

Citation: Tang, M.; Li, C.; Zhang, C.; Cai, Y.; Zhang, Y.; Yang, L.; Chen, M.; Zhu, F.; Li, Q.; Li, K. SWATH-MS-Based Proteomics Reveals the Regulatory Metabolism of Amaryllidaceae Alkaloids in Three *Lycoris* Species. *Int. J. Mol. Sci.* **2023**, *24*, 4495. https://doi.org/10.3390/ijms24054495

Academic Editor: Setsuko Komatsu

Received: 18 January 2023
Revised: 14 February 2023
Accepted: 15 February 2023
Published: 24 February 2023

1. Introduction

Lycoris spp. Amaryllidaceae is found in tropical and temperate regions, especially in China and Japan. It has long been widely used as folk medicine to treat various diseases [1,2]. *Lycoris* bulbs are used as traditional Chinese herbal medicine to treat sore throat, abscess, cancer, suppurative wound, poliomyelitis, mastitis, otitis media, ulcer, and neurodegenerative diseases [3]. According to the "Compendium of Chinese Materia Medica", *Lycoris* also works in detoxifying, diminishing inflammation, pain relief, and diuresis [4]. Alkaloids are the main medicinal chemicals in the bulbs of *Lycoris* plants. Alkaloids are heterocyclic nitrogen compounds and have significant biological activities in human health applications. For example, the application in medicine is morphine, which has a powerful analgesic and sedative effect in clinical practice. Diterpenoid alkaloids isolated from Ranunculaceae were found to have antibacterial properties [5,6]. Carbohydrate alkaloids extracted from *Solanum nigrum* berries are medically used in the prevention of HIV infection and AIDS-related intestinal infections [7,8]. More than 600 alkaloids with various structures have been isolated from *Lycoris* plants including galanthamine, lycoramine, and lycorine thus far and have a variety of biological activities including anticancer, anti-inflammatory, anti-plasmodium, and antibacterial activities [3,9].

Galanthamine is a representative alkaloid of *Lycoris*, which is a selective, long-acting, reversible, and competitive acetylcholinesterase inhibitor. Symptomatic treatment for AD slows the progression of the disease and helps relieve memory loss [10]. Recently, transcriptomic and metabolomic analyses of *L. radiata* were performed to investigate the biosynthesis of galanthamine in different organs. It was found that *LrNNR*, *LrN4OMT*, and *LrCYP96T* were highly expressed in bulbs, which is consistent with the observation that more galanthamine is present in bulbs than in roots and leaves [11]. Functional characterization of the *lycoris* phenylalanine ammonia-lyase gene (*LrPAL*) revealed that *LrPAL* may be the limiting factor for the biosynthesis of galanthamine [12]. On the other hand, NpNBS condenses tyramine and 3,4-DHBA into norbelladine which is the first step in benzylisoquinoline alkaloid biosynthesis [13]. In addition, recombinant LlOMT catalyzes norbelladine to generate 4′-O-methylnorbelladine and overexpression of *LlOMT* in *Lycoris longituba* could increase the galanthamine content [14,15]. To date, previous research on the biosynthesis of galanthamine only focused on a few key genes, and the whole biosynthetic pathway of galanthamine in plants has not been comprehensively investigated particularly for the metabolism regulatory network from the perspective of proteome level.

This study was conducted by SWATH-MS technology to explore the proteome of *Lycoris*. SWATH-MS is a specific data-independent-acquisition (DIA)-based method and is emerging as a technology that combines deep proteome coverage capabilities with quantitative consistency and accuracy [16]. SWATH overcomes the problem of stochasticity in data-dependent acquisition (DDA), making the detection results more reproducible and consistent. The wider coverage and higher detection sensitivity will further provide more ways to verify the function of the target protein in plants, which makes SWATH-MS a huge prospect in plant proteomics research [17]. In this study, the relative quantitative comparison of proteins among three groups of *Lycoris* samples with different alkaloid contents was carried out by quantitative proteomics technology based on SWATH-MS, and a total of 2193 proteins were detected. The differentially expressed proteins among the three groups were enriched in the amino acid metabolism pathway, galanthamine synthesis pathway, and sucrose metabolism pathway. Analysis of bioinformatics data will offer considerable information on the proteome among the three *Lycoris* with different alkaloid contents, and also provide deeper insights into the molecular mechanisms of Amaryllidaceae alkaloid's regulatory metabolism. In summary, our SWATH-MS-based proteomics study can provide new insights into the mechanism of alkaloid-regulated metabolism in Amaryllidaceae at the protein level, and provide new ideas for the yield enhancement of alkaloid biosynthesis in the future.

2. Results

2.1. SWATH-MS-Based Proteomic Analysis among Three Species of Lycoris with Different Alkaloid Content

To further investigate the biosynthesis of Amaryllidaceae alkaloids in *Lycoris* species, we performed a SWATH-MS-based proteomic analysis of *Lycoris longituba* (*Ll*), *Lycoris incarnata* (*Li*), and *Lycoris sprengeri* (*Ls*) with different alkaloid contents. The three *Lycoris* materials were collected from the Institute of Botany, Chinese Academy of Sciences, Jiangsu Province, China. Among them, the total content of alkaloids in *Ll* was the highest, whereas in *Ls* was the lowest (Figure 1B).

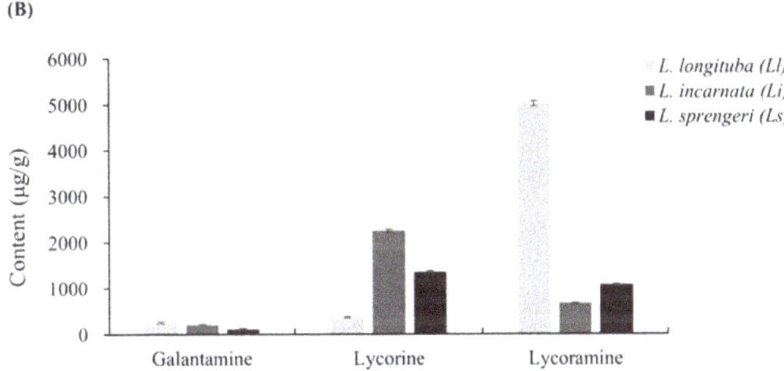

Figure 1. Phenotype of samples and alkaloid contents of three kinds of *Lycoris* species. (**A**) The three kinds of plant materials: *Lycoris longituba*, *Lycoris incarnata*, and *Lycoris sprengeri*. They were collected from the Institute of Botany, Chinese Academy of Sciences, Jiangsu Province, China (32.05° N, 118.83° E) in late March when the leaves were growing vigorously (Bars = 3 cm). (**B**) The total content of three alkaloids in *Lycoris longituba* was the highest, whereas that in *Lycoris sprengeri* was the lowest.

We selected three groups of samples as biological replicates for proteomic analysis. The correlation coefficient of the samples within the group and the principal component clustering results strongly reflected the high reliability of the appropriate sampling and suitability for subsequent analysis (Figure S1A). A total of 2193 proteins were quantified, and the group with the lowest alkaloid content *Ls* served as the control group. Proteins with a fold change above 2 or below 0.5 ($p < 0.05$) were considered differentially expressed proteins (DEPs) in this study. As shown in the volcano diagram, 485 upregulated proteins and 235 downregulated proteins were detected in *Ll* compared with *Ls*, whereas 288 upregulated proteins and 175 downregulated proteins were detected in *Li* (Figure 2A). The cluster heatmap showed significant changes in protein abundance. When we compared *Ll* with *Ls*, red and green colors, respectively, indicated upregulation and downregulation; when we compared *Li* with *Ls*, red and blue colors, respectively, indicated upregulation and downregulation (Figure S1C,D). This suggests that some crucial regulators or pathways have been induced, potentially affecting the biosynthesis of Amaryllidaceae alkaloids in *Lycoris*.

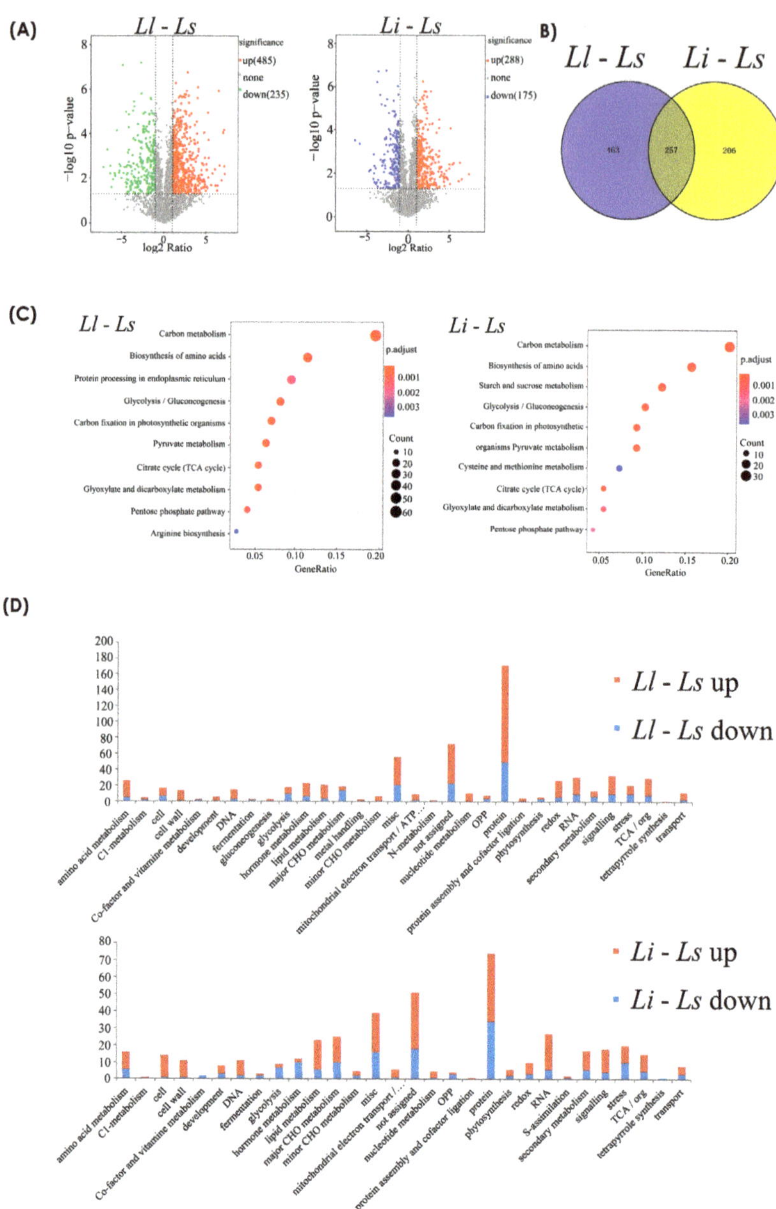

Figure 2. The overview and functional classification by KEGG and MAPMAN BIN.of DEPs. (**A**) The volcano picture shows the distribution of differentially expressed proteins with different fold changes. (**B**) Venn diagram shows the number of differential proteins compared between the two groups. (**C**) KEGG pathway classification was performed on the differential proteins of the 2 groups and enriched in the top 10 pathways. Amino acid biosynthesis, starch, and sucrose metabolic pathways are significantly activated. (**D**) MAPMAN BIN categorization of DEPs; blue and red represent down-regulated and up-regulated proteins, respectively.

2.2. Functional Classification of DEPs

KEGG pathway classification was performed on the two groups' differential proteins and enriched in the top 10 pathways in order to quickly view the pathways that affect the biosynthesis of Amaryllidaceae alkaloids in *Lycoris*. For example, amino acid biosynthesis and starch and sucrose metabolic pathways were significantly activated and may be involved in contributing to the biosynthesis of Amaryllidaceae alkaloids in *Lycoris* (Figure 2C). Subsequently, we used the MapMAN BIN system to functionally classify the differentially expressed proteins into two groups. For example, compared with *Ls*, all differential proteins were assigned to 32 functional categories in *Ll*, including "protein metabolism", "RNA processing and transport", "signal transduction", and other categories which constitute the main part of differential proteins, indicating these proteins have a great impact on the biosynthesis of Amaryllidaceae alkaloids in *Lycoris*. Among these, 21 of the 26 proteins involved in amino acid metabolism were downregulated. In the following sections, these upregulated or downregulated proteins will be mapped to different physiological and biochemical pathways, and the correlation between these biochemical pathways and the biosynthesis of alkaloids will be discussed (Figure 2D).

2.3. Effects of Amino Acid Metabolism on the Biosynthesis of Amaryllidaceae Alkaloids in Lycoris

Amino acids are closely related to the synthesis of plant alkaloids, and the nitrogen in complex alkaloids comes from amines derived from amino acid metabolism [18]. The amines contributing to alkaloid biosynthesis are derived from various amino acids, and are divided into two categories including polyamines and aromatic amines. Tyrosine is the precursor of multiple alkaloid families, including benzylisoquinolines (BIAs), Amaryllidaceae alkaloids, and betalains. *TyrDC* from Amaryllidaceae alkaloid biosynthesis has recently been discovered, it enables the incorporation of tyramine into the structures [15].

Through previous analyses of KEGG and MAPMAN BIN, we integrated the identified proteins embedded into different amino acid biosynthesis pathways by KEGG and MAPMAN BIN analysis (Table S1). Proteins with significant changes in abundance were enriched in the biosynthesis of alanine, aspartate, glutamate, cysteine, methionine, and arginine (Figure 3), of which alanine and arginine are the raw materials for the biosynthesis of plant alkaloids. Conversely, various alkaloids can negatively regulate the biosynthesis of glutamate in cells [19].

We detected a total of 38 differential proteins in the amino acid metabolism pathway, mainly concentrated in the biosynthesis and metabolism pathways of amino acids such as alanine, arginine, and cysteine. Among them, 18 genes were upregulated in *Ll*, 12 genes were upregulated in *Li*, and 8 genes were upregulated in *Ls*. In general, the highly expressed proteins were mainly concentrated in *Ll*, which was consistent with the highest alkaloid content of *Ll*, implicating that amino acid metabolism is closely related to the biosynthesis of Amaryllidaceae alkaloids in *Lycoris*.

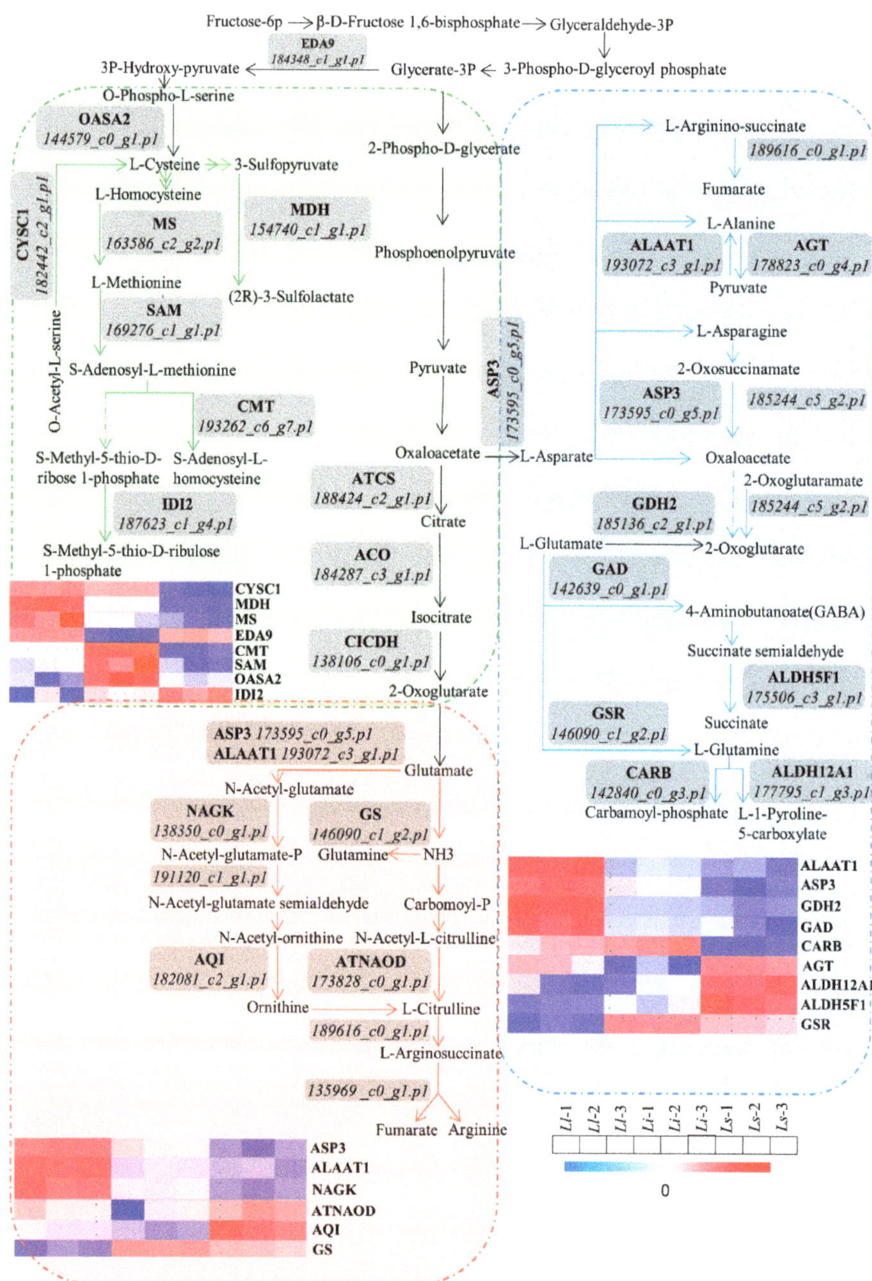

Figure 3. Amino acid metabolism is closely related to the biosynthesis of Amaryllidaceae alkaloids in *Lycoris*. We detected a total of 38 differential proteins in the amino acid metabolism pathway. The highly expressed proteins were mainly concentrated in Ll, which was consistent with the actual detected trend of alkaloid content of the three varieties.

2.4. Important Genes Involved in the Galanthamine Biosynthesis Pathway for Alkaloids Production in Lycoris

Galanthamine is a unique isoquinoline alkaloid and competitive acetylcholinesterase inhibitor that has broad prospects for future medical applications. The difficulty and high cost of synthesizing galanthamine have been the major obstacles to industrial production. The biosynthetic pathway of Gal has recently been elucidated including phenylalanine ammonia lyase (PAL), cinnamic acid-4-hydroxylase (C4H), coumaric acid 3-hydroxylase, tyrosine decarboxylase (TYDC), desmethylbelladine synthase (NBS), desmethylbelladine 40-O-methyltransferase (OMT), desmethoxymalidine synthase (CYP96T1), and N-methyltransferase (NMT) [15,20]. *OMT* is a methyltransferase that is involved in the biosynthesis of many alkaloids. For example, *OMT* has been shown to be responsible for multiple substrate and region-specific methylation in the roots of *G. flavum* and the biosynthesis and metabolism of plant alkaloids [21,22]. *NMT* is also a methyltransferase that has been shown to catalyze phenylisoquinoline alkaloids, one biosynthetic precursor of morphine [23]. In this study, we detected ten differentially expressed proteins among the three *Lycoris* species in the biosynthetic pathway of galanthamine. Among them, three proteins collectively known as OMT probably catalyzed norbelladine into 4′-O-methylnorbelladine and subsequently oxidized to N-demethylnawedine, further reduced to N-demethylgalanthamine. Seven other identified proteins known as NMT are probably responsible for the methylation of N-demethylgalanthamine to galanthamine (Figure 4).

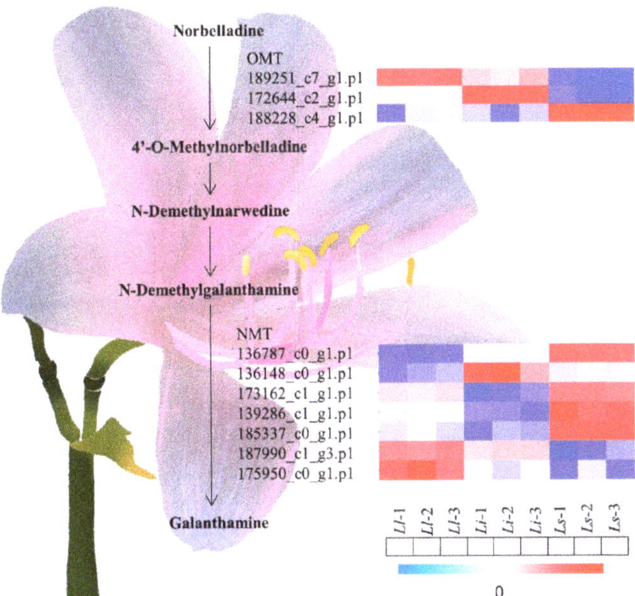

Figure 4. Candidate genes responsible for the biosynthesis of galanthamine in *Lycoris*. 189251_c7_g1.p1 and 175950_c0_g1.p1 and other genes upregulated in *Ll*, which provide the raw material for the biosynthesis of galanthamine, consistent with the highest galanthamine content in Ll.

2.5. Other Pathways Related to the Amaryllidaceae alkaloids in Lycoris

According to our MapMAN BIN system results, some proteins were detected to be enriched in starch and sucrose metabolism pathways suggesting a reciprocal regulation between alkaloids and sugar metabolism on the seasonal variation of alkaloids and total polyphenol contents, showing a contrasting tendency in sugar content and tissue development [24] (Figure 5A). Likewise, some proteins showed differences in abundance in secondary metabolic pathways, which are likely to contribute to the biosynthesis of alkaloids. For example, the upregulated *187456_c3_g1.p1* and *189251_c7_g1.p1* in *Ll* participate in the biosynthesis of sinapyl alcohol, resulting in an increase in the alkaloid raw material sinapyl alcohol, which would eventually lead to a rise in alkaloid content [25] (Figure 5B).

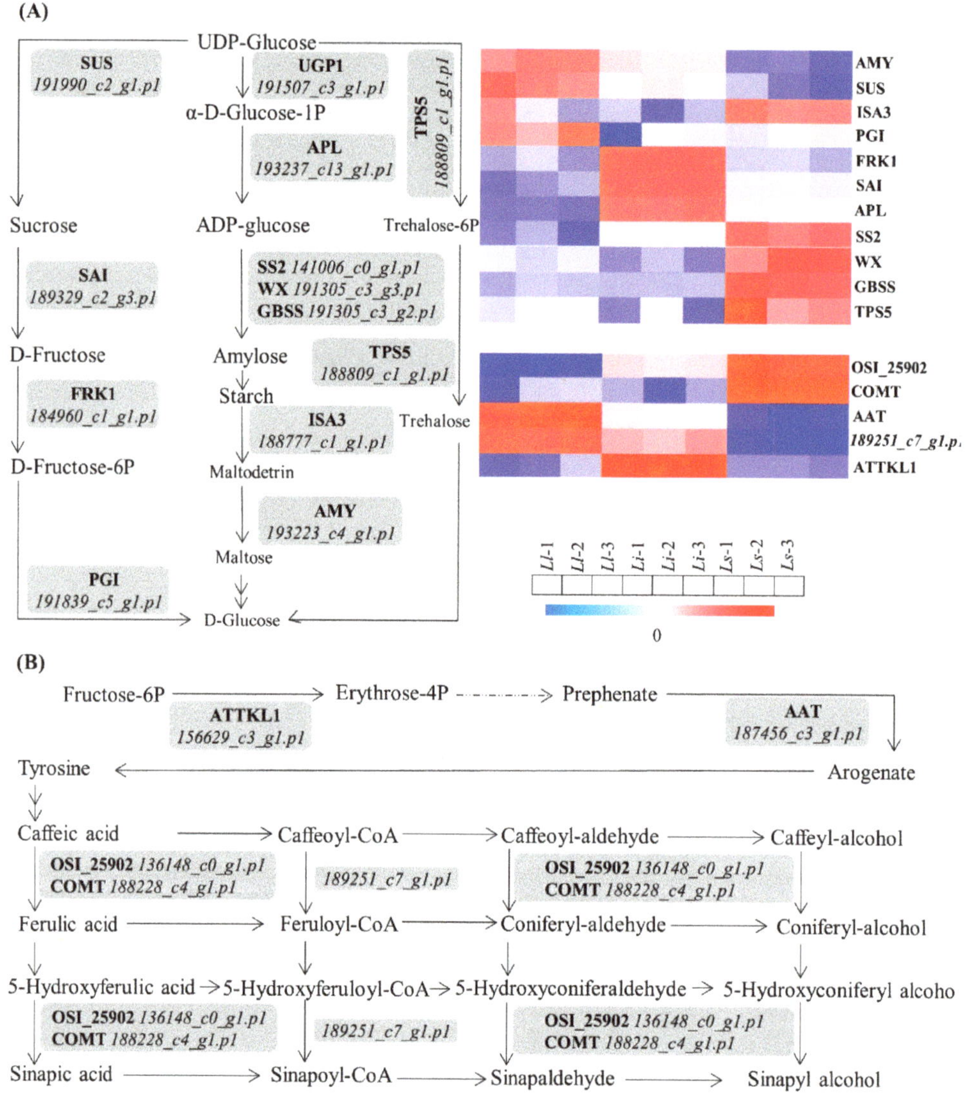

Figure 5. Potenial pathways contribute to the biosynthesis of Amaryllidaceae alkaloids in *Lycoris*. (**A**) Some proteins were detected to be enriched in starch and sucrose metabolism pathways suggesting a reciprocal regulation between alkaloids and sugar metabolism. (**B**) In secondary metabolic pathways, the up-regulated 187456_c3_g1.p1 and 189251_c7_g1.p1 in Ll participate in the biosynthesis of sinapyl alcohol, resulting in an increase in the alkaloid raw material sinapyl alcohol. Eventually, this leads to a rise in alkaloid content.

3. Discussion

3.1. Advanced High-Throughput Omics Technology Facilities the Elucidation on the Regulatory Metabolism of Lycoris Species

Multiple-omics technologies have been extensively applied to various aspects of *Lycoris* research. Transcriptome analysis facilities the identification of key regulatory genes involved in anthocyanin metabolism during flower development in *Lycoris radiata* [26]. Proteomic analysis helps us better understand the molecular mechanism of *L. radiata* development and provides valuable information about the proteins involved in the development and stress response of other *Lycoris* genera [27]. Multi-omics such as transcriptomic and metabolomic analyses reveal that exogenous methyl jasmonate regulates galanthamine biosynthesis in *Lycoris longituba* seedlings; furthermore, anthocyanin biosynthesis, steroid biosynthesis, and R2R3 MYB TFs may play vital regulatory roles in petal color development in *L. sprengeri* [15,28]. Therefore, we used innovative SWATH-MS quantitative proteomics to investigate the molecular regulatory network of Amaryllidaceae alkaloids metabolism in *Lycoris*, which provides a new strategy for the future exploration of the biosynthesis and function of secondary metabolites. Furthermore, single-cell-based omics reveals that alkaloids are localized in *Catharanthus roseus* stem and leaf tissues [29], which would be an effective strategy to determine intercellular localization of alkaloids from different tissues in *Lycoris*, thus understanding the mechanism of alkaloids biosynthesis at the cellular level. A flow cytometry study of the nuclear DNA contents of *Lycoris* species (Amaryllidaceae) with different chromosome numbers revealed that the *Lycoris* genome contains approximately 30 G [30]. Since the *Lycoris* genome is so large and undiscovered, Pacbio's new revolutionary long-read length sequencing system, revio, will significantly reduce sequencing costs, and is expected to be applied to the *Lycoris* genome, thus providing a high-quality map of the *Lycoris* genome and providing a scientific basis for the metabolic regulatory mechanisms in *Lycoris*.

3.2. Candidate Genes Responsible for the Biosynthesis of Amaryllidaceae alkaloids in Lycoris

Seeking some target proteins among the differentially expressed proteins is necessary to illuminate the biosynthesis process of Amaryllidaceae alkaloids in *Lycoris*. Here, we collected five candidate proteins with dramatically increasing abundances (fold changes > 100), as shown in Table 1. Representatively, non-specific lipid-transfer protein 3 (*LTP3: TRIBITY_DN163977_c0_g2.p1*) exhibited 118.2-fold upregulation in *Ll* and 173.7-fold upregulation in *Li*. Plant lipid transfer proteins (LTPs) exhibit the ability to transfer lipids between membranes in vitro and have been shown to promote the production of abscisic acid (ABA) production, thereby stimulating the accumulation of alkaloids in cells [31]. Interestingly, crosstalk was detected between the JA and abscisic acid (ABA) signaling pathways in the regulation of tobacco (Nicotiana tabacum) alkaloid biosynthesis [32]. Therefore, further research is needed to explore secondary metabolites which can regulate the biosynthesis of Amaryllidaceae alkaloids through the ABA signaling pathway. In short, further genetic and molecular strategies with bioinformatics analysis will confirm the roles of these candidate proteins related to the biosynthesis of Amaryllidaceae alkaloids in *Lycoris*.

Table 1. Some candidate genes responsible for the biosynthesis of Amaryllidaceae alkaloids in *Lycoris*.

Protein ID	Homolog-Arabidopsis	Description	Fold Change (*Ll/Ls*) ($p < 0.05$)	Fold Change (*Li/Ls*) ($p < 0.05$)
TRINITY_DN183542_c1_g2.p1	AT3G05500	Rubber elongation factor protein (REF)	153.6	21.0
TRINITY_DN147476_c3_g1.p1	AT1G08080	Alpha carbonic anhydrase 7	145.7	0.5
TRINITY_DN182379_c0_g1.p1	AT1G61490	S-locus lectin protein kinase family protein	142.6	312.8
TRINITY_DN163977_c0_g2.p1	AT5G59320	Non-specific lipid-transfer protein 3	118.2	173.7
TRINITY_DN194079_c6_g2.p1	AT5G59320	Non-specific lipid-transfer protein 3	1.2	330.9

3.3. RNA Processing Related Proteins May Contribute to the Biosynthesis of Amaryllidaceae alkaloids in Lycoris

RNA processing includes mRNA capping, splicing, cleavage, and polyadenylation, which participates in various key cellular processes and probably plays an important role in the biosynthesis of alkaloids. It was recently discovered that two potent anticancer alkaloids SANG and CHEL could directly bind to single-stranded RNAs, which reveals the fundamental structural and calorimetric aspects of the interaction of the natural benzophenanthridine alkaloids with single-stranded RNAs, facilitating the development of next-generation alkaloid therapeutics targeting single-stranded RNA [33]. Coincidentally, the polyadenylate [poly(A)] tail of mRNA was also found to have recently been identified as a potential drug target due to its important role in translational initiation, maturation, and stabilization of mRNA, and production of alternative proteins in eukaryotic cells. Some small molecular alkaloids with isoquinoline groups can bind to poly A with high affinity to form self-structures. It provides a reference for the development of novel bio-base molecules targeting poly(A) structures [34].

Eukaryotic pre-mRNAs are spliced to form mature mRNA. Alternative splicing greatly expands the transcriptomic and proteomic diversities related to physiological and developmental processes in higher eukaryotes. Alternative splicing is a posttranscriptional regulatory mechanism that generates multiple protein isoforms from a single gene through the use of alternative splice sites during splicing. However, the biosynthesis of alkaloids is a special metabolic pathway in plants, in which many genes have alternative splicing at different developmental stages and under stress conditions [35]. The finding of PR3b splicing regulation by JA/ET and NIC loci in Burley 21 is valuable to the genetic studies of low-alkaloid mutants and could provide clues to unravel the mechanisms by which JA/ET-signaling pathways regulate PR protein gene splicing [36]. Homoharringtonine (HHT) is a natural alkaloid with potent antitumor activity, which regulates the alternative splicing of Bel-x and Caspase 9 through a PP1-dependent mechanism, revealing a novel mechanism underlying the antitumor activities of HHT [37]. However, its role in the biosynthesis of galanthamine remains unclear. Interestingly, we found nine differentially expressed proteins related to RNA processing in the tested data in Table 2. This indicates that RNA processing is an important step in the biosynthesis of Amaryllidaceae alkaloids in *Lycoris*, and further experimental investigations are needed for functional validation.

Table 2. List of RNA processing-related proteins in DEPs.

Protein Name	Homolog-Arabidopsis	Fold Change	*p* Value	Functional Classification	Description
TRINITY_DN183625_c1_g1.p1	AT1G09760	2.054	0.005	RNA.processing	U2 small nuclear ribonucleoprotein A (U2A′)
TRINITY_DN183625_c1_g2.p1	AT1G09760	2.106	0.003	RNA.processing	U2 small nuclear ribonucleoprotein A (U2A′)
TRINITY_DN171323_c3_g1.p1	AT4G24770	12.734	0.008	RNA.processing.editing	31-kDa RNA binding protein (RBP31)
TRINITY_DN191152_c4_g3.p1	AT4G34110	7.243	0.035	RNA.processing	poly(A) binding protein 2 (PAB2)
TRINITY_DN171323_c3_g1.p1	AT4G24770	7.100	0.010	RNA.processing.editing	31-kDa RNA binding protein (RBP31)
TRINITY_DN183625_c1_g1.p1	AT1G09760	2.075	0.028	RNA.processing	U2 small nuclear ribonucleoprotein A (U2A′)
TRINITY_DN178056_c1_g3.p1	AT4G25630	2.001	0.002	RNA.processing	fibrillarin 2 (FIB2)
TRINITY_DN178691_c2_g1.p1	AT2G33430	0.018	0.008	RNA.processing.editing	Multiple organellar RNA editing factor 2, chloroplastic (MORF2)
TRINITY_DN188290_c2_g1.p1	AT2G47640	0.423	0.021	RNA.processing	Small nuclear ribonucleoprotein family protein (AT2G47640)
TRINITY_DN188183_c1_g1.p1	AT1G43190	0.086	0.042	RNA.processing	polypyrimidine tract-binding protein 3 (PTB3)
TRINITY_DN183238_c1_g2.p1	AT1G24450	0.269	0.001	RNA.processing.ribonucleases	Ribonuclease III family protein (NFD2)

4. Materials and Methods

4.1. Plant Materials

Plant materials of the three *Lycoris* varieties *Lycoris longituba*, *Lycoris incarnata*, and *Lycoris sprengeri* were collected from the Institute of Botany, Chinese Academy of Sciences, Jiangsu Province, China (32.05° N, 118.83° E) in late March when the leaves were growing vigorously (Figure 1A). Voucher specimens of *Lycoris longituba* (0653969), *Lycoris incarnata* (0653959), and *Lycoris sprengeri* (0653967) were deposited at the herbarium in the Institute of Botany, Chinese Academy of Sciences. The bulb samples with similar diameters and specifications to the three *Lycoris* varieties were selected, chopped, mixed well, and frozen at $-70\,°C$.

4.2. Protein Extraction for Proteomic Analysis

Tissue samples were first ground into a powder with liquid nitrogen and added to an appropriate volume of lysis buffer (2.5% SDS/100 mM Tris-HCl, pH = 8.5), sonicated in an ice-water bath for 15 min and centrifuged at $16,000\times g$ for 20 min until clarified. The protein in the supernatant was precipitated by the acetone method. After washing with acetone and drying, 8 M urea/100 mM Tris-HCl solution (pH 8.0) was added to the protein precipitate to fully dissolve the protein. The samples were centrifuged at $12,000\times g$ for 15 min, the supernatant was collected, dithiothreitol (DTT) was added to a final concentration of 10 mM, and the samples were incubated at 37 °C for 1 h to perform a reduction reaction to open disulfide bonds. Then, iodoacetamide (IAA) was added to a final concentration of 40 mM, and an alkylation reaction was performed at room temperature in the dark to block sulfhydryl groups. An appropriate volume of 100 mM Tris-HCl solution (pH 8.0) was added, the Bradford method was used to quantify the protein concentration, the urea concentration was diluted to below 2 M, and trypsin was added to each of the samples according to the ratio of protein amount trypsin amount =50:1, and incubated at 37 °C overnight with shaking. The next day, TFA was added to terminate the digestion and the pH value of the solution was adjusted to approximately 6.0. The solution was centrifuged at $12,000\times g$ for 15 min and a homemade C18 cartridge was used for desalting. The desalted peptide solution was dried by a centrifugal concentrator and then stored frozen at $-20\,°C$ for on-board detection.

4.3. SWATH-MS Analysis

The SWATH methods were used for subsequent MS-analysis using Triple TOF 5600 (Sciex) LC/MS system. The prepared peptide samples were first bound to the trap column and then separated by the analytical column (45 min gradient, 60 min total duration). Two mobile phases used to establish the analytical gradient were buffer A-0.1% (v/v) formic acid, 5% DMSO in H_2O, buffer B-0.1% (v/v) formic acid, and 5% DMSO in acetonitrile. During SWATH scanning, each scan cycle consisted of one MS1 scan (ion accumulation time 250 ms, scan range 350–1500 m/z) and 100 variable window MS2 scans (ion accumulation time 33 ms, scan range 100–1800 m/z).

4.4. Alkaloid Extraction and Quantification

Galanthamine, lycoramine and lycorine were purchased from Shanghai TCI Development Co. Their characteristics have been described previously [38]. A total of 0.2 g of bulb samples of *Lycoris longituba*, *Lycoris incarnata*, and *Lycoris sprengeri* were taken to be tested, freeze-dried, and ground into powder. Extraction was performed by sonication with 2 mL ethanol (70% high-performance liquid chromatography (HPLC)–grade) for 30 min. After centrifugation at 12,000 rpm for 10 min, the supernatant was taken and dried under vacuum. The samples were redissolved in 1 mL 0.1% formic acid-acetonitrile (v/v = 95/5) for liquid phase analysis [15]. Waters ACQUITY UPLC BEH C18 column (150 mm × 2.1 mm, 1.7 μm) was used as the liquid chromatography column and the separation was conducted using 0.1% formic acid (v/v) (A) and acetonitrile (B) with a 6-min linear gradient of 5–60% B at a flow rate of 0.2 mL/min. Quantification of galanthamine, lycoramine, and lycorine used $288 \rightarrow 231$, $290 \rightarrow 189$, $288 \rightarrow 146$ (m/z) transition reactions respectively. Experiments were conducted with three independent biological replicates. Least-significant difference test (LSD, $p < 0.05$) was used to compare the means. Different letters represent significant differences between groups ($n = 3$, $p < 0.05$).

4.5. Data Analysis

The mass spectrum files obtained by SWATH scanning are processed by DIA-Umpire to obtain secondary mass spectrum files that are used for database search. TPP software, Comet, and X!tandem search engines were used to search the database, and the search results were used as the spectral library for subsequent target extraction. The algorithm used for SWATH targeted extraction quantification is OpenSWATH. The test results were screened with 1% FDR.

The protein quantitative intensity information obtained by SWATH analysis was used for difference comparison and T-test analysis after log2 transformation, data filling (imputation algorithm in Perseus software), and data normalization. Differential proteins were screened by fold difference (Ratio) and BH-corrected p-value (*P*.adjust). Proteins with a fold change above 2 or below 0.5 ($p < 0.05$) were considered differentially expressed proteins (DEPs) in this study. In the bioinformatics analysis, the DEPs identified were used for the Kyoto Encyclopedia of Genes and Genomes (KEGG) enrichment analysis by the clusterProfiler R package, and pathways of KEGG enrichment analysis results were drawn with reference to the KEGG mapper [39]. After the DEPs were compared with the homologous proteins in Arabidopsis thaliana, the MapMAN BIN system was used for functional classification.

5. Conclusions

Amaryllidaceae plants are rich in alkaloids such as galanthamine (Gal), lycoramine (Lycm), and lycorine (Lyc) with a variety of biological activities, including anticancer, anti-inflammatory, anti-plasmodium, and antibacterial activities. In this study, three alkaloids profiling of *Lycoris longituba* (*Ll*), *Lycoris incarnata* (*Li*), and *Lycoris sprengeri* (*Ls*) were determined, and a comprehensive proteomic analysis was carried out with the aim of investigating the regulatory metabolism of Amaryllidaceae alkaloids in *Lycoris* species. The significant proteome changes in amino acid metabolism, starch, and sucrose metabolism, and galanthamine biosynthesis strengthen our understanding of the regulatory metabolism of Amaryllidaceae alkaloids in *Lycoris*. Additionally, we have also identified important candidate genes involved in the galanthamine biosynthesis pathway of alkaloid production in *Lycoris*. Taken together, we have offered new thoughts on the use of SWATH-MS to explore the regulatory metabolism of Amaryllidaceae alkaloids, providing a new strategy for the future exploitation of alkaloids.

Supplementary Materials: The following supporting information can be downloaded at: https://www.mdpi.com/article/10.3390/ijms24054495/s1.

Author Contributions: Conceptualization, Q.L. and K.L.; methodology, C.L.; formal analysis, C.Z. and Y.C.; data curation, Y.Z. and L.Y.; writing—original draft, M.T.; writing—review and editing, M.C. and F.Z.; visualization, M.T. and C.Z.; funding acquisition, Q.L. and K.L. All authors have read and agreed to the published version of the manuscript.

Funding: This work was supported by the Shanghai Rising-Star Program, China (No. 20QB1404100), Jiangsu Agricultural Science and Technology Innovation Fund [CX(21)2023], the National Natural Science Foundation of China (31801889, 32002067), Natural Science Foundation of Jiangsu Province (BK20221334), the Natural Science Foundation of Jiangsu Province (SBK2020042924), and the Shanghai Sailing Program (Project No. 20YF1442800).

Institutional Review Board Statement: Not applicable.

Informed Consent Statement: Informed consent was obtained from all subjects involved in the study.

Data Availability Statement: The data presented in this study are available upon request. The MS data have been deposited to the ProteomeXchange Consortium via the iProX repository with the data set identifier PXD040341.

Conflicts of Interest: The authors declare no conflict of interest.

References

1. Hayashi, A.; Saito, T.; Mukai, Y.; Kurita, S.; Hori, T. Genetic variations in *Lycoris radiata* var. radiata in Japan. *Genes Genet. Syst.* **2005**, *80*, 199–212. [CrossRef] [PubMed]
2. Wang, L.; Yin, Z.-Q.; Cai, Y.; Zhang, X.-Q.; Yao, X.-S.; Ye, W.-C. Amaryllidaceae alkaloids from the bulbs of *Lycoris radiata*. *Biochem. Syst. Ecol.* **2010**, *38*, 444–446. [CrossRef]
3. Cao, Z.; Yang, P.; Zhou, Q. Multiple biological functions and pharmacological effects of lycorine. *Sci. China Chem.* **2013**, *56*, 1382–1391. [CrossRef] [PubMed]
4. Jin, Z. Amaryllidaceae and Sceletium alkaloids. *Nat. Prod. Rep.* **2016**, *33*, 1318–1343. [CrossRef] [PubMed]
5. Attaur, R.; Choudhary, M.I. Diterpenoid and steroidal alkaloids. *Nat. Prod. Rep.* **1997**, *14*, 191–203.
6. Omulokoli, E.; Khan, B.; Chhabra, S.C. Antiplasmodial activity of four Kenyan medicinal plants. *J. Ethnopharmacol.* **1997**, *56*, 133–137. [CrossRef]
7. Sethi, M.L. Inhibition of Reverse Transcriptase Activity by Benzophenanthridine Alkaloids. *J. Nat. Prod.* **1979**, *42*, 187–196. [CrossRef]
8. McMahon, J.B.; Currens, M.J.; Gulakowski, R.J.; Buckheit, R.W., Jr.; Lackman-Smith, C.; Hallock, Y.F.; Boyd, M.R. Michellamine B, a novel plant alkaloid, inhibits human immunodeficiency virus-induced cell killing by at least two distinct mechanisms. *Antimicrob. Agents Chemother.* **1995**, *39*, 484–488. [CrossRef]
9. Nair, J.J.; van Staden, J. Antiplasmodial constituents in the minor alkaloid groups of the Amaryllidaceae. *South Afr. J. Bot.* **2019**, *126*, 362–370. [CrossRef]
10. Heinrich, M. Chapter 4—Galanthamine from Galanthus and Other Amaryllidaceae—Chemistry and Biology Based on Traditional Use. In *The Alkaloids: Chemistry and Biology*; Cordell, G.A., Ed.; Academic Press: Guelph, ON, Canada, 2010; Volume 68, pp. 157–165.
11. Park, C.H.; Yeo, H.J.; Park, Y.E.; Baek, S.-A.; Kim, J.K.; Park, S.U. Transcriptome Analysis and Metabolic Profiling of *Lycoris radiata*. *Biology* **2019**, *8*, 63. [CrossRef]
12. Jiang, Y.; Xia, N.; Li, X.; Shen, W.; Liang, L.; Wang, C.; Wang, R.; Peng, F.; Xia, B. Molecular cloning and characterization of a phenylalanine ammonia-lyase gene (LrPAL) from *Lycoris radiata*. *Mol. Biol. Rep.* **2011**, *38*, 1935–1940. [CrossRef] [PubMed]
13. Singh, A.; Massicotte, M.A.; Garand, A.; Tousignant, L.; Ouellette, V.; Bérubé, G.; Desgagné-Penix, I. Cloning and characterization of norbelladine synthase catalyzing the first committed reaction in Amaryllidaceae alkaloid biosynthesis. *BMC Plant Biol.* **2018**, *18*, 338. [CrossRef] [PubMed]
14. Li, Q.; Xu, J.; Yang, L.; Zhou, X.; Cai, Y.; Zhang, Y. Transcriptome Analysis of Different Tissues Reveals Key Genes Associated With Galanthamine Biosynthesis in *Lycoris longituba*. *Front. Plant Sci.* **2020**, *11*, 519752. [CrossRef] [PubMed]
15. Li, Q.; Xu, J.; Zheng, Y.; Zhang, Y.; Cai, Y. Transcriptomic and Metabolomic Analyses Reveals That Exogenous Methyl Jasmonate Regulates Galanthamine Biosynthesis in *Lycoris longituba* Seedlings. *Front. Plant Sci.* **2021**, *12*, 713–795. [CrossRef] [PubMed]
16. Gillet, L.C.; Navarro, P.; Tate, S.; Röst, H.; Selevsek, N.; Reiter, L.; Bonner, R.; Aebersold, R. Targeted Data Extraction of the MS/MS Spectra Generated by Data-independent Acquisition: A New Concept for Consistent and Accurate Proteome Analysis*. *Mol. Cell. Proteom.* **2012**, *11*, O111.016717. [CrossRef] [PubMed]
17. Ludwig, C.; Gillet, L.; Rosenberger, G.; Amon, S.; Collins, B.C.; Aebersold, R. Data-independent acquisition-based SWATH-MS for quantitative proteomics: A tutorial. *Mol. Syst. Biol.* **2018**, *14*, e8126. [CrossRef] [PubMed]
18. Chen, D.; Shao, Q.; Yin, L.; Younis, A.; Zheng, B. Polyamine Function in Plants: Metabolism, Regulation on Development, and Roles in Abiotic Stress Responses. *Front. Plant Sci.* **2019**, *9*, 1945. [CrossRef] [PubMed]
19. Campisi, A.; Acquaviva, R.; Mastrojeni, S.; Raciti, G.; Vanella, A.; Pasquale, R.D.; Puglisi, S.; Iauk, L. Effect of berberine and Berberis aetnensis C. Presl. alkaloid extract on glutamate-evoked tissue transglutaminase up-regulation in astroglial cell cultures. *Phytother. Res.* **2011**, *25*, 816–820. [CrossRef]
20. Desgagné-Penix, I. Biosynthesis of alkaloids in Amaryllidaceae plants: A review. *Phytochem. Rev.* **2021**, *20*, 409–431. [CrossRef]
21. Hawkins, K.M.; Smolke, C.D. Production of benzylisoquinoline alkaloids in *Saccharomyces cerevisiae*. *Nat. Chem. Biol.* **2008**, *4*, 564–573. [CrossRef]
22. Chang, L.; Hagel, J.M.; Facchini, P.J. Isolation and Characterization of O-methyltransferases Involved in the Biosynthesis of Glaucine in *Glaucium flavum*. *Plant Physiol.* **2015**, *169*, 1127–1140. [CrossRef]
23. Grobe, N.; Ren, X.; Kutchan, T.M.; Zenk, M.H. An (R)-specific N-methyltransferase involved in human morphine biosynthesis. *Arch. Biochem. Biophys.* **2010**, *506*, 42–47. [CrossRef] [PubMed]
24. Kudo, Y.; Umemoto, K.; Obata, T.; Kaneda, A.; Ni, S.-r.; Mikage, M.; Sasaki, Y.; Ando, H. Seasonal variation of alkaloids and polyphenol in Ephedra sinica cultivated in Japan and controlling factors. *J. Nat. Med.* **2022**, *77*, 137–151. [CrossRef] [PubMed]
25. Hua, L.; Chen, J.; Gao, K. A new pyrrolizidine alkaloid and other constituents from the roots of *Ligularia achyrotricha* (Diels) Ling. *Phytochem. Lett.* **2012**, *5*, 541–544. [CrossRef]

26. Wang, N.; Shu, X.; Zhang, F.; Zhuang, W.; Wang, T.; Wang, Z. Comparative Transcriptome Analysis Identifies Key Regulatory Genes Involved in Anthocyanin Metabolism During Flower Development in *Lycoris radiata*. *Front. Plant Sci.* **2021**, *12*, 761862. [CrossRef] [PubMed]

27. Jiang, X.; Chen, H.; Wei, X.; Cai, J. Proteomic analysis provides an insight into the molecular mechanism of development and flowering in *Lycoris radiata*. *Acta Physiol. Plant.* **2021**, *43*, 4. [CrossRef]

28. Yang, F.; Li, C.H.; Das, D.; Zheng, Y.H.; Song, T.; Wang, L.X.; Chen, M.X.; Li, Q.Z.; Zhang, J. Comprehensive Transcriptome and Metabolic Profiling of Petal Color Development in *Lycoris sprengeri*. *Front. Plant Sci.* **2021**, *12*, 2540. [CrossRef]

29. Yamamoto, K.; Takahashi, K.; O'Connor, S.E.; Mimura, T. Imaging MS Analysis in Catharanthus roseus. In *Catharanthus roseus: Methods and Protocols*; Courdavault, V., Besseau, S., Eds.; Springer US: New York, NY, USA, 2022.

30. Jiang, Y.; Xu, S.; Han, X.; Wang, R.; He, J.; Xia, B. Investigation of nuclear DNA contents of *Lycoris* species (Amaryllidaceae) with different chromosome number by flow cytometry. *Pak. J. Bot.* **2017**, *49*, 2197–2200.

31. Gao, S.; Guo, W.; Feng, W.; Liu, L.; Song, X.; Chen, J.; Hou, W.; Zhu, H.; Tang, S.; Hu, J. LTP3 contributes to disease susceptibility in Arabidopsis by enhancing abscisic acid (ABA) biosynthesis. *Mol. Plant Pathol.* **2016**, *17*, 412–426. [CrossRef]

32. Lackman, P.; González-Guzmán, M.; Tilleman, S.; Carqueijeiro, I.; Pérez, A.C.; Moses, T.; Seo, M.; Kanno, Y.; Häkkinen, S.T.; Van Montagu, M.C.E.; et al. Jasmonate signaling involves the abscisic acid receptor PYL4 to regulate metabolic reprogramming in *Arabidopsis* and *tobacco*. *PNAS* **2011**, *108*, 5891–5896. [CrossRef]

33. Basu, P.; Suresh Kumar, G. Small molecule–RNA recognition: Binding of the benzophenanthridine alkaloids sanguinarine and chelerythrine to single stranded polyribonucleotides. *J. Photochem. Photobiol. B Biol.* **2017**, *174*, 173–181. [CrossRef] [PubMed]

34. Kumar, G.S.; Giri, P. Isoquinoline alkaloids and their binding with polyadenylic acid: Potential basis of therapeutic action. *Mini-Rev. Med. Chem.* **2010**, *10*, 568–577.

35. Lam, P.Y.; Wang, L.; Lo, C.; Zhu, F.-Y. Alternative Splicing and Its Roles in Plant Metabolism. *Int. J. Mol. Sci.* **2022**, *23*, 7355. [CrossRef] [PubMed]

36. Ma, H.; Wang, F.; Wang, W.; Yin, G.; Zhang, D.; Ding, Y.; Timko, M.P.; Zhang, H. Alternative splicing of basic chitinase gene PR3b in the low-nicotine mutants of *Nicotiana tabacum L.* cv. Burley 21. *J. Exp. Bot.* **2016**, *67*, 5799–5809. [CrossRef]

37. Sun, Q.; Li, S.; Li, J.; Fu, Q.; Wang, Z.; Li, B.; Liu, S.-S.; Su, Z.; Song, J.; Lu, D. Homoharringtonine regulates the alternative splicing of Bcl-x and caspase 9 through a protein phosphatase 1-dependent mechanism. *BMC Complement. Altern. Med.* **2018**, *18*, 164. [CrossRef]

38. Qing-Zhu, L.; Zhang, Y.; Zheng, Y.; Yang, L.; Cai, Y. Simultaneous Determination of Three Alkaloids in *Lycoris* spp.by UPLC-QTRAP-MS/MS. *J. Instrum. Anal.* **2018**, *37*, 211–216.

39. Yu, G.; Wang, L.-G.; Han, Y.; Han, Y. clusterProfiler: An R package for comparing biological themes among gene clusters. *OMICS J. Integr. Biol.* **2012**, *16*, 284–287. [CrossRef]

International Journal of
Molecular Sciences

Review

Alternative Splicing and Its Roles in Plant Metabolism

Pui Ying Lam [1,*], Lanxiang Wang [2,3], Clive Lo [4] and Fu-Yuan Zhu [2]

[1] Center for Crossover Education, Graduate School of Engineering Science, Akita University, Tegata Gakuen-machi 1-1, Akita City 010-8502, Akita, Japan
[2] Co-Innovation Center for Sustainable Forestry in Southern China, College of Biology and the Environment, Nanjing Forestry University, Nanjing 210037, China; amelia0610@163.com (L.W.); fyzhu@njfu.edu.cn (F.-Y.Z.)
[3] CAS Key Laboratory of Quantitative Engineering Biology, Shenzhen Institute of Synthetic Biology, Shenzhen Institute of Advanced Technology, Chinese Academy of Sciences, Shenzhen 518055, China
[4] School of Biological Sciences, The University of Hong Kong, Pokfulam, Hong Kong, China; clivelo@hku.hk
* Correspondence: lam@gipc.akita-u.ac.jp

Abstract: Plant metabolism, including primary metabolism such as tricarboxylic acid cycle, glycolysis, shikimate and amino acid pathways as well as specialized metabolism such as biosynthesis of phenolics, alkaloids and saponins, contributes to plant survival, growth, development and interactions with the environment. To this end, these metabolic processes are tightly and finely regulated transcriptionally, post-transcriptionally, translationally and post-translationally in response to different growth and developmental stages as well as the constantly changing environment. In this review, we summarize and describe the current knowledge of the regulation of plant metabolism by alternative splicing, a post-transcriptional regulatory mechanism that generates multiple protein isoforms from a single gene by using alternative splice sites during splicing. Numerous genes in plant metabolism have been shown to be alternatively spliced under different developmental stages and stress conditions. In particular, alternative splicing serves as a regulatory mechanism to fine-tune plant metabolism by altering biochemical activities, interaction and subcellular localization of proteins encoded by splice isoforms of various genes.

Keywords: alkaloids; alternative splicing; ascorbate; phytohormones; lipids; metabolism; phenylpropanoids; plants; starch; terpenoids

Citation: Lam, P.Y.; Wang, L.; Lo, C.; Zhu, F.-Y. Alternative Splicing and Its Roles in Plant Metabolism. *Int. J. Mol. Sci.* **2022**, 23, 7355. https://doi.org/10.3390/ijms23137355

Academic Editors: Melvin J. Oliver, Bei Gao and Moxian Chen

Received: 8 June 2022
Accepted: 28 June 2022
Published: 1 July 2022

Publisher's Note: MDPI stays neutral with regard to jurisdictional claims in published maps and institutional affiliations.

1. Introduction

Metabolism is made up of networks of biochemical reactions that produce a diverse array of organic compounds in living organisms. In plants, metabolism can be categorized as primary metabolism and specialized metabolism. Primary metabolism such as tricarboxylic acid cycle, glycolysis, shikimate, lipid and amino acid pathways generates low-molecular-weight organic molecules directly involved in basic life functions and is essential for growth, development and survival of plants [1]. On the other hand, specialized metabolism, which is also known as secondary metabolism, comprises metabolic processes that are not absolutely essential for growth of plants. Specialized metabolites are natural phytochemicals synthesized from primary metabolic pathways [2] and were once considered as byproducts, waste products or detoxification products of primary metabolism [3]. Well-known plant-specialized metabolites include phenolics, alkaloids, terpenoids, saponins, etc.

In response to different growth and developmental stages as well as for adaptation to environmental changes, metabolism in plants is tightly regulated at different molecular levels. These include transcriptional regulation; post-transcriptional modifications such as alternative splicing, RNA methylation [4], RNA editing [5], and mRNA decay; translational regulation; post-translational modifications such as protein phosphorylation, methylation and acetylation [6], ubiquitination, and protein turnover. Understanding these regulations

131

will establish the foundation for molecular breeding or bioengineering to improve plant growth, yield and biomass production as well as enhance plant stress resistance.

Bibliometric analyses through keyword co-occurrence network analysis and burst word detection analysis of the Web of Science Core Collection database reveal that studies of alternative splicing in the last decade are mainly associated with various processes and conditions such as stresses, seed development and circadian clock (Figure 1). However, research on the relationship between alternative splicing and plant metabolism has seldom been reported or explored in detail. In this review, we focus on the roles of alternative splicing, an understudied yet potentially important regulatory mechanism in plant metabolism. We first briefly introduce the occurrence of alternative splicing in plants. We then describe and update the current understanding of the roles of alternative splicing in regulating metabolism of starch, lipid, photorespiration, ascorbate, auxin, jasmonates, terpenoids, alkaloids and phenylpropanoids in plants, with specific examples of how alternative splicing regulates gene functions and metabolism (Table 1), and recommend future research directions.

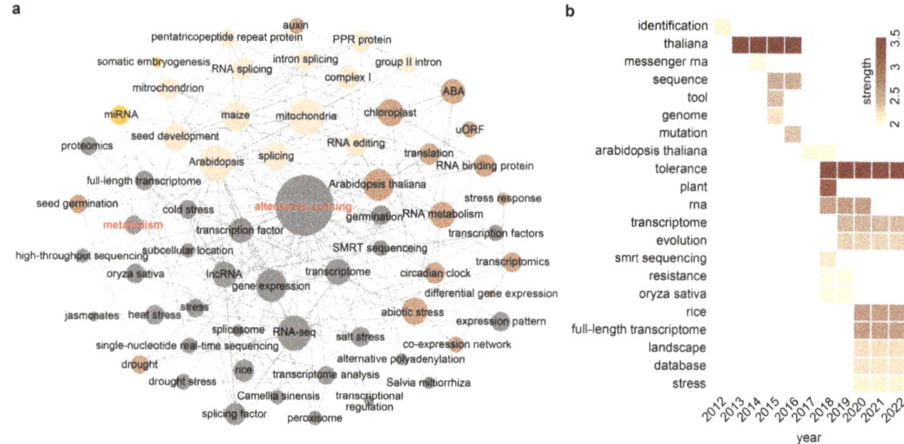

Figure 1. Analysis of keywords in Web of Science Core Collection database from January 2012 to May 2022. (**a**) Keyword co-occurrence network analyzed by BibExcel and Pajek. Nodes represent keywords. Node size represents the frequency of keywords that appear. Node colors represent modularity. (**b**) Burst keyword analysis. Length of colored boxes represents burst status duration. Colors represent burst strength. Bibliometric analysis was carried out by retrieving citation data on topic search using query: "TS = (alternative splicing OR splicing factor) AND plant AND (metabolism OR metabolic OR metabolize)" and was further analyzed by CiteSpace (https://citespace.podia.com, accessed on 1 June 2022).

2. Alternative Splicing and Its Roles in Plants

2.1. Gene Splicing Machinery in Plants

Most plant genes contain introns, which are non-coding sequences interrupting exons that have to be removed from pre-mRNA by a process called splicing. Pre-mRNA splicing is performed and controlled by a ribonucleoprotein complex called spliceosome and other spliceosome-associated proteins [7,8]. So far, two types of spliceosomes have been described [7,9,10]. U2 spliceosomes that are composed of five small nuclear RNA (snRNA) subcomplexes (U1, U2, U4, U5 and U6) represent the major type [7,9,10]. The minor and rare U12 spliceosome also consists of five snRNA, including U11, U12, U4atac, U5 and U6atac [10]. Both types mainly recognize and splice the exon-intron junctions that harbor 5′-GT-AG-3′ sequences [10]. Accordingly, for assembly of spliceosomes, a 5′ splice site with a conserved GT sequence and a 3′ splice site with a conserved AG sequence as well as a conserved A nucleotide at 18 to 40 nucleotides upstream of the 3′ splice site and a polypyrimidine tract after branch point are required [11]. Meanwhile, non-snRNA splicing factors, including serine-/arginine-rich proteins and heterogeneous nuclear ribonucleoproteins (hnRNPs), are also known to assist localization of splicing enhancers and suppressors, thereby regulating the selection of splice sites by spliceosomes [7,12,13]. Numerous studies suggest that mutations of spliceosomal proteins have led to altered alternative splicing as well as growth and developmental defects, including embryonic lethality, abnormal flower and leaf morphology, early flowering, defects in seed maturation and/or hypersensitivity to abscisic acid [14–18]. Apparently, spliceosomes govern accurate splicing, which is essential for normal plant growth and development. During splicing, two transesterification reactions occur, excising introns and joining exons [19]. The detailed process of splicing was reviewed elsewhere previously [7,20].

2.2. Alternative Splicing in Plants

Alternative splicing refers to the generation of multiple splice isoforms (mRNA transcripts) from a single gene due to the use of different splice sites [21]. Such a process largely expands the coding capacity of genomes [22,23] and is believed to be a post-transcriptional regulatory mechanism in response to developmental and environmental changes. The existence of alternative splicing was initially uncovered in some plant species in the early 1980s [24–27] and was later found to be a common phenomenon in different living organisms, including plants [21]. For example, in *Arabidopsis* (*Arabidopsis thaliana*) and rice (*Oryza sativa*), it was estimated that at least 42% and 48% of intron-containing genes are alternatively spliced, respectively [28,29]. Several types of alternative splicing events have been reported in plants. These include exon skipping, intron retention, alternative 5′ splicing (alternative donor site), alternative 3′ splicing (alternative acceptor site), alternative 5′ and 3′ splicing (alternative position), mutually exclusive exon, alternative first exon and alternative last exon (Figure 2a) [21,22,30]. Among these, intron retention is the most prevalent type of alternative splicing event that takes place in plants [21,28,30].

In plants, alternative splicing is regulated by several factors. For example, the concentration and compositional ratio of splicing factors was suggested to mediate alternative splicing [31]. In addition, structure of chromatin, including DNA methylation and histone modifications, may affect accessibility and elongation speed of RNA polymerase II as well as the recruitment of splicing factors, thus mediating splicing outcomes [31–33]. Alternative splicing mediates gene expression or functions at both RNA and protein levels. At RNA level, alternative splicing may affect mRNA stability and result in nonsense-mediated mRNA decay due to frame shift mutations and formation of premature stop codons [30,34,35]. At protein level, alternative splicing may impact translation efficiency [36], subcellular location [37], biological and/or biochemical functions and interaction [38] of proteins translated. Alternative splicing is a regulatory mechanism during stress defense and development of plants [39]. With the development of high throughput RNA sequencing techniques and proteomic approaches such as sequential window acquisition of all theoretical mass spectra (SWATH-MS) and alternative splicing-related bioinformatic platforms such as PlantSPEAD for alternative splicing analysis and validation [40,41], numerous genes in plants have been uncovered to be alternatively spliced when subject to stresses such as pathogen infection [38,42,43], herbivore attack [44], high-temperature stress [45–47], cold stress [48], salt stress [49,50], iron and phosphate deficiencies [51], flooding [52,53] and drought [54]. They are also detected in various developmental stages, functioning in the regulation of circadian rhythm [55], wood formation [56], flowering [57,58] and fruit ripening (Figure 2b) [59].

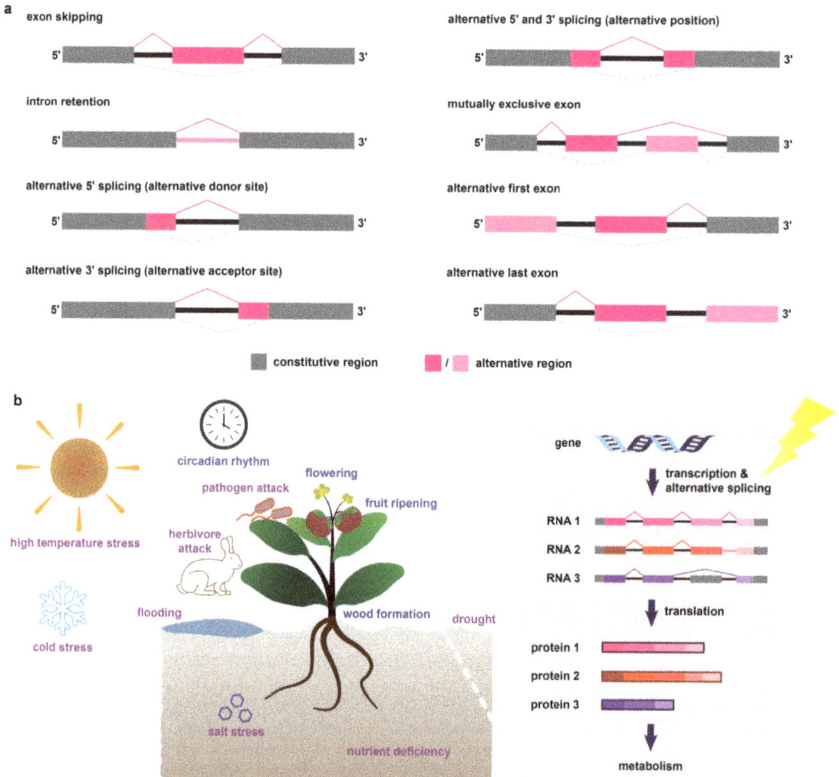

Figure 2. Alternative splicing and its regulation on metabolism in plants during stress responses and development. (**a**) Common types of alternative splicing events in plants. Boxes: exons; horizontal lines: introns; (**b**) regulation of alternative splicing by stresses and developmental stages. Stresses that regulate alternative splicing are indicated in purple. Developmental stages that regulate alternative splicing are indicated in navy blue.

3. Current Understanding of the Roles of Alternative Splicing in Plant Metabolism

3.1. Overview

Regarding our knowledge on alternative splicing in plant metabolism, recent studies have mostly focused on global evaluation of the extent of alternative splicing, particularly on detecting alternative splicing events and types. From these research studies, numerous genes of plant metabolisms have been reported to be alternatively spliced under different developmental stages and stress conditions. However, the precise physiological and biochemical meanings of the presence of different splice isoforms of these genes have only been partially understood.

3.2. Primary Metabolism

3.2.1. Starch Metabolism

Starch is a primary carbon source in human and animal diets [60]. Together with sucrose, starch serves as the major product of photosynthesis and the primary energy storage in plants. Starch is composed of amylose and amylopectin, which are both formed by polymerization of glucose [60]. Starch metabolism, including its biosynthesis and degradation, controls the storage and release of carbon in plants [61]. Thus, starch metabolism defines the quality and yield of cereal grains and other food crops, serving as an important constituent in agriculture and biorefineries.

Cold stresses induce alternative splicing of genes related to starch metabolism. For example, under cold stress, genes related to starch and sugar metabolism in tea plants (*Camellia sinensis*) are alternatively spliced and show correlation with various sugar accumulation, likely conferring resistance towards cold stress [62].

Alternative splicing alters subcellular localization and catalytic activities of starch-branching enzyme (SBE), which functions in determining chain length and branch point frequency of amylopectin in *Phaseolus vulgaris L.* [37]. This is achieved by using alternative first exons and translation start sites of the pre-mRNA of SBE for protein translation [37]. Consequently, an isoform (LF-PvSBE2) harboring an N-terminal plastid targeting sequence that is targeted to both starch granule and cytosol and another isoform (PvSBE2) with a truncated N-terminus that leads to cytosolic localization are generated [37]. Interestingly, the altered N-terminus also affects substrate binding affinity and catalytic efficiency of the enzyme [37].

Alternative splicing has been found to mediate starch metabolism in plants by controlling the expression abundances of different splice isoforms of transcriptional factors. For example, in *Arabidopsis*, the transcription factor indeterminate domain 14 (IDD14) mediates starch accumulation by activating *Qua-Quine Starch* (*QQS*) [63], which is involved in starch degradation [64]. Under cold stress, a splice isoform IDD14β that lacks DNA binding domains was induced [63]. IDD14β binds to the full-length functional isoform IDD14α and inhibits its DNA binding ability, resulting in reduced transcription of *QQS* and altered starch accumulation [63]. On the other hand, rice *OsbZIP58*, a basic leucine zipper transcription factor that is highly and specifically expressed in endosperm, mediates grain filling by regulating the expression of starch biosynthetic and hydrolyzing genes [65,66]. The truncated isoform OsbZIP58β is induced under heat stress and displays lower transactivation activity than the full-length isoform OsbZIP58α, thus fine-tuning starch accumulation and grain filling during heat stress [65].

Alternative splicing regulates fruit ripening associated with starch metabolism in banana (*Musa acuminate*) (Figure 3a) [67]. An R1-type MYB transcription factor, namely *MaMYB16*, is spliced through alternative 5′ splicing, exon skipping and alternative 3′ splicing, generating two isoforms [67]. The full-length isoform MaMYB16L binds to the promotors of genes involved in starch degradation, including isoamylases (*MaISA2*), β-amylase (*MaBAM7*) and α-amylases (*MaAMY3*), as well as positive master ripening regulator dehydration-responsive element-binding factor (*MsDREB2*), suppressing their expressions [67]. By contrast, the short isoform MaMYB16S could not bind to these promoters due to the lack of a DNA binding domain, but could bind to MaMYB16L and inhibit its DNA binding and transactivation activities [67]. During fruit ripening, *MaMYB16L* is downregulated, while *MaMYB16S* is upregulated, hence promoting starch degradation and fruit softening [67].

3.2.2. Lipid Metabolism

Lipids are primary metabolites essential for plants. As a major constituent of membranes, lipids make up 5–10% dry weight of plant vegetative cells [68]. In seeds, lipid storage is an energy reserve for securing the survival of young seedlings after germination [69]. They are also consumed as food by humans and animals as well as used as biofuels in the biorefinery industry [70].

Phosphatidic acid is an intermediate for the generation of membrane lipids and storage lipids [71]. It is also involved in various cellular processes such as signal transduction in response to stimuli, secretion and membrane trafficking [71]. In plants, diacylglycerol kinase (DGK) catalyzes the conversion of diacylglycerol to phosphatidic acid [72]. Alternative splicing generates two splice isoforms of DGK in tomato (*Solanum lycopersicum*), both harboring DGK catalytic activity [72]. The full-length isoform delineated as LeCBDGK harbors a calmodulin-binding domain at the C-terminus and could bind to calmodulin [72], which is a calcium ion binding regulatory protein that could be activated by calcium ion [73]. In response to calcium ion, LeCBDGK is translocated from soluble cell fraction to membrane fraction [72]. On the other hand, the truncated isoform LeDGK1 that lacks the calmodulin-binding domain is insensitive to calcium ion [72]. Thus, alternative splicing regulates the generation of calcium-sensitive and -insensitive DGK isoforms, providing flexibility in response to calcium ion [72].

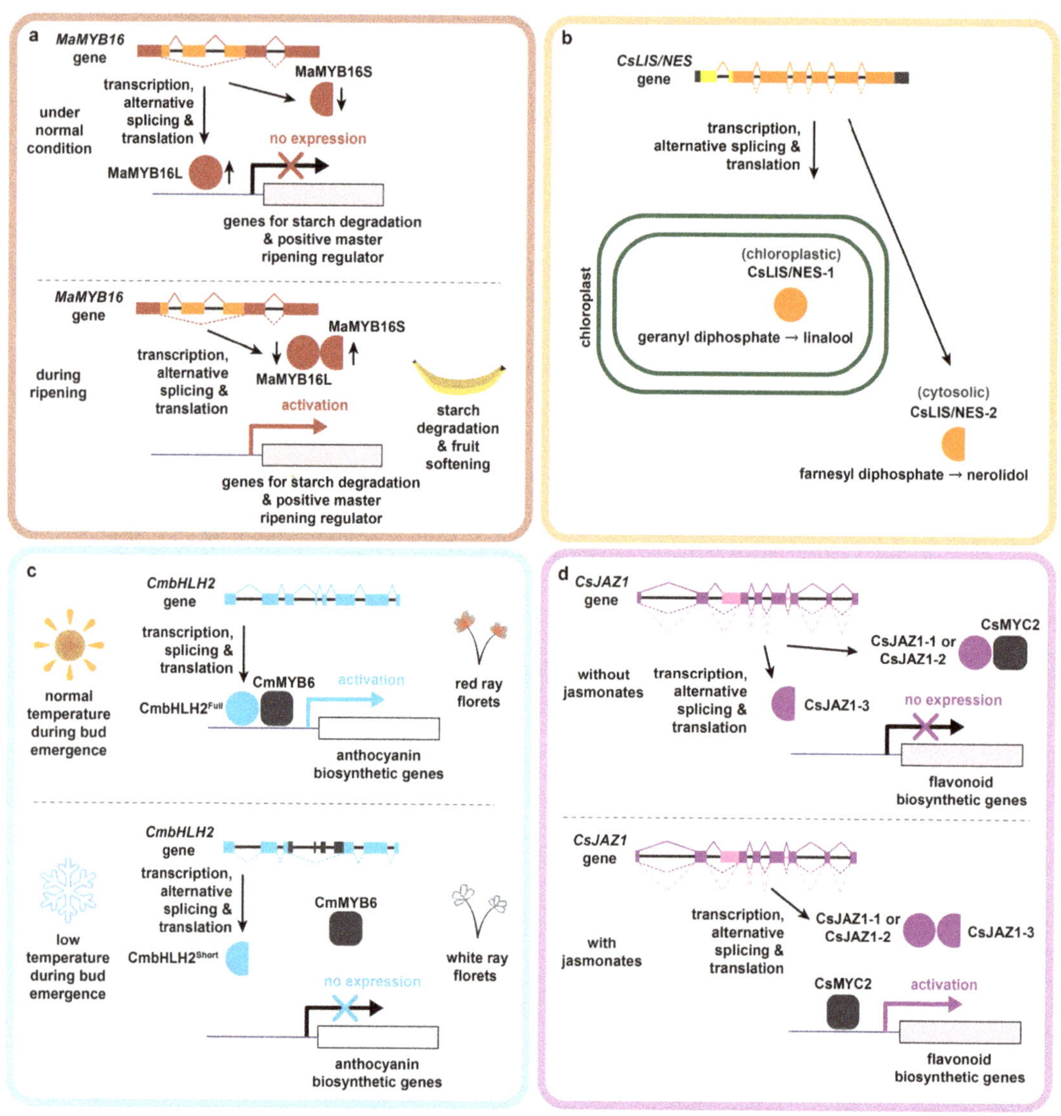

Figure 3. Examples of plant metabolism regulated by alternative splicing. (**a**) Regulation of transcription factor MaMYB16 in banana by alternative splicing and its roles in starch metabolism; (**b**) regulation of CsLIS/NES in tea plants by alternative splicing and its roles in linalool and nerolidol biosynthesis; (**c**) regulation of CmbHLH2 in chrysanthemums by alternative splicing and its roles in anthocyanin biosynthesis and floret coloration; (**d**) regulation of JASMONATE ZIM-domain (JAZ) repressor in tea plants by alternative splicing and its roles in jasmonate-mediated flavonoid biosynthesis.

Biosynthesis of triacylglycerol, the major form of energy storage in seed oil crops, requires diacylglycerol acyltransferase (DGAT) in the Kennedy pathway [74]. In peanuts (*Arachis hypogaea*), AhDGAT1 is regulated by alternative splicing in an organ-dependent manner [75]. Except the two truncated splice isoforms AhDGAT1.2 and AhDGAT1.4, all the other five isoforms are functional and could complement the free fatty acid lethality phenotype of a triacylglycerol-deficient *Saccharomyces cerevisiae* strain [75]. However, the actual contribution and biochemical meanings of these isoforms on triacylglycerol production and storage in peanuts remain unclear.

WRINKLED1 (WRI1), which belongs to the APETALA2/ethylene-responsive element binding protein transcriptional factor family, acts as a master regulator for triacylglycerol biosynthesis in plants [76]. In castor bean (*Ricinus connunis* L.), both of the two splice isoforms, RcWRI1-A and RcWRI1-B, which differ by three amino acids in length, are functional, with RcWRI1-B appearing to be more active [77]. RcWRI1-A is expressed in all plant tissues, whereas RcWRI1-B expression is specific to seeds [77]. Thus, alternative splicing likely plays a regulatory role to improve lipid biosynthesis in castor bean seeds.

3.2.3. Photorespiration

Photorespiration is a pathway to recycle the toxic product, 2-phosphoglycolate, formed when ribulose-1,5-bisphosphate carboxylase/oxygenase (RuBisCO) utilizes oxygen instead of carbon dioxide [78]. As a competitive pathway for carbon dioxide fixation that causes carbon, nitrogen and energy loss, photorespiration is a major target for bioengineering to improve plant growth and yield [78]. As the intermediates of photorespiration are also toxic, glycolate-glyoxylate metabolism exists for detoxification to maintain normal plant growth [78].

Hydroxypyruvate reductase (HPR) converts hydroxypyruvate to glycerate in glycolate-glyoxylate metabolism [79]. In pumpkin (*Cucurbita* sp.), two splice isoforms of HPR (HPR1 and HPR2) are generated by alternative $5'$ splicing [79]. Compared with HPR1, HPR2 lacks a targeting sequence for peroxisome localization [79]. Accordingly, HPR2 is cytosolic, whereas HPR1 is localized in peroxisomes [79]. In darkness, HPR1 and HPR2 are both weakly expressed [79]. Under light, HPR2 but not HPR1 is strongly induced [79]. Therefore, light regulates alternative splicing of HPR with specific cellular localization. It remains unclear whether the two HPRs harbor the same catalytic ability.

3.2.4. Ascorbate Metabolism

Ascorbate, a downstream metabolite of D-glucose, D-mannose and/or *myo*-inositol, is the most abundant water-soluble antioxidant in plants [80]. It scavenges and regulates the level of hydrogen peroxide, a reactive oxygen species, in plant cells by reduction in the ascorbate/glutathione cycle (or Asada–Halliwell pathway) through the action of ascorbate peroxidase (APX) [81]. Consequently, ascorbate and hydrogen peroxide are converted into monodehydroascorbate and water, respectively, and recycled in the pathway [73,74]. Due to its antioxidant activities and health-promoting effects, ascorbate is considered an important food component in human diets [82].

Early studies suggested that chloroplastic APXs are regulated by alternative splicing developmentally in some plant species [83,84]. In pumpkin, the stromal APX that lacks a putative membrane-spanning domain in the C-terminus is a splice isoform of the thylakoid-bound APX, which is formed as a consequence of alternative $3'$ splicing [83]. Similarly, stromal and thylakoid-bound APXs are produced from a single gene by alternative splicing in spinach (*Spinacia oleracea*), but these isoforms are generated by intron retention and/or alternative last exon [84–86]. A putative splicing regulatory *cis*-element upstream of the acceptor site of intron 12 in this spinach APX was found to be crucial for the selection of splice sites, but the exact nuclear protein(s) that interact with this *cis*-element and the regulatory mechanism remain unknown [87]. Interestingly, the stromal and thylakoid-bound APXs in *Arabidopsis*, rice and tomato are encoded by separate genes [80,88,89].

Alternative splicing regulates wild emmer wheat's (*Triticum turgidum* ssp. *Dicoccoides*) wheat kinase start1 (*WKS1*) resistance gene that confers resistance to *Puccinia striiformis* f. sp. *tritici* (*Pst*), the causal agent of stripe rust, through catalyzing phosphorylation of thylakoid-bound APX [90,91]. When inoculated with *Pst* under high temperatures, the full-length isoform WKS1 that harbors a START domain at the C-terminus is upregulated, whereas the major splice isoform WKS2 that lacks the START domain is downregulated [90,91]. WKS1 can be translocated to chloroplasts where it binds, phosphorylating the thylakoid-bound APX, reducing its activity and ability to detoxify hydrogen peroxide [90]. It was proposed that this would eventually lead to cell death after several days, considerably longer than *R*-genes-triggered hypersensitive responses which restrict pathogen growth in the host [90]. By contrast, WKS2 appears to be non-functional and is unable to bind or phosphorylate APX [90]. Hence, WKS1, but not WKS2, is a candidate isoform that confers resistance towards *Pst* and could be introduced to wheat (*Triticum aestivum*), a close relative of wild emmer wheat and a major food crop, by transgenic approaches [90].

3.3. Phytohormones

3.3.1. Auxin Metabolism

Auxin is a major phytohormone that regulates plant growth and development by mediating cell division, elongation and differentiation [92]. Its roles in apical-basal polarity and responses to tropisms have been extensively studied [93,94]. Auxin metabolism includes biosynthesis, conjugation and degradation [92]. Our current understanding of auxin biosynthesis suggests the existence of multiple pathways in plants [95].

In the proposed auxin biosynthetic pathways, a family of YUCCA proteins, which are flavin-dependent mono-oxygenases, catalyzes the conversion of tryptamine to *N*-hydroxytryptamine in the tryptophan-dependent pathway [95,96]. As a result of alternative splicing, two isoforms of YUCCA4 are formed in *Arabidopsis* [96]. YUCCA4-1 is cytosolic and is detectable in all tissues [96]. By contrast, YUCCA4-2, which harbors a predicted C-terminal hydrophobic transmembrane domain, is inserted into the endoplasmic reticulum membrane and is specific in flowers [96]. Both isoforms harbor the expected YUCCA4 catalytic activities [96]. Taken together, alternative splicing regulates subcellular localization of YUCCA4, which may lead to compartmentation of auxin biosynthesis [96].

3.3.2. Jasmonate Metabolism

Jasmonates, including jasmonic acid as well as its precursor and derivatives, are a group of phytohormones that regulate plant growth and development, especially under biotic and abiotic stresses [97]. For example, jasmonates mediate senescence and leaf abscission and inhibit seed germination [98]. Jasmonates are derived from linolenic acid and are structurally similar to eicosanoids in mammals [98].

In poplar (*Populus tomentosa*), alternative splicing fine-tunes the molecular mechanism of leaf senescence [99]. PtRD26 is a NAC transcription factor that acts as a positive regulator for leaf senescence [99]. The downstream genes that PtRD26 regulates include several senescence-associated NAC family transcription factors, proteins related to chlorophyll degradation and lysine catabolism as well as lipoxygenase 2 (LOX2) for jasmonate biosynthesis and 1-aminocyclopropane-1-carboxylic acid synthase 6 (ACS6) for ethylene biosynthesis [99]. Alternative splicing of PtRD26 by intron retention occurs during leaf senescence, generating the truncated splice isoform PtRD26[IR] [92]. PtRD26[IR] could interact with several senescence-associated NAC family transcription factors and repress their DNA binding affinity, resulting in the delay of age-, dark- and PtRD26-induced leaf senescence [99].

Lipoxygenase (LOX) catalyzes the oxygenation of polyunsaturated fatty acids for further jasmonate biosynthesis [100]. In tea plants, six out of eleven LOX genes are differentially regulated by alternative splicing [100]. During feeding by tea geometrids, infection by *Glomerella cingulate*, cold stress and jasmonate treatment, the corresponding CsLOX truncated isoforms are induced [100]. It was suggested that these splice isoforms might regulate LOXs by competing or compensating with the full-length isoforms [100].

3.4. Primary and Specialized Metabolism

Terpenoid Metabolism

Terpenoids, which include all compounds derived from isopentenyl diphosphate (IPP) and dimethylallyl diphosphate (DMAPP), are a large family of primary and specialized metabolites found in all living organisms [101,102]. They are structurally diverse compounds commonly categorized as hemiterpene (C_5), monoterpenes (C_{10}), sesquiterpenes (C_{15}), diterpenes (C_{20}), sesterterpenes (C_{25}), triterpenes (C_{30}) and tetraterpenes (C_{40}) [103,104]. Terpenoids have diverse functions in plants. For example, they act as phytohormones [105] and also serve as signals for attracting pollinators [106], avoiding herbivores [107] and mediating interactions among plants [108]. Terpenoids have a wide range of applications in human diets and the food industry. Volatile terpenoids such as linalool and nerolidol contribute to the odor of teas [109,110]. Terpenoids including bixin, lycopene ,and astaxanthin are widely used as colorants in the food industry [111]. Terpenoids in herbs and spices help in preserving food due to their microbicidal and insecticidal activities [112]. Some terpenoids also harbor pharmaceutical and health-promoting effects against cancer, inflammation and infectious diseases [103,104].

Various transcriptomic studies have revealed that terpenoid biosynthetic genes are regulated by alternative splicing in different tissues and stress conditions. For instance, terpenoid biosynthetic genes are regulated by alternative splicing in different tissues of *Ginkgo biloba* [113], Sichuan pepper (*Zanthoxylum armatum*) [114], *Artemisia argyi* [115] and *Lindera glauca* [116]. They are also alternatively spliced under drought and heat stress in tea plants [117].

Terpene synthases are a family of enzymes catalyzing the committed steps for generating isoprene, monoterpenes, sesquiterpenes, diterpenes and triterpenes [102,118]. In Dong Ling Cao (*Isodon rubescens*), alternative 3′ splicing of a type I terpene synthase *IrKSL3* generates two splice isoforms with different biochemical activities [119]. Using copalyl diphosphate as a substrate, the full-length IrKSL3 produces miltiradiene as the sole product, whereas the splice isoform IrKSL3a that harbors a deletion of six amino acids could simultaneously generate isopimaradiene and miltiradiene [119]. These results illustrate that alternative splicing could influence product outcomes in enzyme-catalyzed reactions [119].

Biosynthesis of linalool and nerolidol also requires terpene synthase [120]. Two splice isoforms of terpene synthase could be detected in tea plants (Figure 3b) [120]. The full-length isoform CsLIS/NES-1 is localized in chloroplasts, functioning as a linalool synthase [120]. The splice isoform CsLIS/NES-2 with an N-terminal truncation is cytosolic, acting as a nerolidol synthase [120]. As both isoforms are bifunctional in catalyzing the in vitro generation of linalool and nerolidol from geranyl diphosphate and farnesyl diphosphate, respectively, the difference in their subcellular localizations likely contributes to the discrepancy of their *in planta* biochemical functions [120]. In addition, expression of *CsLIS/NES-1* and *CsLIS/NES-2* was differentially regulated [120]. *CsLIS/NES-1* is induced by jasmonates, whereas *CsLIS/NES-2* expression level is higher in flowers than in leaves [120].

Phylloquinone, which is also known as vitamin K_1, is a prenylated naphthoquinone [121]. In most photosynthetic plants, phylloquinone serves as an electron acceptor in photosystem I [122] and an electron carrier for disulfide bond formation in proteins essential for photosystem II assembly [123]. Dietary consumption of phylloquinone is beneficial for human health due to its roles in maintenance of bones [124], blood coagulation [125] and prevention of cardiovascular diseases [126]. Isochorismate synthase (ICS) converts chorismate from the shikimate pathway to isochorismate, a key intermediate for the biosynthesis of phylloquinone [127]. Poplar *(Populus trichocarpa) ICS* undergoes extensive alternative splicing, producing at least 37 splice isoforms that represent approximately 50% of total *ICS* transcripts [128]. Most splice isoforms are formed from intron retention and/or alternative 3′ splicing and harbor premature stop codons [128]. This is in contrast to *Arabidopsis AtICS1* that predominantly generates a full-length transcript [128]. Accordingly, it was proposed that alternative splicing of *ICS* was recruited independently during evolution [128]. *Populus* ICS mainly functions in phylloquinone biosynthesis, which can be maintained at a low functional transcript level [128], whereas *Arabidopsis* AtICS1 is predominantly involved in the biosynthesis of stress-induced salicylic acid, which could also be synthesized from isochorismate [129].

3.5. Specialized Metabolism

3.5.1. Alkaloid Metabolism

Alkaloids are specialized metabolites that harbor at least one nitrogen atom in their heterocyclic rings [130,131]. They could be further classified into different groups according to their backbone structures [131]. Most of them are generated from amino acids: phenylalanine, tyrosine, tryptophan and ornithine [132]. In plants, some alkaloids are toxins involved in defense against pathogens [133], insects [134] and herbivores [135]. Owing to their excellent toxicity and pharmaceutical activities, alkaloids have been extensively exploited for poisoning and medical uses [136,137]. Well-known examples of alkaloids include caffeine, morphine, strychnine and nicotine.

Monoterpene indole alkaloids, a group of alkaloids generated from the combination of tryptophan and terpenoid precursors, are mainly distributed in Apocynaceae, Loganiaceae and Rubiaceae [138]. To generate the active substrate for biosynthesis of monoterpene indole alkaloids, strictosidine β-D-glucosidase (SGD) catalyzes deglycosylation of strictosidine to release the highly reactive aglycone [139]. In Madagascar periwinkle (*Catharanthus roseus*), alternative last exon splicing generates a full-length isoform SGD with glucosidase activities and another isoform shSGD harboring a truncated C-terminus and lacking glucosidase activities [139]. shSGD interacts with SGD and disrupts multimerization of SGDs, which, in turn, inhibits the SGD deglycosylation activities [139]. Thus, generation of a pseudo-enzyme by alternative splicing of SGD serves as a regulatory mechanism to fine-tune monoterpene indole alkaloid biosynthesis [139].

Allantoin is a nitrogen-rich ureide compound generated from degradation of purines [140]. In plants, its biosynthesis requires transthyretin-like (TTL) protein [140]. The two isoforms of TTL in *Arabidopsis*, which are generated by alternative 3′ splicing, harbor similar in vitro catalytic activities with different subcellular localizations [140]. An internal peroxisomal targeting signal present in the long isoform TTL^{1-} is missing from the short isoform TTL^{2-} [140]. Exploration of TTLs in other plant species suggests that alternative splicing of internal peroxisomal targeting signal appears to be a conserved regulatory mechanism in angiosperm [140].

3.5.2. Phenylpropanoid Metabolism

Phenylpropanoids and their downstream metabolites such as flavonoids, monolignols/lignin, stilbenoids, lignans and suberin are phenolic compounds derived from the amino acids phenylalanine and/or tyrosine [141,142]. Their biosynthesis and regulation have been extensively studied due to their functions in cell wall structure [143,144], anti-oxidation [145,146], UV protection [147,148], determination of fruit and flower colors [149,150], defense against pathogens [151,152] and herbivores [153,154] and fertility [155,156], as well as their contribution to human activities such as biomass utilization [157], nutrition [158] and breeding [159]. Depending on the types of phenylpropanoids and plant species, some of the phenylpropanoids are constitutively accumulated, while some of them are induced by stresses [151]. Their accumulation is often correlated to stress tolerances [151].

Biosynthetic genes in phenylpropanoid and its downstream pathways are regulated by alternative splicing during plant development. For example, in kiwifruit (*Actinidia chinensis*), several structural genes for anthocyanin biosynthesis, including chalcone synthase (*CHS*), flavanone 3-hydroxylase (*F3H*), dihydroflavonol-4-reductase (*DFR*), anthocyanidin synthase (*ANS*) and uridine diphosphate (UDP)-glucosyltransferase (*UGT*) are regulated by alternative splicing during fruit development and ripening [160]. In tea plants, a series of genes involved in phenylpropanoid, anthocyanin and monolignol biosynthesis is also alternatively spliced in cultivars that harbor purple or green tender shoots [161], throughout leaf development [162] and in different tissues [81]. Similar observations were found in *G. biloba*, in which alternative splicing regulates flavonoid biosynthetic genes and their transcription regulators in different tissues [163].

Alternative splicing also regulates phenylpropanoid and its downstream pathway genes during stress conditions. In sorghum (*Sorghum bicolor*), upregulation and alternative splicing of flavonoid and phenylpropanoid biosynthetic genes occur upon infection of *Colletotrichum sublineola*, the causal agent of anthracnose [65]. In *Arabidopsis*, phenylpropanoid biosynthetic genes such as phenylalanine ammonia lyases (*PAL*) and 4-coumarate CoA ligases (*4CL*) as well as the downstream monolignol biosynthetic genes such as hydroxycinnamoyl-CoA shikimate/quinate hydroxycinnamoyltransferase (*HCT*) and coniferaldehyde 5-hydroxylase (*CAld5H*) are alternatively spliced during iron deficiency [51]. These are expected to be related to the excretion of phenylpropanoids, which could chelate iron, to the rhizosphere [51]. Collectively, these works suggest that alternative splicing potentially plays crucial roles in mediating phenylpropanoid biosynthesis in plants under both biotic and abiotic stresses.

Alternative splicing is a regulatory mechanism that governs anthocyanin production and floret colors in some plant species under different conditions. For instance, temperatures during flower bud emergence affect anthocyanin biosynthesis and floret coloration in chrysanthemums (*Chrysanthemum morifolium*) [164]. The temperature-dependent coloration was found to be attributed by alternative splicing of a basic helix–hoop–helix transcription factor gene *CmbHLH2* (Figure 3c) [164]. The red ray florets generate a full-length functional CmbHLH2^{Full}, whereas the white ray florets produce a truncated CmbHLH2^{Short} due to alternative position and exon skipping [164]. Unlike CmbHLH2^{Full}, CmbHLH2^{Short} fails to interact with the MYB transcription factor CmMYB6 or activate anthocyanin biosynthetic genes [164]. A similar example was reported in peach (*Prunus persica*), which produces flowers of different colors on the same tree [165]. It was found that white flowers generate a truncated non-functional ANS by alternative splicing [165]. Hence, different cultivation conditions induce alternative splicing of the key genes for anthocyanin biosynthesis, thereby impacting floret colors.

Anthocyanin biosynthesis in higher plants is typically activated by the MBW (MYB–basic helix-loop-helix protein–WD40 repeat; MYB–bHLH–WDR) protein complexes [166]. In rapeseed (*Brassica napus* L.), BnaPAP2 is an MYB transcription factor required for regulating anthocyanin biosynthesis [167]. Alternative splicing generates different BnaPAP2.A7 isoforms with opposite functions [167]. The full-length BnaPAP2.A7-744 harbors all the essential domains of MYB and could interact with a bHLH protein in vitro [167]. The splice isoforms BnaPAP2.A7-910 and BnaPAP2.A7-395 are truncated and cannot interact with bHLH proteins [167]. Although the exact molecular mechanism is still unknown, BnaPAP2.A7-910 and BnaPAP2.A7-395 downregulate flavonoid biosynthetic genes when overexpressed in *Arabidopsis*, suggesting that their roles as suppressors are in opposition to the function of the full-length BnaPAP2.A7-744 as an activator [167]. Thus, alternative splicing provides a mechanism to balance the positive and negative regulations of anthocyanin biosynthesis in rapeseed.

JASMONATE ZIM-domain (*JAZ*) repressor, a negative regulator of diverse jasmonate responses including flavonoid biosynthesis [168], is regulated by alternative splicing in tea plants which produces three JAZ splice isoforms (Figure 3d) [169]. The full-length isoform CsJAZ1-1 is localized in nucleus, whereas the truncated isoforms CsJAZ1-2 and CsJAZ1-3, which lack the 3' coding sequences, are localized in both nucleus and cytoplasm [169]. In the absence of jasmonates, CsJAZ1-1 and CsJAZ1-2, but not CsJAZ1-3, competitively bind to CsMYC2, a positive regulator of flavonoid biosynthesis [170], thereby inactivating CsMYC2 and repressing flavonoid biosynthetic genes such as dihydroflavonol reductase (*DFR*) and leucoanthocyanidin dioxygenase (*LDOX*) [169]. In the presence of jasmonates,

CsJAZ1-3 interacts with CsJAZ1-1 and CsJAZ1-2, preventing their binding to CsMYC2 and eventually leading to their degradation. Consequently, repression of CsMYC2 and flavonoid biosynthesis was released [169]. Collectively, alternative splicing coordinately regulates jasmonate-mediated flavonoid biosynthesis in tea plants [169].

Table 1. Genes related to plant metabolism that are regulated by alternative splicing.

Type of Metabolism	Metabolic Pathways	Species	Gene Alternatively Spliced	Spliced Isoforms and Their Functions	References
Primary metabolism	Starch metabolism	*Phaseolus vulgaris* L.	Starch-branching enzyme (SBE)	LF-PvSBE2: long form, targeted to starch granule and cytosol PvSBE: truncated, targeted to cytosol	[37]
		Arabidopsis (*Arabidopsis thaliana*)	Indeterminate domain 14 (IDD14)	IDD14α: full-length, activates *Qua-Quine Starch* (*QQS*) IDD14β: truncated, lacks DNA binding domains, inhibits DNA binding ability of IDD14α	[63]
		Rice (*Oryza sativa*)	OsbZIP58	OsbZIP58: full-length, mediates grain filling by regulating the expression of starch biosynthetic and hydrolyzing genes OsbZIP58β: induced under heat stress, displayed a lower transactivation activity than the full-length isoform OsbZIP58α	[65]
		Banana (*Musa acuminate*)	MaMYB16L	MaMYB16L: full-length, binds to the promotors and activates genes involved in starch degradation MaMYB16S: truncated, binds to MaMYB16L, inhibits its DNA binding and transactivation activities	[67]
	Lipid metabolism	Tomato (*Solanum lycopersicum*)	Diacylglycerol kinase (DGK)	LeCBDGK: full-length, harbors DGK catalytic activity, harbors a calmodulin-binding domain, could bind to calmodulin LeDGK1: truncated, harbors DGK catalytic activity, lacks a calmodulin-binding domain, could not bind to calmodulin	[72]
		Peanuts (*Arachis hypogaea*)	Diacylglycerol acyltransferase (DGAT)	AhDGAT1.1, AhDGAT1.3, AhDGAT1.5, AhDGAT1.6 and AhDGAT1.7: harbor DGAT activities AhDGAT1.2 and AhDGAT1.4: truncated, lack DGAT activities	[75]
		Castor bean (*Ricinus connunis* L.)	WRINKLED1 (WRI1)	RcWRI1-A: functional, less active, is expressed in all tissues RcWRI1-B: functional, more active, expression specific to seeds	[77]
	Photorespiration	Pumpkin (*Cucurbita* sp.)	Hydroxypyruvate reductase (HPR)	HPR1: full-length, harbors a targeting sequence for peroxisome localization, localized in peroxisomes, induced under light HPR1: truncated, lacks a targeting sequence for peroxisome localization, localized in cytosol, weakly expressed in dark and under light	[79]
	Ascorbate metabolism	Pumpkin (*Cucurbita* sp.)	Ascorbate peroxidase (APX)	Thylakoid-bound APX: harbors a putative membrane-spanning domain in the C-terminus, localized in thylakoid Stromal APX: lacks a putative membrane-spanning domain in the C-terminus, localized in stroma	[83]
		Spinach (*Spinacia oleracea*)	Ascorbate peroxidase (APX)	Thylakoid-bound APX: harbors a putative membrane-spanning domain in the C-terminus, localized in thylakoid Stromal APX: lacks a putative membrane-spanning domain in the C-terminus, localized in stroma	[84–86]
		Wheat (*Triticum turgidum* ssp. *Dicoccoides*)	Wheat kinase start1 (WKS1) resistance gene	WKS1: full-length, harbors a START domain at the C-terminus, upregulated under high temperature and when inoculated with *Pst*, translocated to chloroplast, binds, phosphorylates and reduces the activity of thylakoid-bound APX WKS2: lacks the START domain, downregulated under high temperature and when inoculated with *Pst*, non-functional, unable to bind or phosphorylate APX	[90]
Phytohormone	Auxin metabolism	*Arabidopsis* (*Arabidopsis thaliana*)	Flavin-dependent mono-oxygenase (YUCCA4)	YUCCA4-1: lacks a predicted C-terminus hydrophobic transmembrane domain cytosolic, expressed in all tissues YUCCA4-2: harbors a predicted C-terminus hydrophobic transmembrane domain, inserted into endoplasmic reticulum membrane; expressed in flowers	[96]
	Jasmonate metabolism	Poplar (*Populus tomentosa*)	NAC transcription factor (PtRD26)	PtRD26: full-length, activates several senescence-associated NAC family transcription factors, proteins related to chlorophyll degradation, lysine catabolism, lipoxygenase 2 (LOX2) for jasmonate biosynthesis and 1-aminocyclopropane-1-carboxylic acid synthase 6 (ACS6) for ethylene biosynthesis PtRD26: truncated, interacts with several senescence-associated NAC family transcription factors and represses their DNA binding affinity	[99]
		Tea plants (*Camellia sinensis*)	Lipoxygenase (LOX)	Full-length isoform: predominant during normal conditions Truncated splice isoforms: induced during feeding by tea geometrids, infection by *Glomerella cingulate*, cold stress and jasmonate treatment	[100]

Table 1. *Cont.*

Type of Metabolism	Metabolic Pathways	Species	Gene Alternatively Spliced	Spliced Isoforms and Their Functions	References
Primary and Specialized metabolism	Terpenoid metabolism	Dong Ling Cao (*Isodon rubescens*)	Terpene synthase (IrKSL3)	IrKSL3: full-length, produces miltiradiene as the sole product from copalyl diphosphate; IrKSL3a: shorter, simultaneously generates isopimaradiene and miltiradiene from copalyl diphosphate	[119]
		Tea plants (*Camellia sinensis*)	Terpene synthase (LIS/NES)	CsLIS/NES-1: full-length, localized in chloroplast, functions as a linalool synthase, induced by jasmonates; CsLIS/NES-2: harbors a truncated N-terminus, localized in cytosol, functions as a nerolidol synthase, expression is higher in flowers than in leaves	[120]
		Poplar (*Populus trichocarpa*)	Isochorismate synthase (ICS)	*Populus ICS* undergoes extensive alternative splicing, produces at least 37 splice isoforms that represent approximately 50% of total *ICS* transcripts	[128]
Specialized metabolism	Alkaloid metabolism	Madagascar periwinkle (*Catharanthus roseus*)	Stictosidine β-D-glucosidase (SGD)	SGD: full-length, harbors glucosidase activities; shSGD: harbors a truncated C-terminus, lacks glucosidase activities, interacts with SGD, disrupts multimerization of SGD, inhibits deglycosylation activities of SGD	[139]
		Arabidopsis (*Arabidopsis thaliana*)	Transthyretin-like (TTL) protein	TTL^{1-}: long isoform, harbors an internal peroxisomal targeting signal; TTL^{2-}: short isoform, lacks an internal peroxisomal targeting signal	[140]
	Phenylpropanoid metabolism	Chrysanthemum (*Chrysanthemum morifolium*)	Basic helix–hoop–helix transcription factor (CmbHLH2)	$CmbHLH2^{Full}$: full-length, expressed in red ray florets, interacts with CmMYB6 and activates anthocyanin biosynthetic genes; $CmbHLH2^{Short}$: truncated, expressed in white ray florets, cannot interact with CmMYB6 or activate anthocyanin biosynthetic genes	[164]
		Peach (*Prunus persica*)	Anthocyanidin synthase (ANS)	Full-length ANS: functional, generates red flowers; Truncated ANS: non-functional, generates white flowers	[165]
		Rapeseed (*Brassica napus* L.)	MYB transcription factor (BnaPAP2)	BnaPAP2.A7-744: full-length, harbors all the essential domains of MYB, could interact with bHLH protein, activates flavonoid biosynthetic genes; BnaPAP2.A7-910 and BnaPAP2.A7-395: truncated, cannot interact with bHLH protein, downregulates flavonoid biosynthetic genes	[167]
		Tea plants (*Camellia sinensis*)	JASMONATE ZIM-domain (JAZ) repressor	CsJAZ1-1 and CsJAZ1-2: full-length (CsJAZ1-1) and truncated (CsJAZ1-2), bind to CsMYB2, resulting in inactivation of flavonoid biosynthetic genes; CsJAZ1-3: truncated, binds to CsJAZ1-1 and CsJAZ1-2 in the presence of jasmonates and prevents their binding to CsMYB2, resulting in activation of flavonoid biosynthetic genes	[169]

4. Conclusions and Future Directions

Owing to the advancement of high-throughput third-generation RNA sequencing technique, omics analyses and related technology in the past decades, full-length transcripts can be efficiently obtained, providing valuable information regarding regulation of genes by alternative splicing [171,172]. Thus far, a diverse array of genes in plant metabolism that are regulated by alternative splicing has been determined in different developmental stages and/or stress conditions. Some specific examples of how alternative splicing fine-tunes various metabolic pathways in plants by altering biochemical activities, interaction and/or subcellular locations of proteins encoded by different splice isoforms of genes have also been provided. Altogether, these findings could advance our understanding of post-transcriptional regulation of plant metabolism for coping with stresses and modulating growth and development.

Remaining tasks ahead include functional characterization of all alternatively spliced isoforms for each gene and determination of their involvement in regulating different metabolic processes. Meanwhile, conservation of alternative splicing of a particular gene in metabolic processes in the plant kingdom, if any, and to what extent, may indicate evolutionary significance of metabolic pathways and reveal mechanisms underlying alternative splicing. Furthermore, it is known that epigenetic modifications, such as DNA methylation, regulate alternative splicing in animals [173]. It has also been shown recently that mRNA methylation, such as methylation of adenosine at the N6 position (m^6A), alters the occurrence of alternative splicing events and expression of splice isoforms in *Arabidopsis* [174].

Currently, it remains to be examined whether epigenetic modification serves as another level of regulation for plant metabolism through mediating alternative splicing.

Studies of the roles of alternative splicing in regulating plant metabolism will provide essential foundational information that could open up new avenues for bioengineering. Current approaches of overexpression, downregulation and knockout mutation usually only consider the predominant splice isoforms of target genes available in public databases. With the knowledge of different splice isoforms, including their differences in biological and biochemical functions and/or subcellular localization, the potential of bioengineering could be largely extended through overexpression, downregulation or knockout via CRISPR/cas9-mediated mutagenesis of specific isoforms of a gene of interest, thus fine-tuning the desired metabolic processes in plants. In fact, specific suppression of targeted splice isoforms of SGD and shSGD was successfully achieved in *C. roseus* by virus-induced gene silencing previously [139]. Overall, these will provide new insights into improvement of plant performance, yield and utility through bioengineering.

Author Contributions: P.Y.L.: writing—original draft preparation; L.W., C.L. and F.-Y.Z.: writing—review and editing. All authors have read and agreed to the published version of the manuscript.

Funding: This study is supported by Reiwa 4 Akita University manuscript submission support project, the Jiangsu Agricultural Science and Technology Innovation Fund (CX (21) 2023) and Natural Science Foundation of Jiangsu Province (SBK2020042924).

Institutional Review Board Statement: Not applicable.

Informed Consent Statement: Not applicable.

Data Availability Statement: Not applicable.

Conflicts of Interest: The authors declare no conflict of interest.

References

1. Pott, D.M.; Osorio, S.; Vallarino, J.G. From central to specialized metabolism: An overview of some secondary compounds derived from the primary metabolism for their role in conferring nutritional and organoleptic characteristics to fruit. *Front. Plant Sci.* **2019**, *10*, 835. [CrossRef] [PubMed]
2. Sato, F. Plant secondary metabolism. *eLS* **2014**. [CrossRef]
3. Hartmann, T. From waste products to ecochemicals: Fifty years research of plant secondary metabolism. *Phytochemistry* **2007**, *68*, 2831–2846. [CrossRef] [PubMed]
4. Zhao, B.S.; Roundtree, I.A.; He, C. Post-transcriptional gene regulation by mRNA modifications. *Nat. Rev. Mol. Cell Biol.* **2017**, *18*, 31–42. [CrossRef]
5. Glisovic, T.; Bachorik, J.L.; Yong, J.; Dreyfuss, G. RNA-binding proteins and post-transcriptional gene regulation. *FEBS Lett.* **2008**, *582*, 1977–1986. [CrossRef]
6. Perkins, N.D. Post-translational modifications regulating the activity and function of the nuclear factor kappa B pathway. *Oncogene* **2006**, *25*, 6717–6730. [CrossRef]
7. Will, C.L.; Lührmann, R. Spliceosome structure and function. *Cold Spring Harb. Perspect. Biol.* **2011**, *3*, a003707. [CrossRef]
8. Liu, L.; Tang, Z.; Liu, F.; Mao, F.; Yujuan, G.; Wang, Z.; Zhao, X. Normal, novel or none: Versatile regulation from alternative splicing. *Plant Signal. Behav.* **2021**, *16*, 1917170. [CrossRef]
9. Reddy, A.S.N.; Rogers, M.F.; Richardson, D.N.; Hamilton, M.; Ben-Hur, A. Deciphering the plant splicing code: Experimental and computational approaches for predicting alternative splicing and splicing regulatory elements. *Front. Plant Sci.* **2012**, *3*, 18. [CrossRef]
10. Simpson, C.; Brown, J. U12-dependent intron splicing in plants. In *Nuclear Pre-mRNA Processing in Plants*; Reddy, A.S.N., Golovkin, M., Eds.; Springer: Berlin/Heidelberg, Germany, 2008; pp. 61–82.
11. Reddy, A.S.N.; Marquez, Y.; Kalyna, M.; Barta, A. Complexity of the alternative splicing landscape in plants. *Plant Cell* **2013**, *25*, 3657–3683. [CrossRef]
12. Zhang, D.; Chen, M.-X.; Zhu, F.-Y.; Zhang, J.; Liu, Y.-G. Emerging functions of plant Serine/Arginine-Rich (SR) proteins: Lessons from animals. *Crit. Rev. Plant Sci.* **2020**, *39*, 173–194. [CrossRef]
13. Chen, M.-X.; Zhang, K.-L.; Gao, B.; Yang, J.-F.; Tian, Y.; Das, D.; Fan, T.; Dai, L.; Hao, G.-F.; Yang, G.-F.; et al. Phylogenetic comparison of plant *U1-70K* gene family, central regulators on 5′ splice site determination, in response to developmental cues and stress conditions. *Plant J.* **2020**, *103*, 357–378. [CrossRef] [PubMed]

14. Jang, Y.H.; Park, H.-Y.; Lee, K.C.; Thu, M.P.; Kim, S.-K.; Suh, M.C.; Kang, H.; Kim, J.-K. A homolog of splicing factor SF 1 is essential for development and is involved in the alternative splicing of pre-m RNA in *Arabidopsis thaliana*. *Plant J.* **2014**, *78*, 591–603. [CrossRef]
15. Kim, W.Y.; Jung, H.J.; Kwak, K.J.; Kim, M.K.; Oh, S.H.; Han, Y.S.; Kang, H. The *Arabidopsis* U12-type spliceosomal protein U11/U12-31K is involved in U12 intron splicing via RNA chaperone activity and affects plant development. *Plant Cell* **2010**, *22*, 3951–3962. [CrossRef]
16. Carvalho, R.F.; Carvalho, S.D.; Duque, P. The plant-specific SR45 protein negatively regulates glucose and ABA signaling during early seedling development in Arabidopsis. *Plant Physiol.* **2010**, *154*, 772–783. [CrossRef] [PubMed]
17. Aki, S.; Nakai, H.; Aoyama, T.; Oka, A.; Tsuge, T. *AtSAP130/AtSF3b-3* function is required for reproduction in *Arabidopsis thaliana*. *Plant Cell Physiol.* **2011**, *52*, 1330–1339. [CrossRef]
18. Zhang, X.-N.; Mount, S.M. Two alternatively spliced isoforms of the Arabidopsis SR45 protein have distinct roles during normal plant development. *Plant Physiol.* **2009**, *150*, 1450–1458. [CrossRef]
19. Shang, X.; Cao, Y.; Ma, L. Alternative splicing in plant genes: A means of regulating the environmental fitness of plants. *Int. J. Mol. Sci.* **2017**, *18*, 432. [CrossRef]
20. Wahl, M.C.; Will, C.L.; Lührmann, R. The spliceosome: Design principles of a dynamic RNP machine. *Cell* **2009**, *136*, 701–718. [CrossRef]
21. Barbazuk, W.B.; Fu, Y.; McGinnis, K.M. Genome-wide analyses of alternative splicing in plants: Opportunities and challenges. *Genome Res.* **2008**, *18*, 1381–1392. [CrossRef]
22. Szakonyi, D.; Duque, P. Alternative splicing as a regulator of early plant development. *Front. Plant Sci.* **2018**, *9*, 1174. [CrossRef] [PubMed]
23. Nilsen, T.W.; Graveley, B.R. Expansion of the eukaryotic proteome by alternative splicing. *Nature* **2010**, *463*, 457–463. [CrossRef] [PubMed]
24. Early, P.; Rogers, J.; Davis, M.; Calame, K.; Bond, M.; Wall, R.; Hood, L. Two mRNAs can be produced from a single immunoglobulin μ gene by alternative RNA processing pathways. *Cell* **1980**, *20*, 313–319. [CrossRef]
25. Fornace, A.J., Jr.; Cummings, D.E.; Comeau, C.M.; Kant, J.A.; Crabtree, G.R. Structure of the human gamma-fibrinogen gene. Alternate mRNA splicing near the 3′end of the gene produces gamma A and gamma B forms of gamma-fibrinogen. *J. Biol. Chem.* **1984**, *259*, 12826–12830. [CrossRef]
26. Kornblihtt, A.R.; Vibe-Pedersen, K.; Baralle, F.E. Human fibronectin: Cell specific alternative mRNA splicing generates polypeptide chains dfffering in the number of internal repeats. *Nucleic Acids Res.* **1984**, *12*, 5853–5868. [CrossRef]
27. Rosenfeld, M.G.; Lin, C.R.; Amara, S.G.; Stolarsky, L.; Roos, B.A.; Ong, E.S.; Evans, R.M. Calcitonin mRNA polymorphism: Peptide switching associated with alternative RNA splicing events. *Proc. Natl. Acad. Sci. USA* **1982**, *79*, 1717–1721. [CrossRef]
28. Filichkin, S.A.; Priest, H.D.; Givan, S.A.; Shen, R.; Bryant, D.W.; Fox, S.E.; Wong, W.-K.; Mockler, T.C. Genome-wide mapping of alternative splicing in *Arabidopsis thaliana*. *Genome Res.* **2010**, *20*, 45–58. [CrossRef]
29. Lu, T.; Lu, G.; Fan, D.; Zhu, C.; Li, W.; Zhao, Q.; Feng, Q.; Zhao, Y.; Guo, Y.; Li, W.; et al. Function annotation of the rice transcriptome at single-nucleotide resolution by RNA-seq. *Genome Res.* **2010**, *20*, 1238–1249. [CrossRef]
30. Wang, L.; Zhou, J. Splicing and alternative splicing in rice and humans. *BMB Rep.* **2013**, *46*, 439–447.
31. Jabre, I.; Reddy, A.S.N.; Kalyna, M.; Chaudhary, S.; Khokhar, W.; Byrne, L.J.; Wilson, C.M.; Syed, N.H. Does co-transcriptional regulation of alternative splicing mediate plant stress responses? *Nucleic Acids Res.* **2019**, *47*, 2716–2726. [CrossRef]
32. Luco, R.F.; Pan, Q.; Tominaga, K.; Blencowe, B.J.; Pereira-Smith, O.M.; Misteli, T. Regulation of alternative splicing by histone modifications. *Science* **2010**, *327*, 996–1000. [CrossRef]
33. Ullah, F.; Hamilton, M.; Reddy, A.S.N.; Ben-Hur, A. Exploring the relationship between intron retention and chromatin accessibility in plants. *BMC Genom.* **2018**, *19*, 21. [CrossRef] [PubMed]
34. Drechsel, G.; Kahles, A.; Kesarwani, A.K.; Stauffer, E.; Behr, J.; Drewe, P.; Rätsch, G.; Wachter, A. Nonsense-mediated decay of alternative precursor mRNA splicing variants is a major determinant of the *Arabidopsis* steady state transcriptome. *Plant Cell* **2013**, *25*, 3726–3742. [CrossRef] [PubMed]
35. Kalyna, M.; Simpson, C.G.; Syed, N.H.; Lewandowska, D.; Marquez, Y.; Kusenda, B.; Marshall, J.; Fuller, J.; Cardle, L.; McNicol, J.; et al. Alternative splicing and nonsense-mediated decay modulate expression of important regulatory genes in *Arabidopsis*. *Nucleic Acids Res.* **2012**, *40*, 2454–2469. [CrossRef] [PubMed]
36. Reddy, A.S.N. Alternative splicing of pre-messenger RNAs in plants in the genomic era. *Annu. Rev. Plant Biol.* **2007**, *58*, 267–294. [CrossRef]
37. Hamada, S.; Ito, H.; Hiraga, S.; Inagaki, K.; Nozaki, K.; Isono, N.; Yoshimoto, Y.; Takeda, Y.; Matsui, H. Differential characteristics and subcellular localization of two starch-branching enzyme isoforms encoded by a single gene in *Phaseolus vulgaris* L. *J. Biol. Chem.* **2002**, *277*, 16538–16546. [CrossRef]
38. Liu, J.; Chen, X.; Liang, X.; Zhou, X.; Yang, F.; Liu, J.; He, S.Y.; Guo, Z. Alternative splicing of rice *WRKY62* and *WRKY76* transcription factor genes in pathogen defense. *Plant Physiol.* **2016**, *171*, 1427–1442. [CrossRef]
39. Chen, M.-X.; Zhang, K.-L.; Zhang, M.; Das, D.; Fang, Y.-M.; Dai, L.; Zhang, J.; Zhu, F.-Y. Alternative splicing and its regulatory role in woody plants. *Tree Physiol.* **2020**, *40*, 1475–1486. [CrossRef]
40. Chen, M.-X.; Zhang, Y.; Fernie, A.R.; Liu, Y.-G.; Zhu, F.-Y. SWATH-MS-based proteomics: Strategies and applications in plants. *Trends Biotechnol.* **2021**, *39*, 433–437. [CrossRef]

41. Chen, M.-X.; Mei, L.-C.; Wang, F.; Dewayalage, I.K.W.B.; Yang, J.-F.; Dai, L.; Yang, G.-F.; Gao, B.; Cheng, C.-L.; Liu, Y.-G.; et al. PlantSPEAD: A web resource towards comparatively analysing stress-responsive expression of splicing-related proteins in plant. *Plant Biotechnol. J.* **2021**, *19*, 227–229. [CrossRef]

42. Bedre, R.; Irigoyen, S.; Schaker, P.D.C.; Monteiro-Vitorello, C.B.; Da Silva, J.A.; Mandadi, K.K. Genome-wide alternative splicing landscapes modulated by biotrophic sugarcane smut pathogen. *Sci. Rep.* **2019**, *9*, 8876. [CrossRef]

43. Wang, L.; Chen, M.; Zhu, F.; Fan, T.; Zhang, J.; Lo, C. Alternative splicing is a *Sorghum bicolor* defense response to fungal infection. *Planta* **2020**, *251*, 14. [CrossRef] [PubMed]

44. Ling, Z.; Zhou, W.; Baldwin, I.T.; Xu, S. Insect herbivory elicits genome-wide alternative splicing responses in *Nicotiana attenuata*. *Plant J.* **2015**, *84*, 228–243. [CrossRef] [PubMed]

45. Jiang, J.; Liu, X.; Liu, C.; Liu, G.; Li, S.; Wang, L. Integrating omics and alternative splicing reveals insights into grape response to high temperature. *Plant Physiol.* **2017**, *173*, 1502–1518. [CrossRef] [PubMed]

46. Kannan, S.; Halter, G.; Renner, T.; Waters, E.R. Patterns of alternative splicing vary between species during heat stress. *AoB Plants* **2018**, *10*, ply013. [CrossRef] [PubMed]

47. Wu, Z.; Liang, J.; Wang, C.; Ding, L.; Zhao, X.; Cao, X.; Xu, S.; Teng, N.; Yi, M. Alternative splicing provides a mechanism to regulate LlHSFA3 function in response to heat stress in lily. *Plant Physiol.* **2019**, *181*, 1651–1667. [CrossRef]

48. Seo, P.J.; Park, M.-J.; Park, C.-M. Alternative splicing of transcription factors in plant responses to low temperature stress: Mechanisms and functions. *Planta* **2013**, *237*, 1415–1424. [CrossRef]

49. Ding, F.; Cui, P.; Wang, Z.; Zhang, S.; Ali, S.; Xiong, L. Genome-wide analysis of alternative splicing of pre-mRNA under salt stress in Arabidopsis. *BMC Genom.* **2014**, *15*, 431. [CrossRef]

50. Zhu, G.; Li, W.; Zhang, F.; Guo, W. RNA-seq analysis reveals alternative splicing under salt stress in cotton, *Gossypium davidsonii*. *BMC Genom.* **2018**, *19*, 73. [CrossRef]

51. Li, W.; Lin, W.-D.; Ray, P.; Lan, P.; Schmidt, W. Genome-wide detection of condition-sensitive alternative splicing in Arabidopsis roots. *Plant Physiol.* **2013**, *162*, 1750–1763. [CrossRef]

52. Van Veen, H.; Vashisht, D.; Akman, M.; Girke, T.; Mustroph, A.; Reinen, E.; Hartman, S.; Kooiker, M.; van Tienderen, P.; Schranz, M.E.; et al. Transcriptomes of eight *Arabidopsis thaliana* accessions reveal core conserved, genotype-and organ-specific responses to flooding stress. *Plant Physiol.* **2016**, *172*, 668–689. [CrossRef] [PubMed]

53. Lee, J.S.; Gao, L.; Guzman, L.M.; Rieseberg, L.H. Genome-wide expression and alternative splicing in domesticated sunflowers (*Helianthus annuus* L.) under flooding stress. *Agronomy* **2021**, *11*, 92. [CrossRef]

54. Ding, Y.; Wang, Y.; Qiu, C.; Qian, W.; Xie, H.; Ding, Z. Alternative splicing in tea plants was extensively triggered by drought, heat and their combined stresses. *PeerJ* **2020**, *8*, e8258. [CrossRef] [PubMed]

55. Filichkin, S.A.; Mockler, T.C. Unproductive alternative splicing and nonsense mRNAs: A widespread phenomenon among plant circadian clock genes. *Biol. Direct* **2012**, *7*, 20. [CrossRef]

56. Zhao, Y.; Sun, J.; Xu, P.; Zhang, R.; Li, L. Intron-mediated alternative splicing of WOOD-ASSOCIATED NAC TRANSCRIPTION FACTOR1B regulates cell wall thickening during fiber development in *Populus* species. *Plant Physiol.* **2014**, *164*, 765–776. [CrossRef]

57. Eckardt, N.A. Alternative splicing and the control of flowering time. *Plant Cell* **2002**, *14*, 743–747. [CrossRef]

58. Khan, M.R.G.; Ai, X.-Y.; Zhang, J.-Z. Genetic regulation of flowering time in annual and perennial plants. *Wiley Interdiscip. Rev. RNA* **2014**, *5*, 347–359. [CrossRef]

59. Gupta, V.; Estrada, A.D.; Blakley, I.; Reid, R.; Patel, K.; Meyer, M.D.; Andersen, S.U.; Brown, A.F.; Lila, M.A.; Loraine, A.E. RNA-Seq analysis and annotation of a draft blueberry genome assembly identifies candidate genes involved in fruit ripening, biosynthesis of bioactive compounds, and stage-specific alternative splicing. *Gigascience* **2015**, *4*, s13742-015. [CrossRef]

60. James, M.G.; Denyer, K.; Myers, A.M. Starch synthesis in the cereal endosperm. *Curr. Opin. Plant Biol.* **2003**, *6*, 215–222. [CrossRef]

61. Tetlow, I.J.; Morell, M.K.; Emes, M.J. Recent developments in understanding the regulation of starch metabolism in higher plants. *J. Exp. Bot.* **2004**, *55*, 2131–2145. [CrossRef]

62. Li, Y.; Mi, X.; Zhao, S.; Zhu, J.; Guo, R.; Xia, X.; Liu, L.; Liu, S.; Wei, C. Comprehensive profiling of alternative splicing landscape during cold acclimation in tea plant. *BMC Genom.* **2020**, *21*, 65. [CrossRef] [PubMed]

63. Seo, P.J.; Kim, M.J.; Ryu, J.-Y.; Jeong, E.-Y.; Park, C.-M. Two splice variants of the IDD14 transcription factor competitively form nonfunctional heterodimers which may regulate starch metabolism. *Nat. Commun.* **2011**, *2*, 303. [CrossRef] [PubMed]

64. Li, L.; Foster, C.M.; Gan, Q.; Nettleton, D.; James, M.G.; Myers, A.M.; Wurtele, E.S. Identification of the novel protein QQS as a component of the starch metabolic network in Arabidopsis leaves. *Plant J.* **2009**, *58*, 485–498. [CrossRef] [PubMed]

65. Xu, H.; Li, X.; Zhang, H.; Wang, L.; Zhu, Z.; Gao, J.; Li, C.; Zhu, Y. High temperature inhibits the accumulation of storage materials by inducing alternative splicing of *OsbZIP58* during filling stage in rice. *Plant Cell Environ.* **2020**, *43*, 1879–1896. [CrossRef]

66. Wang, J.-C.; Xu, H.; Zhu, Y.; Liu, Q.-Q.; Cai, X.-L. OsbZIP58, a basic leucine zipper transcription factor, regulates starch biosynthesis in rice endosperm. *J. Exp. Bot.* **2013**, *64*, 3453–3466. [CrossRef]

67. Jiang, G.; Zhang, D.; Li, Z.; Liang, H.; Deng, R.; Su, X.; Jiang, Y.; Duan, X. Alternative splicing of *MaMYB16L* regulates starch degradation in banana fruit during ripening. *J. Integr. Plant Biol.* **2021**, *63*, 1341–1352. [CrossRef]

68. Ohlrogge, J.; Browse, J. Lipid biosynthesis. *Plant Cell* **1995**, *7*, 957–970.

69. Vanhercke, T.; Tahchy, A.E.; Liu, Q.; Zhou, X.-R.; Shrestha, P.; Divi, U.K.; Ral, J.-P.; Mansour, M.P.; Nichols, P.D.; James, C.N.; et al. Metabolic engineering of biomass for high energy density: Oilseed-like triacylglycerol yields from plant leaves. *Plant Biotechnol. J.* **2014**, *12*, 231–239. [CrossRef]

70. Lu, C.; Napier, J.A.; Clemente, T.E.; Cahoon, E.B. New frontiers in oilseed biotechnology: Meeting the global demand for vegetable oils for food, feed, biofuel, and industrial applications. *Curr. Opin. Biotechnol.* **2011**, *22*, 252–259. [CrossRef]

71. Wang, X.; Devaiah, S.P.; Zhang, W.; Welti, R. Signaling functions of phosphatidic acid. *Prog. Lipid Res.* **2006**, *45*, 250–278. [CrossRef]

72. Snedden, W.A.; Blumwald, E. Alternative splicing of a novel diacylglycerol kinase in tomato leads to a calmodulin-binding isoform. *Plant J.* **2000**, *24*, 317–326. [CrossRef] [PubMed]

73. Trewavas, A.J.; Malhó, R. Ca^{2+} signalling in plant cells: The big network! *Curr. Opin. Plant Biol.* **1998**, *1*, 428–433. [CrossRef]

74. Kennedy, E.P. Biosynthesis of Complex Lipids. *Fed. Proc.* **1961**, *20*, 934–940.

75. Zheng, L.; Shockey, J.; Guo, F.; Shi, L.; Li, X.; Shan, L.; Wan, S.; Peng, Z. Discovery of a new mechanism for regulation of plant triacylglycerol metabolism: The peanut *diacylglycerol acyltransferase-1* gene family transcriptome is highly enriched in alternative splicing variants. *J. Plant Physiol.* **2017**, *219*, 62–70. [CrossRef] [PubMed]

76. Cernac, A.; Benning, C. *WRINKLED1* encodes an AP2/EREB domain protein involved in the control of storage compound biosynthesis in Arabidopsis. *Plant J.* **2004**, *40*, 575–585. [CrossRef] [PubMed]

77. Ji, X.-J.; Mao, X.; Hao, Q.-T.; Liu, B.-L.; Xue, J.-A.; Li, R.-Z. Splice variants of the castor *WRI1* gene upregulate fatty acid and oil biosynthesis when expressed in tobacco leaves. *Int. J. Mol. Sci.* **2018**, *19*, 146. [CrossRef]

78. Dellero, Y.; Jossier, M.; Schmitz, J.; Maurino, V.G.; Hodges, M. Photorespiratory glycolate–glyoxylate metabolism. *J. Exp. Bot.* **2016**, *67*, 3041–3052. [CrossRef]

79. Mano, S.; Hayashi, M.; Nishimura, M. Light regulates alternative splicing of hydroxypyruvate reductase in pumpkin. *Plant J.* **1999**, *17*, 309–320. [CrossRef]

80. Ishikawa, T.; Shigeoka, S. Recent advances in ascorbate biosynthesis and the physiological significance of ascorbate peroxidase in photosynthesizing organisms. *Biosci. Biotechnol. Biochem.* **2008**, *72*, 1143–1154. [CrossRef]

81. Zhu, J.; Wang, X.; Xu, Q.; Zhao, S.; Tai, Y.; Wei, C. Global dissection of alternative splicing uncovers transcriptional diversity in tissues and associates with the flavonoid pathway in tea plant (*Camellia sinensis*). *BMC Plant Biol.* **2018**, *18*, 266. [CrossRef]

82. Frei, B.; England, L.; Ames, B.N. Ascorbate is an outstanding antioxidant in human blood plasma. *Proc. Natl. Acad. Sci. USA* **1989**, *86*, 6377–6381. [CrossRef] [PubMed]

83. Mano, S.; Yamaguchi, K.; Hayashi, M.; Nishimura, M. Stromal and thylakoid-bound ascorbate peroxidases are produced by alternative splicing in pumpkin 1. *FEBS Lett.* **1997**, *413*, 21–26. [CrossRef]

84. Yoshimura, K.; Yabuta, Y.; Tamoi, M.; Ishikawa, T.; Shigeoka, S. Alternatively spliced mRNA variants of chloroplast ascorbate peroxidase isoenzymes in spinach leaves. *Biochem. J.* **1999**, *338*, 41–48. [CrossRef] [PubMed]

85. Ishikawa, T.; Yoshimura, K.; Tamoi, M.; Takeda, T.; Shigeoka, S. Alternative mRNA splicing of 3′-terminal exons generates ascorbate peroxidase isoenzymes in spinach (*Spinacia oleracea*) chloroplasts. *Biochem. J.* **1997**, *328*, 795–800. [CrossRef]

86. Ishikawa, T.; Sakai, K.; Yoshimura, K.; Takeda, T.; Shigeoka, S. cDNAs encoding spinach stromal and thylakoid-bound ascorbate peroxidase, differing in the presence or absence of their 3′-coding regions. *FEBS Lett.* **1996**, *384*, 289–293. [CrossRef]

87. Yoshimura, K.; Yabuta, Y.; Ishikawa, T.; Shigeoka, S. Identification of a *cis* element for tissue-specific alternative splicing of chloroplast ascorbate peroxidase pre-mRNA in higher plants. *J. Biol. Chem.* **2002**, *277*, 40623–40632. [CrossRef]

88. Najami, N.; Janda, T.; Barriah, W.; Kayam, G.; Tal, M.; Guy, M.; Volokita, M. Ascorbate peroxidase gene family in tomato: Its identification and characterization. *Mol. Genet. Genom.* **2008**, *279*, 171–182. [CrossRef]

89. Jespersen, H.M.; Kjærsgård, I.V.H.; Østergaard, L.; Welinder, K.G. From sequence analysis of three novel ascorbate peroxidases from *Arabidopsis thaliana* to structure, function and evolution of seven types of ascorbate peroxidase. *Biochem. J.* **1997**, *326*, 305–310. [CrossRef]

90. Gou, J.-Y.; Li, K.; Wu, K.; Wang, X.; Lin, H.; Cantu, D.; Uauy, C.; Dobon-Alonso, A.; Midorikawa, T.; Inoue, K.; et al. Wheat stripe rust resistance protein WKS1 reduces the ability of the thylakoid-associated ascorbate peroxidase to detoxify reactive oxygen species. *Plant Cell* **2015**, *27*, 1755–1770. [CrossRef]

91. Fu, D.; Uauy, C.; Distelfeld, A.; Blechl, A.; Epstein, L.; Chen, X.; Sela, H.; Fahima, T.; Dubcovsky, J. A kinase-START gene confers temperature-dependent resistance to wheat stripe rust. *Science* **2009**, *323*, 1357–1360. [CrossRef]

92. Ljung, K. Auxin metabolism and homeostasis during plant development. *Development* **2013**, *140*, 943–950. [CrossRef] [PubMed]

93. Estelle, M. Plant tropisms: The ins and outs of auxin. *Curr. Biol.* **1996**, *6*, 1589–1591. [CrossRef]

94. Friml, J.; Vieten, A.; Sauer, M.; Weijers, D.; Schwarz, H.; Hamann, T.; Offringa, R.; Jürgens, G. Efflux-dependent auxin gradients establish the apical–basal axis of *Arabidopsis*. *Nature* **2003**, *426*, 147–153. [CrossRef]

95. Cao, X.; Yang, H.; Shang, C.; Ma, S.; Liu, L.; Cheng, J. The roles of auxin biosynthesis YUCCA gene family in plants. *Int. J. Mol. Sci.* **2019**, *20*, 6343. [CrossRef] [PubMed]

96. Kriechbaumer, V.; Wang, P.; Hawes, C.; Abell, B.M. Alternative splicing of the auxin biosynthesis gene YUCCA4 determines its subcellular compartmentation. *Plant J.* **2012**, *70*, 292–302. [CrossRef] [PubMed]

97. Ruan, J.; Zhou, Y.; Zhou, M.; Yan, J.; Khurshid, M.; Weng, W.; Cheng, J.; Zhang, K. Jasmonic acid signaling pathway in plants. *Int. J. Mol. Sci.* **2019**, *20*, 2479. [CrossRef]

98. Creelman, R.A.; Mullet, J.E. Jasmonic acid distribution and action in plants: Regulation during development and response to biotic and abiotic stress. *Proc. Natl. Acad. Sci. USA* **1995**, *92*, 4114–4119. [CrossRef]

99. Wang, H.-L.; Zhang, Y.; Wang, T.; Yang, Q.; Yang, Y.; Li, Z.; Li, B.; Wen, X.; Li, W.; Yin, W.; et al. An alternative splicing variant of PtRD26 delays leaf senescence by regulating multiple NAC transcription factors in *Populus. Plant Cell* **2021**, *33*, 1594–1614. [CrossRef]

100. Zhu, J.; Wang, X.; Guo, L.; Xu, Q.; Zhao, S.; Li, F.; Yan, X.; Liu, S.; Wei, C. Characterization and alternative splicing profiles of the lipoxygenase gene family in tea plant (*Camellia sinensis*). *Plant Cell Physiol.* **2018**, *59*, 1765–1781. [CrossRef]

101. Pichersky, E.; Raguso, R.A. Why do plants produce so many terpenoid compounds? *New Phytol.* **2018**, *220*, 692–702. [CrossRef]

102. Chen, F.; Tholl, D.; Bohlmann, J.; Pichersky, E. The family of terpene synthases in plants: A mid-size family of genes for specialized metabolism that is highly diversified throughout the kingdom. *Plant J.* **2011**, *66*, 212–229. [CrossRef] [PubMed]

103. Wang, G.; Tang, W.; Bidigare, R.R. Terpenoids as therapeutic drugs and pharmaceutical agents. In *Natural Products*; Zhang, L., Demain, A.L., Eds.; Humana Press: Totowa, NJ, USA, 2005; pp. 197–227.

104. Huang, M.; Lu, J.-J.; Huang, M.-Q.; Bao, J.-L.; Chen, X.-P.; Wang, Y.-T. Terpenoids: Natural products for cancer therapy. *Expert Opin. Investig. Drugs* **2012**, *21*, 1801–1818. [CrossRef] [PubMed]

105. Bowers, W.S. Juvenile hormone: Activity of aromatic terpenoid ethers. *Science* **1969**, *164*, 323–325. [CrossRef] [PubMed]

106. Pichersky, E.; Gershenzon, J. The formation and function of plant volatiles: Perfumes for pollinator attraction and defense. *Curr. Opin. Plant Biol.* **2002**, *5*, 237–243. [CrossRef]

107. Dicke, M.; Baldwin, I.T. The evolutionary context for herbivore-induced plant volatiles: Beyond the 'cry for help'. *Trends Plant Sci.* **2010**, *15*, 167–175. [CrossRef]

108. Godard, K.-A.; White, R.; Bohlmann, J. Monoterpene-induced molecular responses in *Arabidopsis thaliana. Phytochemistry* **2008**, *69*, 1838–1849. [CrossRef]

109. Schuh, C.; Schieberle, P. Characterization of the key aroma compounds in the beverage prepared from Darjeeling black tea: Quantitative differences between tea leaves and infusion. *J. Agric. Food Chem.* **2006**, *54*, 916–924. [CrossRef]

110. Ma, C.; Qu, Y.; Zhang, Y.; Qiu, B.; Wang, Y.; Chen, X. Determination of nerolidol in teas using headspace solid phase microextraction–gas chromatography. *Food Chem.* **2014**, *152*, 285–290. [CrossRef]

111. Riaz, M.; Zia-Ul-Haq, M.; Dou, D. Chemistry of Carotenoids. In *Carotenoids: Structure and Function in the Human Body*; Zia-Ul-Haq, M., Dewanjee, S., Riaz, M., Eds.; Springer: Cham, Switzerland, 2021; pp. 43–76.

112. Bhavaniramya, S.; Vishnupriya, S.; Al-Aboody, M.S.; Vijayakumar, R.; Baskaran, D. Role of essential oils in food safety: Antimicrobial and antioxidant applications. *Grain Oil Sci. Technol.* **2019**, *2*, 49–55. [CrossRef]

113. Sun, S.; Li, Y.; Chu, L.; Kuang, X.; Song, J.; Sun, C. Full-length sequencing of ginkgo transcriptomes for an in-depth understanding of flavonoid and terpenoid trilactone biosynthesis. *Gene* **2020**, *758*, 144961. [CrossRef]

114. Liu, X.; Tang, N.; Xu, F.; Chen, Z.; Zhang, X.; Ye, J.; Liao, Y.; Zhang, W.; Kim, S.-U.; Wu, P. SMRT and Illumina RNA sequencing reveal the complexity of terpenoid biosynthesis in *Zanthoxylum armatum. Tree Physiol.* **2022**, *42*, 664–683. [CrossRef] [PubMed]

115. Mao, L.; Jin, B.; Chen, L.; Tian, M.; Ma, R.; Yin, B.; Zhang, H.; Guo, J.; Tang, J.; Chen, T.; et al. Functional identification of the terpene synthase family involved in diterpenoid alkaloids biosynthesis in *Aconitum carmichaelii. Acta Pharm. Sin. B* **2021**, *11*, 3310–3321. [CrossRef] [PubMed]

116. Niu, J.; Hou, X.; Fang, C.; An, J.; Ha, D.; Qiu, L.; Ju, Y.; Zhao, H.; Du, W.; Qi, J.; et al. Transcriptome analysis of distinct *Lindera glauca* tissues revealed the differences in the unigenes related to terpenoid biosynthesis. *Gene* **2015**, *559*, 22–30. [CrossRef] [PubMed]

117. Ding, Y.Q.; Fan, K.; Wang, Y.; Fang, W.P.; Zhu, X.J.; Chen, L.; Sun, L.T.; Qiu, C.; Ding, Z.T. Drought and heat stress-mediated modulation of alternative splicing in the genes involved in biosynthesis of metabolites related to tea quality. *Mol. Biol.* **2022**, *56*, 257–268. [CrossRef]

118. Thimmappa, R.; Geisler, K.; Louveau, T.; O'Maille, P.; Osbourn, A. Triterpene biosynthesis in plants. *Annu. Rev. Plant Biol.* **2014**, *65*, 225–257. [CrossRef]

119. Jin, B.; Guo, J.; Tang, J.; Tong, Y.; Ma, Y.; Chen, T.; Wang, Y.; Shen, Y.; Zhao, Y.; Lai, C.; et al. An alternative splicing alters the product outcome of a class I terpene synthase in *Isodon rubescens. Biochem. Biophys. Res. Commun.* **2019**, *512*, 310–313. [CrossRef]

120. Liu, G.-F.; Liu, J.-J.; He, Z.-R.; Wang, F.-M.; Yang, H.; Yan, Y.-F.; Gao, M.-J.; Gruber, M.Y.; Wan, X.-C.; Wei, S. Implementation of CsLIS/NES in linalool biosynthesis involves transcript splicing regulation in *Camellia sinensis. Plant Cell Environ.* **2018**, *41*, 176–186. [CrossRef]

121. J Basset, G.; Latimer, S.; Fatihi, A.; Soubeyrand, E.; Block, A. Phylloquinone (vitamin K1): Occurrence, biosynthesis and functions. *Mini Rev. Med. Chem.* **2017**, *17*, 1028–1038. [CrossRef]

122. Brettel, K.; Setif, P.; Mathis, P. Flash-induced absorption changes in photosystem I at low temperature: Evidence that the electron acceptor A_1 is vitamin K1. *FEBS Lett.* **1986**, *203*, 220–224. [CrossRef]

123. Furt, F.; van Oostende, C.; Widhalm, J.R.; Dale, M.A.; Wertz, J.; Basset, G.J. A bimodular oxidoreductase mediates the specific reduction of phylloquinone (vitamin K_1) in chloroplasts. *Plant J.* **2010**, *64*, 38–46. [CrossRef]

124. Shearer, M.J.; Bach, A.; Kohlmeier, M. Chemistry, nutritional sources, tissue distribution and metabolism of vitamin K with special reference to bone health. *J. Nutr.* **1996**, *126*, 1181S–1186S. [CrossRef] [PubMed]

125. Uchida, K.; Nomura, Y.; Takase, H.; Harauchi, T.; Yoshizaki, T.; Nakano, H. Effects of vitamin K-deficient diets and fasting on blood coagulation factors in conventional and germ-free rats. *Jpn. J. Pharmacol.* **1986**, *40*, 115–122. [CrossRef] [PubMed]

126. Erkkilä, A.T.; Booth, S.L.; Hu, F.B.; Jacques, P.F.; Lichtenstein, A.H. Phylloquinone intake and risk of cardiovascular diseases in men. *Nutr. Metab. Cardiovasc. Dis.* **2007**, *17*, 58–62. [CrossRef] [PubMed]

127. Walsh, C.T.; Liu, J.; Rusnak, F.; Sakaitani, M. Molecular studies on enzymes in chorismate metabolism and the enterobactin biosynthetic pathway. *Chem. Rev.* **1990**, *90*, 1105–1129. [CrossRef]

128. Yuan, Y.; Chung, J.-D.; Fu, X.; Johnson, V.E.; Ranjan, P.; Booth, S.L.; Harding, S.A.; Tsai, C.-J. Alternative splicing and gene duplication differentially shaped the regulation of isochorismate synthase in *Populus* and *Arabidopsis*. *Proc. Natl. Acad. Sci. USA* **2009**, *106*, 22020–22025. [CrossRef]

129. Wildermuth, M.C.; Dewdney, J.; Wu, G.; Ausubel, F.M. Isochorismate synthase is required to synthesize salicylic acid for plant defence. *Nature* **2001**, *414*, 562–565. [CrossRef]

130. Hashimoto, T.; Yamada, Y. Alkaloid biogenesis: Molecular aspects. *Annu. Rev. Plant Biol.* **1994**, *45*, 257–285. [CrossRef]

131. Ziegler, J.; Facchini, P.J. Alkaloid biosynthesis: Metabolism and trafficking. *Annu. Rev. Plant Biol.* **2008**, *59*, 735–769. [CrossRef]

132. De Luca, V.; St Pierre, B. The cell and developmental biology of alkaloid biosynthesis. *Trends Plant Sci* **2000**, *5*, 168–173. [CrossRef]

133. Powell, R.G.; Petroski, R.J. Alkaloid toxins in endophyte-infected grasses. *Nat. Toxins* **1993**, *1*, 163–170. [CrossRef]

134. Saunders, J.A.; O'neill, N.R.; Romeo, J.T. Alkaloid chemistry and feeding specificity of insect herbivores. In *Alkaloids: Chemical and Biological Perspectives*; Pelletier, S.W., Ed.; Springer: New York, NY, USA, 1992; pp. 151–196.

135. Shymanovich, T.; Saari, S.; Lovin, M.E.; Jarmusch, A.K.; Jarmusch, S.A.; Musso, A.M.; Charlton, N.D.; Young, C.A.; Cech, N.B.; Faeth, S.H. Alkaloid variation among epichloid endophytes of sleepygrass (*Achnatherum robustum*) and consequences for resistance to insect herbivores. *J. Chem. Ecol.* **2015**, *41*, 93–104. [CrossRef] [PubMed]

136. Ng, Y.P.; Or, T.C.T.; Ip, N.Y. Plant alkaloids as drug leads for Alzheimer's disease. *Neurochem. Int.* **2015**, *89*, 260–270. [CrossRef] [PubMed]

137. Cheeke, P.R. Toxicity and metabolism of pyrrolizidine alkaloids. *J. Anim. Sci.* **1988**, *66*, 2343–2350. [CrossRef] [PubMed]

138. O'Connor, S.E.; Maresh, J.J. Chemistry and biology of monoterpene indole alkaloid biosynthesis. *Nat. Prod. Rep.* **2006**, *23*, 532–547. [CrossRef] [PubMed]

139. Carqueijeiro, I.; Koudounas, K.; de Bernonville, T.D.; Sepúlveda, L.J.; Mosquera, A.; Bomzan, D.P.; Oudin, A.; Lanoue, A.; Besseau, S.; Cruz, P.L.; et al. Alternative splicing creates a pseudo-strictosidine β-D-glucosidase modulating alkaloid synthesis in *Catharanthus roseus*. *Plant Physiol.* **2021**, *185*, 836–856. [CrossRef]

140. Lamberto, I.; Percudani, R.; Gatti, R.; Folli, C.; Petrucco, S. Conserved alternative splicing of *Arabidopsis* transthyretin-like determines protein localization and S-allantoin synthesis in peroxisomes. *Plant Cell* **2010**, *22*, 1564–1574. [CrossRef]

141. Dixon, R.A.; Achnine, L.; Kota, P.; Liu, C.-J.; Reddy, M.S.S.; Wang, L. The phenylpropanoid pathway and plant defence—A genomics perspective. *Mol. Plant Pathol.* **2002**, *3*, 371–390. [CrossRef]

142. Deng, Y.; Lu, S. Biosynthesis and regulation of phenylpropanoids in plants. *Crit. Rev. Plant Sci.* **2017**, *36*, 257–290. [CrossRef]

143. Ralph, J.; Lapierre, C.; Boerjan, W. Lignin structure and its engineering. *Curr. Opin. Biotechnol.* **2019**, *56*, 240–249. [CrossRef]

144. Vanholme, R.; Demedts, B.; Morreel, K.; Ralph, J.; Boerjan, W. Lignin biosynthesis and structure. *Plant Physiol.* **2010**, *153*, 895–905. [CrossRef]

145. Boz, H. *p*-Coumaric acid in cereals: Presence, antioxidant and antimicrobial effects. *Int. J. Food Sci. Technol.* **2015**, *50*, 2323–2328. [CrossRef]

146. Bors, W.; Michel, C.; Stettmaier, K. Antioxidant effects of flavonoids. *Biofactors* **1997**, *6*, 399–402. [CrossRef] [PubMed]

147. Ryan, K.G.; Swinny, E.E.; Winefield, C.; Markham, K.R. Flavonoids and UV photoprotection in Arabidopsis mutants. *Z. Naturforsch. C* **2001**, *56c*, 745–754. [CrossRef] [PubMed]

148. Solovchenko, A.; Schmitz-Eiberger, M. Significance of skin flavonoids for UV-B-protection in apple fruits. *J. Exp. Bot.* **2003**, *54*, 1977–1984. [CrossRef] [PubMed]

149. Fukusaki, E.-I.; Kawasaki, K.; Kajiyama, S.; An, C.-I.; Suzuki, K.; Tanaka, Y.; Kobayashi, A. Flower color modulations of *Torenia hybrida* by downregulation of chalcone synthase genes with RNA interference. *J. Biotechnol.* **2004**, *111*, 229–240. [CrossRef] [PubMed]

150. Tanaka, Y.; Brugliera, F. Flower colour and cytochromes P450. *Philos. Trans. R. Soc. B Biol. Sci.* **2013**, *368*, 20120432. [CrossRef] [PubMed]

151. Agati, G.; Azzarello, E.; Pollastri, S.; Tattini, M. Flavonoids as antioxidants in plants: Location and functional significance. *Plant Sci.* **2012**, *196*, 67–76. [CrossRef]

152. Du, Y.; Chu, H.; Wang, M.; Chu, I.K.; Lo, C. Identification of flavone phytoalexins and a pathogen-inducible flavone synthase II gene (*SbFNSII*) in sorghum. *J. Exp. Bot.* **2010**, *61*, 983–994. [CrossRef]

153. Elliger, C.A.; Chan, B.G.; Waiss, A.C., Jr.; Lundin, R.E.; Haddon, W.F. C-glycosylflavones from *Zea mays* that inhibit insect development. *Phytochemistry* **1980**, *19*, 293–297. [CrossRef]

154. Waiss, A.C.W.; Chan, B.G.; Elliger, C.A.; Wiseman, B.R.; McMillian, W.W.; Widstrom, N.W.; Zuber, M.S.; Keaster, A.J. Maysin, a flavone glycoside from corn silks with antibiotic activity toward corn earworm. *J. Econ. Entomol.* **1979**, *72*, 256–258. [CrossRef]

155. Wang, L.; Lam, P.Y.; Lui, A.C.W.; Zhu, F.-Y.; Chen, M.-X.; Liu, H.; Zhang, J.; Lo, C. Flavonoids are indispensable for complete male fertility in rice. *J. Exp. Bot.* **2020**, *71*, 4715–4728. [CrossRef] [PubMed]

156. van der Meer, I.M.; Stam, M.E.; van Tunen, A.J.; Mol, J.N.; Stuitje, A.R. Antisense inhibition of flavonoid biosynthesis in petunia anthers results in male sterility. *Plant Cell* **1992**, *4*, 253–262. [PubMed]

157. Guerriero, G.; Hausman, J.-F.; Strauss, J.; Ertan, H.; Siddiqui, K.S. Lignocellulosic biomass: Biosynthesis, degradation, and industrial utilization. *Eng. Life Sci.* **2016**, *16*, 1–16. [CrossRef]

158. Yao, L.H.; Jiang, Y.-M.; Shi, J.; Tomás-Barberán, F.; Datta, N.; Singanusong, R.; Chen, S.S. Flavonoids in food and their health benefits. *Plant Foods Hum. Nutr.* **2004**, *59*, 113–122. [CrossRef]

159. Hong, L.; Qian, Q.; Tang, D.; Wang, K.; Li, M.; Cheng, Z. A mutation in the rice chalcone isomerase gene causes the *golden hull and internode 1* phenotype. *Planta* **2012**, *236*, 141–151. [CrossRef]

160. Tang, W.; Zheng, Y.; Dong, J.; Yu, J.; Yue, J.; Liu, F.; Guo, X.; Huang, S.; Wisniewski, M.; Sun, J.; et al. Comprehensive transcriptome profiling reveals long noncoding RNA expression and alternative splicing regulation during fruit development and ripening in kiwifruit (*Actinidia chinensis*). *Front. Plant Sci.* **2016**, *7*, 335. [CrossRef]

161. Chen, L.; Shi, X.; Nian, B.; Duan, S.; Jiang, B.; Wang, X.; Lv, C.; Zhang, G.; Ma, Y.; Zhao, M. Alternative splicing regulation of anthocyanin biosynthesis in *Camellia sinensis* var. *assamica* unveiled by PacBio Iso-Seq. *G3 Gene Genomes Genet.* **2020**, *10*, 2713–2723. [CrossRef]

162. Qiao, D.; Yang, C.; Chen, J.; Guo, Y.; Li, Y.; Niu, S.; Cao, K.; Chen, Z. Comprehensive identification of the full-length transcripts and alternative splicing related to the secondary metabolism pathways in the tea plant (*Camellia sinensis*). *Sci. Rep.* **2019**, *9*, 2709. [CrossRef]

163. Ye, J.; Cheng, S.; Zhou, X.; Chen, Z.; Kim, S.U.; Tan, J.; Zheng, J.; Xu, F.; Zhang, W.; Liao, Y.; et al. A global survey of full-length transcriptome of *Ginkgo biloba* reveals transcript variants involved in flavonoid biosynthesis. *Ind. Crops and Prod.* **2019**, *139*, 111547. [CrossRef]

164. Lim, S.-H.; Kim, D.-H.; Jung, J.-A.; Lee, J.-Y. Alternative splicing of the basic helix–loop–helix transcription factor gene *CmbHLH2* affects anthocyanin biosynthesis in ray florets of chrysanthemum (*Chrysanthemum morifolium*). *Front. Plant Sci.* **2021**, *12*, 669315. [CrossRef]

165. Yin, P.; Zhen, Y.; Li, S. Identification and functional classification of differentially expressed proteins and insight into regulatory mechanism about flower color variegation in peach. *Acta Physiol. Plant.* **2019**, *41*, 95. [CrossRef]

166. Xu, W.; Dubos, C.; Lepiniec, L. Transcriptional control of flavonoid biosynthesis by MYB–bHLH–WDR complexes. *Trends Plant Sci.* **2015**, *20*, 176–185. [CrossRef] [PubMed]

167. Chen, D.; Liu, Y.; Yin, S.; Qiu, J.; Jin, Q.; King, G.J.; Wang, J.; Ge, X.; Li, Z. Alternatively spliced *BnaPAP2.A7* isoforms play opposing roles in anthocyanin biosynthesis of *Brassica napus* L. *Front. Plant Sci.* **2020**, *19*, 983. [CrossRef] [PubMed]

168. Qi, T.; Song, S.; Ren, Q.; Wu, D.; Huang, H.; Chen, Y.; Fan, M.; Peng, W.; Ren, C.; Xie, D. The jasmonate-ZIM-domain proteins interact with the WD-Repeat/bHLH/MYB complexes to regulate jasmonate-mediated anthocyanin accumulation and trichome initiation in *Arabidopsis thaliana*. *Plant Cell* **2011**, *23*, 1795–1814. [CrossRef] [PubMed]

169. Zhu, J.; Yan, X.; Liu, S.; Xia, X.; An, Y.; Xu, Q.; Zhao, S.; Liu, L.; Guo, R.; Zhang, Z.; et al. Alternative splicing of CsJAZ1 negatively regulates flavan-3-ols biosynthesis in tea plants. *Plant J.* **2022**, *110*, 243–261. [CrossRef]

170. Dombrecht, B.; Xue, G.P.; Sprague, S.J.; Kirkegaard, J.A.; Ross, J.J.; Reid, J.B.; Fitt, G.P.; Sewelam, N.; Schenk, P.M.; Manners, J.M.; et al. MYC2 differentially modulates diverse jasmonate-dependent functions in *Arabidopsis*. *Plant Cell* **2007**, *19*, 2225–2245. [CrossRef]

171. Egan, A.N.; Schlueter, J.; Spooner, D.M. Applications of next-generation sequencing in plant biology. *Am. J. Bot.* **2012**, *99*, 175–185. [CrossRef]

172. Martin, L.B.B.; Fei, Z.; Giovannoni, J.J.; Rose, J.K.C. Catalyzing plant science research with RNA-seq. *Front. Plant Sci.* **2013**, *4*, 66. [CrossRef]

173. Maor, G.L.; Yearim, A.; Ast, G. The alternative role of DNA methylation in splicing regulation. *Trends Genet.* **2015**, *31*, 274–280. [CrossRef]

174. Parker, M.T.; Knop, K.; Sherwood, A.V.; Schurch, N.J.; Mackinnon, K.; Gould, P.D.; Hall, A.J.W.; Barton, G.J.; Simpson, G.G. Nanopore direct RNA sequencing maps the complexity of Arabidopsis mRNA processing and m^{6}A modification. *Elife* **2020**, *9*, e49658. [CrossRef]

International Journal of
Molecular Sciences

Review

Emerging Function of Ecotype-Specific Splicing in the Recruitment of Commensal Microbiome

Yue-Han Li [1,2,3], Yuan-You Yang [1], Zhi-Gang Wang [2,3,*] and Zhuo Chen [1,*]

[1] State Key Laboratory Breeding Base of Green Pesticide and Agricultural Bioengineering, Key Laboratory of Green Pesticide and Agricultural Bioengineering, Ministry of Education, Research and Development Center for Fine Chemicals, Guizhou University, Guiyang 550025, China; leeenxiuu@163.com (Y.-H.L.); gs.yangyy16@gzu.edu.cn (Y.-Y.Y.)
[2] School of Life Science and Agriculture Forestry, Qiqihar University, Qiqihar 161006, China
[3] Heilongjiang Provincial Technology Innovation Center of Agromicrobial Preparation Industrialization, Qiqihar 161006, China
* Correspondence: wangzhigang@qqhru.edu.cn (Z.-G.W.); gychenzhuo@aliyun.com (Z.C.)

Abstract: In recent years, host–microbiome interactions in both animals and plants has emerged as a novel research area for studying the relationship between host organisms and their commensal microbial communities. The fitness advantages of this mutualistic interaction can be found in both plant hosts and their associated microbiome, however, the driving forces mediating this beneficial interaction are poorly understood. Alternative splicing (AS), a pivotal post-transcriptional mechanism, has been demonstrated to play a crucial role in plant development and stress responses among diverse plant ecotypes. This natural variation of plants also has an impact on their commensal microbiome. In this article, we review the current progress of plant natural variation on their microbiome community, and discuss knowledge gaps between AS regulation of plants in response to their intimately related microbiota. Through the impact of this article, an avenue could be established to study the biological mechanism of naturally varied splicing isoforms on plant-associated microbiome assembly.

Keywords: alternative splicing; plant genotype; plant–microbe interaction; microbiome assembly; post-transcriptional regulation; proteogenomics

Citation: Li, Y.-H.; Yang, Y.-Y.; Wang, Z.-G.; Chen, Z. Emerging Function of Ecotype-Specific Splicing in the Recruitment of Commensal Microbiome. *Int. J. Mol. Sci.* **2022**, *23*, 4860. https://doi.org/10.3390/ijms23094860

Academic Editors: Melvin J. Oliver, Bei Gao, Moxian Chen, Toshifumi Tsukahara and Alfonso Gutiérrez-Adán

Received: 19 March 2022
Accepted: 25 April 2022
Published: 27 April 2022

Publisher's Note: MDPI stays neutral with regard to jurisdictional claims in published maps and institutional affiliations.

1. Introduction

The interaction between host plants and their commensal microbiome has become a research hotspot in recent years. As sessile organisms, plants have evolved a series of mechanisms to cope with changing environments [1,2]. In fact, the interaction between plant immunity and the plant microbiota is bidirectional, with plants, microbiota, and the environment forming a complex system and working together to coordinate the plant microbiota. An increasing number of researchers have suggested that those plant-associated microorganisms can be studied together with their host plants as one combined holobiont in the field [3]. Benefits from adequate mutualistic interactions will improve plant growth and soil productivity, inhibit pathogen propagation, enhance plant tolerance to adverse environments, and accelerate nutrient recycling in fields [4], ultimately providing an effective way for agriculture to increase crop yield in a sustainable manner [5].

Influencing factors that affect this mutualistic interaction have been extensively reported, including internal and external factors in both plants and microorganisms [6]. Specifically, there is a growing body of evidence that plant genetics information is crucial in response to microbial colonization, extending the knowledge of plant biology to other disciplines such as microbiology and ecology [7]. Studies using different plant ecotypes or cultivars have been conducted to unravel the underlying mechanism of this interaction [3]. For example, plants decode genetic information to determine the composition of root exudates [8], root architecture [9], or ingredients for the microbiome assembly to respond at its surface or rhizosphere. Current reports aim to understand the diversity and dynamics of indigenous microbial communities coupled with the identification of core microbiota or keystone species [3]. However, most of the studies are descriptive, paying little attention to the underlying mechanism to mediate this interaction [1,3].

In addition, plants can translate their genetic differences by adjusting the repertoire of proteins they express in response to surrounding abiotic and biotic factors. One post-transcriptional regulatory mechanism, notably alternative splicing (AS), has emerged as a pivotal process for plants, to adapt to changing environments by significantly increasing proteome diversity or fine-tuning transcript abundance [10]. To date, approximately 95% of human intron-containing genes undergo AS regulation and up to 35% of human diseases are related to mutations that can affect splicing [11,12]. Importantly, single nucleotide polymorphisms (SNPs) observed at a splice site will affect the expression of certain transcript isoforms, thus resulting in phenotypic diversity in research on human diseases [13]. In plants, 60–83% of intron-containing genes undergo AS [14]. Natural variation of AS, identified as splicing quantitative trait loci (functions of sQTLs) [15], and its contribution to plant traits has been observed by using population-level transcriptome analyses, providing evidence for linking genetic variants to plant trait variation [16–18]. However, which function of sQTLs contribute to plant–microbe interactions is poorly understood. To this end, we first introduce the current progress of these research questions in this review article. Then, we discuss the possible mechanism to connecting diverse plant ecotype–phenotype linkages, splicing isoforms, and commensal microbiomes. In addition, we discuss further investigation using modern high-throughput omics technology to identify target plant alleles that can modulate interactions between host plants and their intimately related microbiota.

2. Ecotype-Specific Microbial Community in Plants

2.1. Host Maintanence of Commensal Microbiome

The microbial community is a crucial environmental component, as approximately 10 billion microorganisms are present in every gram of soil [19]. The interaction between plants and microorganism communities is a vital driver of ecosystem functioning [20] and can affect the maintenance of agricultural ecosystems [21]. These microorganisms will form complex co-associations with plants and have necessary roles in promoting the productivity and health of the plant type in natural environments. In particular, plant-associated microbiota and plants from a 'holobiont', and evolutionary selection among microbes and plants contributes to the steadiness of an ecosystem [22,23]. Host-associated microbial communities have a crucial role in shaping the health and fitness of plants [24]. Diverse microbial communities colonize plant surfaces and tissues, and beneficial microbial groups provide plants with a large array of life supporting functions such as resilience to organic phenomena and abiotic stresses, growth promotion, and nutrient acquisition [25,26]. It is well known that many beneficial microorganisms in the soil after planting plants will enrich the roots of the plants [27]. Plants and microorganisms work together as an important part of plant growth and illustrate the importance of plant hosts; however, there is a need to better understand the governing plant–microorganism interactions at the community level and explore their potential agricultural application value. It was found that the composition of microbial communities among different plants is diverse in bulk soils under different vegetation covers [28]. Furthermore, the field bacterial community structure in the bulk and inter-root soils with different plant diversity treatments was examined, and significant differences in bacterial communities were observed in both bulk and inter-root soils [29]. In recent years, culture-independent high-throughput sequencing has greatly expanded the repertoire of microorganisms known to reside in and on plants [30–32]. Plant growth-promoting rhizobacteria (PGPR) are gradually attracting public research attention. If we visualize an ideal agricultural picture, crops produced should be equipped with disease resistance, salt tolerance, drought tolerance, serious metal stress tolerance, and higher nutritional value. One way is to use soil microorganisms (bacteria, fungi, algae, etc.) to achieve this goal [33]. Soil microbial populations are immersed in a framework of interactions known to have an effect on plant fitness and soil quality. Cooperative microbial activities may be exploited as a low-input biotechnology to assist sustainable, environmentally friendly, agro-technological practices. Among these potential soil microorganisms, PGPR bacteria are the most promising research targets. PGPR could also be used to enhance plant health and promote the plant growth rate without environmental contamination [34]. There have also been deeper developments in recent years regarding the separation of PGPR, which also has a positive effect on plants. The most remarkably studied bacteria in relation to biocontrol are members of the genera *Pseudomonas* spp., *Bacillus* spp., *Azospirillum* spp., and *Streptomyce* spp. [35–37]. For example, *Bacillus thuringiensis* is renowned as a good bioinsecticidal bacterium, and acyl homoserine lactone lactonase created by *B. thuringiensis* can open the lactone ring of N-acyl homoserine lactone, a signature molecule within the bacterial quorum-sensing system [38]. The capacity of microorganisms to rescue plants' capacity of stress has been recently studied [39–41]. For example, wheat microbiome bacteria can relieve plant stress, and a compound secreted by the bacteria (phenazine-1-carboxamide) directly affects the activity of fungal protein FgGcn5, a histone acetyltransferase. This results in deregulation of simple protein acetylation at H2BK11, H3K14, H3K18, and H3K27 in *F. graminearum*, as suppression of fungal growth, virulence, and mycotoxin biogenesis [41]. Therefore, the associated degree of antagonistic bacteria can inhibit the growth and virulence of a plant pathogenic fungus by manipulating fungi simple protein modification. Strains of *Chitinophaga* spp., *Chryseobacterium* spp., *Flavobacterium* spp., *Microbacterium* spp., *Pseudomonas* spp., *Sphingomonas* spp., *Stenotrophomonas* spp., and *Xanthomonas* spp. are typically found to complement plant responses to totally different pathogen/pest attacks. However, the benefits of microorganisms on plants may occur via two modes. For example, a change in the structure of the inter-root microbiome would give the common tomato the property of resistance to wilt. The model dicot plant *Arabidopsis thaliana* specifically promotes three bacterial species within the rhizosphere upon foliar defence activation by the mildew pathogen *Hyaloperonospora arabidopsidis* [42]. 2,4-diacetylphloroglucinol (Phl) produced by *Pseudomonas* on roots of wheat grown in a soil suppressive to take-all of wheat. Phl-producing fluorescent *Pseudomonas* are key elements of the natural biological control that operates in take-all-suppressive soils [43]. Recruitment of microbes, via plant hosts, could have evolved to prime plant defence communication pathways and inhibit the expansion and virulence of pathogens that consequently ameliorate plant stresses [44–46]. This suggests that plant hosts largely determine the changes in plant microbial communities. Therefore, we believe that plants depend on microorganisms to bring them positive effects, and they have unique ways to 'communicate'. These works need to be explored continuously and need more research to explore important research implications.

2.2. Function of Microbiome in Plant Defense and Immune Systems, Corollary of the Relationship between Plant Active and Passive Immunity

The plant microbiome is not static: its structure and therefore the provided host functions in response to stresses and environmental stimuli [47–50]. Recent studies suggest that the changes within the microbiome are not merely passive responses of plants, but rather a consequence of millions of years of coevolution [39–41,45,51].

Emerging evidence has indicated that plant-associated microbiomes are closely related to plant health [52] and the helpful features of plant-associated microbes will boost the immune responses in plants against biotic/abiotic environmental constraints [53,54]. In high abundance, beneficial microbes directly inhibit pathogens by producing antimicrobial compounds. However, beneficial microbes can also inhibit pathogens indirectly by stimulating the immune system of plants, a development referred to as induced systemic resistance (ISR) [55]. Two forms of systemic immunity triggered by plant–microbe interactions, systemic acquired resistance (SAR) and ISR, are classified due to differences in the site of induction and the manner in which the microbes are induced. Both SAR and ISR influence growth regulator crosstalk towards enhanced defence against pathogens, which affects the composition of the plant microbiome. ISR inducers include PGPR, such as *Pseudomonas* spp., *Bacillus* spp., *streptomyces* spp., and plant growth-promoting fungi (PGPF), as well as *Trichoderma* spp. and *Serendipita indica* (formerly *Piriformospora indica*) [56–59]. In plants, interaction with beneficial microbes of the microbiome will additionally result in the activation of the plant's system, not least in the form of ISR. Reciprocally, the composition of the plant microbiome is influenced by plant immunity, which we call active immunity. This appears to be only determined by two main mechanisms thus far: direct microbe–microbe interactions and stimulation or priming of the plant immune system. For instance, a molecule secreted by the *Pseudomonas piscium* ZJU60 strain, which was isolated from infected wheat head, antagonizes the flora *Fusarium graminearum* by inhibiting its simple protein acetyltransferases [41]. The ability to antagonize other microbes, together with pathogens, may be a common attribute in bacteria isolated from the *Arabidopsis* leaf microbiome [60]. These studies all suggest that plant microbiota are a rich source of pathogen antagonists that act through direct inhibition [61].

Plant–pathogen interactions are mediated by the interplay of multifaceted processes, which are expedited by pathogen- and plant-oriented molecules [62,63]. Pathogens and commensal microbes that survive competition with different soil and plant-associated microbes then encounter the plant innate system, principally microbe-associated molecular patterns (MAMPs), triggered immunity (MTI), and effector triggered immunity (ETI), two layers of molecular defence referred to within the advanced zigzag defence model [64]. PTI and ETI both depend on SA, and each induce a systemic, SA-dependent defence response that is known as systemic acquired resistance (SAR) [65–67]. SAR may be a durable style of resistance against a broad spectrum of (hemi-)biotrophic pathogens [68].

The vital next step, currently, is to check these molecular mechanisms and unravel how commensal microbes move with plant immunity under various plant conditions. Such interactions between the system and microbiome are likely critical for plant defence against diverse stresses and influence host microbiota associations. Evidently, internal secretion signalling pathways of jasmonic acid (JA), salicylic acid (SA), and ethylene (ET) and their interactions are crucial in regulating plant defences and their associated microbiota [69,70]. Biotrophic and hemi-biotrophic pathogens that depend upon living host tissues and cells are fended off chiefly through SA-dependent immune mechanisms. Other findings show that the diversity of the endophytic microorganism community in *Arabidopsis* leaves is in response to the associated degree of activation of the SA signalling pathway [70] and SA signalling is involved in the modulation of root microbiota.

Taken together, unravelling the interactions between plant defence systems, microbiomes, and environmental factors will be an essential next step for understanding plant selection of microbes and immune modulation shifts of plant microbiota of various environmental conditions.

2.3. Plant Hosts Selectively Recruit Microbes via Ecotype Specificity

In light of growing concern over the threat of water and nutrient stress facing terrestrial ecosystems, particularly those used for agricultural production, increased emphasis has been placed on understanding how abiotic stress conditions influence the composition and functioning of the root microbiome and its ultimate consequences for plant health. However, under abiotic stress conditions, the composition of root microorganisms will change accordingly, which will be reflected in changes in root secretions. Although there are many conclusions about abiotic stresses on microbial community alteration, in recent years, attention has gradually shifted to the genotype differences of plant hosts. Following the appearance of next-generation sequencing, many studies have characterized the rhizospheric wheat microbiome and investigated the influence of the compartment (rhizoplane vs. endosphere), crop management, or wheat genotypes on the diversity and structure of those complicated microbial communities [71–74]. For many crops, studying soil microbial structure is indisputably the most necessary issue in crop management and plant genotype research. A number of studies have demonstrated that plants confirm changes of the microbial community under the same environmental conditions. There are many studies on the effect of different genotypes of the same plant on microorganisms (Table 1). The importance of genotype–environment interactions is in the structural assembly of plant microbiomes. For example, the reference plant *Arabidopsis thaliana* demonstrates that host cultivars (genotypes) mediate a weak but measurable impact on the root-associated microbiota [75]. Plant genotypes and soil have specific effects on the wheat rhizosphere microbial community [76]. The effect of plant genotypes on the inter-root microbial community of maize showed that the two genotypes had a significant effect on the inter-root microbial community of maize [77]. Lamit investigated different cottonwood genotypes supporting different aboveground fungal communities [78]. Jiang et al. investigated the inter-root bacterial community of 12 Rabbiteye blueberry (RB) cultivars and demonstrated that inter-rooting from plant cultivars affects the bacterial association network [79]. Microbial community (SMC) structure and root turnover were assessed in two contrasting *Lolium perenne* cultivars (AberDove and S23), and microbial communities with differences were discovered [80]. Hou et al. found significant differences in the composition of inter-root bacterial communities of different rice varieties under metal contamination conditions [81]. A significant genotypic variation in rhizosphere microbial communities in rice plants was reported [82]. Huang et al. showed significant differences in the inter-root microbial communities of two varieties of kale-type oilseed rape [83]. By analysing the soil samples of three different peony species, it was found that the microbial community structure was greatly influenced by the peony species [84]. Zhou et al. found significant differences in the inter-root microbial communities of four cucumber varieties [85]. A comparison of different genotypes of soybean [5], winter wheat [86], and maize [4] also revealed that the bacterial community structure differed among cultivars.

All of this evidence suggests that microbial communities change in response to changes in hosts, so a hypothesis has been proposed that plants will recruit different microorganisms by releasing different root secretions based on their own demand recruitment to relieve stress on their own growth or when they encounter stressful environments. Specifically, a mechanism employed by plants is the 'cry for help' strategy [87], a phenomenon whereby plants experiencing abiotic, pest-induced, or pathogen-induced stresses recruit helpful microbes/traits from the environment by employing a range of chemical stimuli to boost their capacity to combat stresses [88,89]. For example, some soils conditioned by take-all disease-infected wheat may lead to less take-all in future generations [43]. By studying the composition and metabolic potential of inter-root bacterial communities of different common bean (*Phaseolus vulgaris*) cultivars with different levels of resistance to the fungal root pathogen Fusarium (Fox), differences in microbial composition between susceptible and nonsusceptible cultivars and higher network complexity of resistant cultivars were found. Pseudomonadaceae and Bacillaceae had a high abundance in the rhizosphere of the Fox-resistant cultivar. Breeding for Fox resistance in common bean could have co-selected other unknown plant traits that support the next abundance of specific helpful microorganism families in the rhizosphere with functional traits that reinforce the primary line of defence [90]. Microbiome structures differed between the two tomato cultivars and transplantation of rhizosphere microbiota from resistant plants suppressed disease symptoms in susceptible plants. Comparative analyses of rhizosphere metagenomes from resistant and susceptible plants enabled the identification and assembly of a *flavobacterial* genome that was much more abundant in the resistant plant, which could suppress *R. solanacearum*-disease development in a susceptible plant [45]. Thus, on the far side of classical 'adapt or migrate' strategies, accumulating evidence suggests that plants use the 'cry for help' strategy as a full-of-life method that enables them to learn from microorganism associations under stresses. The host is the subject, and it is unclear in what way they recruit microorganisms; therefore, mining the mechanism of plant–microbe interactions will be an important part of future plant breeding. However, this theory still needs more evidence to prove it.

Table 1. A comparison of some representative microbial communities of different genotypes and a comparison of alternative splicing of plants of different ecotypes is summarized.

Plant Varieties	Genotype Varieties	Region	Reproductive Stages	Result
Wheat	Apache (AP), Bermude (BM), Carstens (CT), Champlein (CH), Cheyenne (CY), Rubisko (RB), Soissons (SS), Terminillo TM	Senegal, Cameroon, France, and Italy	Harvest season	Plant genotypes and soil have specific effects on the wheat rhizosphere microbial community.
Rice	Zhefu No. 7, hereafter: HA, Xiangzaoxian No. 45, hereafter: LA	Zhejiang Province, China	Harvest season	Significant differences in the composition of inter-root bacterial communities of different rice varieties.
Soybean	Kwangan (KA), Poongsannamul (PS), Poongwon (PW), and Taekwang (TK)	Jeonju, South Korea	R1, R3, R5, and R7	Influenced by the variety and growth stage.
Winter Wheat	Eltan, Finch, Hill81,Lewjain, Madsen, PI561722, PI561725, PI561726 and PI561727	Pullman, WA, USA.	Late May to early June, 2010 and 2014	Wheat cultivars are involved in shaping the rhizosphere by differentially altering the bacterial OTUs consistently across different sites.
Maize	Zhengdan 958 (ZD), Gaoneng 1(G1), and Gaoneng 2 (G2)	Hebei province, China	18 September 2017	The bacterial community structure in bulk soil of different cultivars was significantly different.
Rabbiteye blueberry	Clearly clustered into three groups (BRI, BRII, and RBIII).	Lishui, Nanjing province, China	Harvest Season	The rhizosphere from the plant cultivars exerted substantial effects on bacterial association networks and putative keystone species.
Tomato	Resistant varieties and non-resistant varieties	Seoul, Republic of Korea.	Harvest season	More abundant in the resistant plant rhizosphere microbiome than in that of the susceptible plant.
Bean	Resistant varieties and non-resistant varieties	Anhembi municipality, Sao Paulo, Brazil	Harvest season	Different levels of resistance to the fungal root pathogen Fusarium (Fox), differences in microbial composition.
Plant Varieties	Genotype Varieties	Unique Splicing quantitative trait loci (sQTL)	Reproductive Stages	Result
Arabidopsis thaliana	666 geographically distributed diverse ecotypes	6406	/	A role for differential isoform usage in regulating these important processes in diverse ecotypes of *Arabidopsis*.
Maize	Maize kernels from 368 inbred lines	19,554	/	The importance of AS in diversifying gene function and regulating phenotypic variation.

2.4. Emerging Factors That Are Responsible for Microbe Recruitment

It is not surprising that various active genes of the soil microbiota play important roles in competition or cooperation with other microbes. Microorganisms synthesize completely different products that affect microbe–microbe interactions. Distinct and numerous gene clusters for biosynthesis of natural products have been identified in plant-associated microorganisms [91,92]. Additionally, the organic chemistry diversity of root exudates was analysed for alterations within the presence or concentrations of metabolites (metabolomics) to determine how root exudation drives stress-induced microbiome assembly. The key metabolic compounds are isolated from the rhizosphere for potential and tested in vitro for interactions with plant microorganism symbionts and pathogens and, in turn, plant health. The addition of various exudate mixtures to plant monocultures increased microorganism diversity. Some studies also found that higher plant diversity was associated with higher microbial diversity. Major challenges remain, such as analysis of root exudation in natural settings, mostly due to the chemical quality of various soil types. A few plant products that affect the rhizosphere microbiome are known. For example, 2,4-diacetylphloroglucinol [50], peroxidases and oxylipins [93], benzoxazinoids [94], phenylpropanoids [95], flavonoids [93], coumarins [96], triterpenes, mucilage [97], and aromatic compounds [98] are reported to attract microbes that benefit plant defence and nutrition. However, how these compounds are synthesized and how they regulate the microbe interplay, along with their underlying biochemical mechanisms, remains to be explored. Although each plant produces exudates, the number and composition of root exudates vary. First, exudation is outlined by the genotype of the host. Nineteen *Arabidopsis* root metabolic patterns and their variability between plants and naturally occurring germplasm were analysed, and for some secondary metabolites, the ratio of total plant-to-plant variability was high and significant [99]. Exudation is modulated by abiotic stresses: the amounts of exuded amino acids, sugars, and organic acids modified in maize fully grown in phosphate-, iron-, nitrogen-, or potassium-deficient conditions [100]. Additionally, phosphate-deficient *Arabidopsis* plants exhibit hyperbolic coumarin and oligolignol exudation [101], and coumarin exudation is significantly reduced in ABCG37-deficient plants [102], serious metal-treated poplar (*Populus tremula*) iatrogenic organic acid exudation, and zinc-deficient wheat hyperbolic phytosiderophore exudation [103]. Differential exudation could be a plausible mechanism that plants might modulate their interaction with microbes. All of this evidence suggests that under different stresses the root secretions of plants are altered, and it is not difficult to speculate that the microbial community is altered accordingly. However, research on the secretion mechanism of plant inter-root secretions and the discovery of key loci for the recruitment of microorganisms remains a challenge.

3. Ecotype-Prone Splicing Events Are Good Candidates to Study Plant-Microbe Interactions

3.1. Pre-mRNA Alternative Splicing and Splicing Regulators of Plant Immunity

Most eukaryotic genes are interrupted by introns. Therefore, an important step in gene expression is the removal of introns through the splicing of precursor mRNA transcripts (premRNAs). Emerging evidence has indicated that plant-associated microbiomes are engaged with plant health and the helpful features of plant-associated microbes will boost the immune responses in plants against biotic/abiotic environmental constraints. Recent progress in high outturn sequencing of ribonucleic acid and bioinformatics tools to research AS events [104–107] at a genome-wide scale have shown that AS is a vital part of host transcriptome reprogramming in response to microorganism and viral infection in several plant species. In plants, varied reports have primarily focused on AS analysis in model plant species or nonwoody plants, resulting in a notable lack of research on AS in woody plants [108]. Alternative splicing (AS) enhances transcriptome malleability and proteome diversity in response to various growth and stress cues, which places AS at the crossroads of adaptation and environmental stress response [109,110]. However, some people are also working on different ways to improve these bottlenecks regarding AS. For example, Chen et al. presented the plant splicing-related protein expression and annotation database PlantSPEAD, introducing the information on their annotations, sequence information, functional domains, protein interaction partners, and expression patterns in response to abiotic stresses, from a dozen existing databases and the literature [111]. Furthermore, the SWATH-MS approach is characterized by a data-independent acquisition (DIA) method followed by a unique targeted information extraction approach [112]. SWATH-MS proteomics is applied for a range of large-scale profiling studies in plants, moving from model plant species to various plants without reference ordination annotation [113]. Within the field of discovery proteomics, alternative splicing is an emerging research area associated with posttranscriptional regulation. SWATH-MS could be applied to specifically establish peptides translated from splicing junctions [114]. This is also a direction worthy of further study in the future.

Plant immune receptors belonging to the receptor-like kinase (RLK) family play important roles in the recognition of microbial pathogens and activation of downstream defence responses. The function of RLKs in plant immune responses involves specific phosphorylation events within and outside the structural domain of the kinase, which leads to altered kinase activity and consequently to immune signalling. The function of several RLKs in pathogen resistance has been extensively studied in model plants with simpler genomes (e.g., rice and *Arabidopsis*). Examples include XA21 from rice [115,116] and PR5K from *Arabidopsis* [117]. The contribution of resistance to powdery mildew infection in durum wheat has been found by studying the expression and role of TtdLRK10 L-1 in wheat defence against powdery mildew infection and by identifying the intron putative MYB binding site (MYB-BS) in the positive role in the expression and function of TtdLRK10 L-1. Nucleotide binding site/leucine-rich repeat (NBS-LRR) and serine/threonine kinase (STK) genes are unit two of the known categories of resistance (R-) genes in plants and occur in massive multigene families. Some receptor-like kinases (RLKs) with serine/threonine were similar to both cytoplasmic RLKs, such as Pto, and RLKs with LRR, S-locus, lectin-like, and thaumatin-like extracellular binding-domains [118]. By analysing kinase domain-targeted isolation of defence-related receptor-like kinases (RLK/Pelle) in *Platanus × acerifolia*: phyletic and structural, we see that some RLKs have indeed been involved in the expression of phenotypic plasticity and are therefore a decent candidate for investigations into pathogen resistance [119].

PremRNA splicing plays an important role in the regulation of plant immunity mediated by the RLKs SNC4 and CERK1. Plant surface pattern receptors are variably spliced to produce different transcripts that respond to pathogen infestation by influencing downstream signalling. Bacterial flagellin 2 (FLS2) is a classical PRR receptor protein that detects conserved bacterial flagellin 2 (FLS2) in plants, sensing external bacterial infestation. Flagellin perception in Arabidopsis works through recognition of its extremely conserved N-terminal epitope (flg22) by flagellin-sensitive 2 (FLS2) to initiate a series of immune responses. In nine families of dicotyledons, FLS2 was shown to undergo AS of its first exon. Point mutations and gene swaps indicated that the position and potency of exitron splicing primarily relied on the nucleotide sequences of FLS2 genes. The position and efficiency of splicing depend mainly on the FLS2 nucleotide sequence of FLS2; exon of FLS2 is spliced via an intron-mediated enhancement (IME), which regulates the accumulation of transcriptional products. Some ATs have the potential to encode suppressors for the FLS2 pathway, and transformed transcripts can encode FLS2 pathway repressors that affect FLS2-mediated reactive oxygen species production. This study reveals that alternative splicing finely regulates the pattern of receptor recognition and thus influences the downstream disease resistance response [120]. The receptor-like cytoplasmic kinase BIK1, a component of the FLS2 immune receptor advanced, not solely positively regulates flg22-triggered calcium influx; it conjointly directly phosphorylates the NADPH oxidase RbohD at specific sites in an exceedingly calcium-independent manner to enhance ROS generation. However, at present, the relationship between some cytoplasmic kinases such as BIK1 and AS is unknown [121].

The *Arabidopsis thaliana* NPR1 loss-of-function mutant PR gene expression is reduced, and the ability to develop SAR loses the capacity to develop SAR. The backfill npr1 mutant to obtain the SNC (suppressor of npr1-1, constitutional) series of mutants can restore both abilities to some extent [122]. The *Arabidopsis thaliana* mutant SNC4 1D contains an RLK SNC4 (the NPR1-1 suppressor, CONSTITUTIVE4) gain-of-function mutation, which leads to constitutive activation of the defence response. Identification of two conserved shearing factors, SUA (ABI3-5 repressor gene) and RSN2 (SNC4-1D essential gene), are necessary to construct a defence response by mutating the snc4 1D mutant.

In SUA and RSN2 mutants, SNC4 shearing is altered, and further analysis showed that SUA and RSN2 are important receptors for shearing CERK1 (an essential receptor for PAMP, titin inducible receptor kinase 1). The precursor mRNA is sheared in the receptor-like kinase protein kinase SNC4 and CERK1-mediated regulation of plant immunity [123]. All these directly or indirectly express the key role played by alternative splicing in plant defence. Alternatively, spliced transcripts in plants result in profound changes in their gene expression patterns throughout developmental growth.

Interestingly, we found that AS events seem to have specificity across various plant species or in response to a variety of stressful environments. For resistance to different pathogenic bacteria, independent splice variants are also formed. IR was the most common AS pattern observed in *Paulownia tomentosa* [124]. Exon skipping was the primary AS pattern in *Populus* throughout salt stress, suggesting a completely different tree species produce different AS types to cope with constant stress. Apparently, an alternative acceptor is the major AS pattern when camellia sinensis is treated with drought and heat stress [125].

Arabidopsis RPS4 senses and is resistant to the Avr Rps4-expressing strain Pst.DC3000, and RPS4 undergoes variable splicing to produce six different transcripts. After infestation, RPS4 expression was upregulated in intact transcripts, whereas clonally incomplete transcripts could not backfill the resistance of the deficient RPS4 plants to *Pst. DC3000*, suggesting that only intact RPS4 transcripts are resistant to Pst [126]. In tobacco, the N gene specifically recognizes a 50 kD decapping enzyme protein (p50) of Tobaccomosaic virus (TMV) and selectively shears the N gene [127,128]. Different tree species manufacture completely different AS varieties to cope with a similar stresses. Curiously, an alternate acceptor is the major AS pattern when *C. sinensis* is treated with drought and heat stress [125]. Comparisons of gene structures from 67 plant species, protein domains, promoter regions, and conserved splicing patterns indicated that plant U1-70Ks are unable to preserve their preserved molecular function across plant lineages and play a very important practical role in response to environmental stresses [129]. A total of 4388 unique proteins were identified and quantified, among which 542 proteins showed vital abundance changes upon Pb(II) exposure, and differentially expressed proteins (DEPs) that were primarily distributed in the lignin and flavonoid synthesis pathways were powerfully activated upon Pb exposure, indicating their potential roles in Pb detoxification in poplar [130].

There are fewer reports related to alternative splicing, but it is easy to see that conserved AS events seem to be consistently present, including the feature that alternative splicing is specific under different stresses as well as under different species. We have also previously summarized the changes in microbial communities under different plants and different stresses, and it is unclear how the release of phytohormones relates to AS, including the recruitment of their preferred microbes by plants, so we speculate that AS may have its own unique splicing.

3.2. Genome-Wide Association Analyses of Splicing Quantitative Trait Loci (sQTLs) in Plants

Although AS is pervasive, the genetic basis for differential isoform usage in plants is increasing, widespread natural variation in AS has been observed in plants, and how AS is regulated and contributes to phenotypic variation is poorly understood. One study performed genome-wide analysis in 666 geographically distributed diverse ecotypes of *Arabidopsis thaliana* to spot genomic regions (splicing quantitative trait loci (sQTLs)) that regulated differential AS, and observed enrichment for trans-sQTLs (trans-sQTL hotspots) on chromosome regions. Many sQTLs were enriched, including the circadian clock, flowering, and stress-responsive genes, suggesting the potential role of differential isoform usage in controlling these necessary processes among diverse ecotypes of *Arabidopsis* [15]. The presence of widespread variation in diverse ecotypes of Arabidopsis at genetic level has been widely reported. For example, epithiospecifier protein (ESP) is responsible for diverting glucosinolate hydrolysis from the generation of isothiocyanates to that of epithionitriles or nitriles and thereby negatively affecting the capacity of the plant to defend itself against certain insects. Some studies have shown that ESP expression is regulated differently between the two *A. thaliana* ecotypes [131]. Comparison of AS in three phenotypic variants of maize revealed the importance of AS in diversifying gene function and regulating phenotypic variation as well [132] (Table 1).

3.3. Crosstalk between Plant Ecotype-Specific Splicing and Their Commensal Microbiome

Increasing evidence has suggested that the splice sites were also differentially selected among various plant ecotypes. Although the effect of genotypic variation on splice site determination among plant ecotypes is less reported, the sQTLs identified in several studies provide fundamental evidence to support this finding. Subsequently, an intriguing hypothesis has been proposed that the splicing variation among plant ecotypes may further influence the recruitment of the plant commensal microbiome by affecting crucial factors involved in secondary metabolism, root exudate secretion, and plant immune responses, etc. However, the role of these differential splice sites present among various plant ecotypes during the interaction of plants and their microbiomes remains unclear (Figure 1).

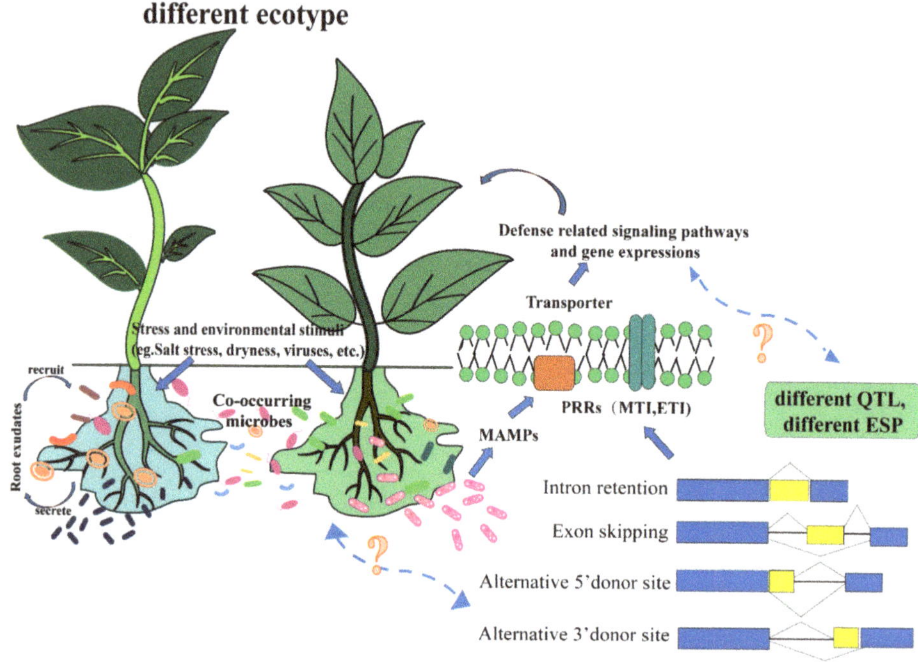

Figure 1. Summary model and research gaps among plant ecotypes, commensal microbiota, and alternative splicing-associated quantitative trait loci. Ecotype-specific microbial recruitment, and hypothesized relationship between plant induced alternatively spliced genes. Different plants, different genotypes, and different stresses will all release different root secretions and recruit microorganisms according to their needs to build their own unique communities in response to certain biotic or abiotic stresses. There is much evidence that PRRs in many plants are associated with AS. However, with different stresses, different plants have different preferences for AS, and different ecotypes of plants have differences in sQTLs that are enriched in flowering and stress response genes. Therefore, we speculate that alternative splicing is somehow associated with expression shape regulation and microbial recruitment, and plant-induced gene expression.

4. Conclusions and Future Perspective

In recent years, the study of plant microbiome interactions has become a hotspot in scientific research. Increasing evidence suggested that alternative splicing plays a pivotal role in plant–microbiome interactions, a dynamic process controlled by both factors generated by plants and microorganisms. Specifically, in this review article, we discussed the recruitment mechanism of commensal microbiome by plant hosts and proposed that the differential recruitment of the microbiome observed among plant ecotypes is closely linked to their splicing quantitative trait loci (sQTLs) identified from these plant ecotypes. For instance, a deep understanding of the genetic linkage between AS and secretion of root exudates for microbiome recruitment will be the key to mastering changes in plant ecology and plant breeding. However, further investigations are required to validate this hypothesis and proof of its value to form the new research direction in this field.

To this end, studies of both host and microbiome aspects are necessary to unravel the underlying mechanism between plant–microbiome interactions. In the host aspect particularly, the propensity for AS varies under different stress conditions, and unique splice variants are developed to fight against different pathogens. AS triggers intracellular defence signals and mass expression of defence genes, yet the interactions between different plant-associated symbionts and the plant immune system are unknown. Plants recruit microbes in a 'call for help' strategy, with different species recruiting different microbes and susceptible versus resistant species recruiting different microbial communities. One major challenge is going to be to research root exudation in natural settings. Due to the chemical complexity of soil, exudation is historically analysed in aquacultural culture, an environment distant from many natural settings of plant microbiome studies. In addition, new technologies are needed for high-throughput screening of functional microorganisms that can be targeted to reveal the impact of core microorganisms on the inter-root biota and, in turn, on plant health. With further understanding of root morphology and secretions, the core strains involved may alter the engineering or breeding of plants with altered talent to act against pathogens. This must be complemented with an improved understanding of the substrate preferences of plant-associated microbes, their interactions, and also the mechanisms which profit the plant. Although research on alternative splicing in plant disease resistance has been ongoing, there are still many unanswered questions about the role of alternative splicing in plant immunity.

Regarding commensal microbes, with the evolution of host–microbe interactions, microorganisms have evolved evolutionary strategies for manipulating host AS machinery to disrupt host immunity, and alternative splicing is involved in the mechanism of R-Avr interactions. Whether this mechanism is widespread in multiple species and in disease-resistance systems in which different microorganisms interact with plants has not been fully explained. The mechanism of alternative splicing in the R–Avr interaction, and whether it is widespread in multiple species and in different microbe–plant interactions in the disease resistance system, is also not fully understood; nor are the functional mechanisms of R proteins and subcellular localization. In the future, we need to investigate the functional mechanism, subcellular localization, and interaction with the target protein Avr to further expand our understanding of the molecular mechanism of plant resistance.

Author Contributions: Conceptualization, Z.C. and Z.-G.W.; writing original draft preparation, Y.-H.L. and Y.-Y.Y.; writing review and editing, Z.C. and Z.-G.W.; funding, Z.C.; The version of the manuscript was agreed by all authors. All authors have read and agreed to the published version of the manuscript.

Funding: This work was supported by National Key Research Development Program of China (2017YFD0200308) and Post-subsidy project of National Key Research Development Program of China (2018-5262), the National Natural Science Foundation of China (numbers 31860515 and 21977023), Program of Introducing Talents to Chinese Universities (111 Program, D20023) and the China Agriculture Research System (CARS-23-D09).

Institutional Review Board Statement: Not applicable.

Informed Consent Statement: Not applicable.

Data Availability Statement: Not applicable.

Conflicts of Interest: The authors declare no conflict of interest.

References

1. Köberl, M.; Schmidt, R.; Ramadan, E.M.; Bauer, R.; Berg, G. The microbiome of medicinal plants: Diversity and importance for plant growth, quality and health. *Front. Microbiol.* **2013**, *4*, 400. [CrossRef] [PubMed]
2. Bouffaud, M.; Poirier, M.; Muller, D.; Moënne-Loccoz, Y. Root microbiome relates to plant host evolution in maize and other Poaceae. *Environ. Microbiol.* **2014**, *16*, 2804–2814. [CrossRef] [PubMed]
3. Fitzpatrick, C.R.; Salas-González, I.; Conway, J.M.; Finkel, O.M.; Gilbert, S.; Russ, D.; Teixeira, P.; Dangl, J.L. The Plant Microbiome: From Ecology to Reductionism and Beyond. *Annu. Rev. Microbiol.* **2020**, *74*, 81–100. [CrossRef] [PubMed]
4. Kong, X.; Han, Z.; Tai, X.; Jin, D.; Ai, S.; Zheng, X.; Bai, Z. Maize (*Zea mays* L. Sp.) varieties significantly influence bacterial and fungal community in bulk soil, rhizosphere soil and phyllosphere. *FEMS Microbiol. Ecol.* **2020**, *96*, fiaa020. [CrossRef]
5. Sohn, S.I.; Ahn, J.H.; Pandian, S.; Oh, Y.J.; Shin, E.K.; Kang, H.J.; Cho, W.S.; Cho, Y.S.; Shin, K.S. Dynamics of Bacterial Community Structure in the Rhizosphere and Root Nodule of Soybean: Impacts of Growth Stages and Varieties. *Int. J. Mol. Sci.* **2021**, *22*, 5577. [CrossRef]
6. Zipfel, C.; Oldroyd, G.E. Plant signalling in symbiosis and immunity. *Nature* **2017**, *543*, 328–336. [CrossRef]
7. Suwabe, K.; Suzuki, G.; Watanabe, M. Achievement of genetics in plant reproduction research: The past decade for the coming decade. *Genes Genet. Syst.* **2010**, *85*, 297–310. [CrossRef]
8. Huang, X.F.; Chaparro, J.; Reardon, F.; Zhang, R.; Shen, Q.; Vivanco, J. Rhizosphere interactions: Root exudates, microbes and microbial communities. *Botany* **2014**, *92*, 4. [CrossRef]
9. Saleem, M.; Law, A.D.; Sahib, M.R.; Pervaiz, Z.H.; Zhang, Q. Impact of root system architecture on rhizosphere and root microbiome. *Rhizosphere* **2018**, *6*, 47–51. [CrossRef]
10. Rigo, R.; Bazin, J.R.M.; Crespi, M.; Charon, C.L. Alternative Splicing in the Regulation of Plant-Microbe Interactions. *Plant Cell Physiol.* **2019**, *60*, 1906–1916. [CrossRef]
11. Kornblihtt, A.R.; Schor, I.E.; Alló, M.; Dujardin, G.; Petrillo, E.; Muñoz, M.J. Alternative splicing: A pivotal step between eukaryotic transcription and translation. *Nat. Rev. Mol. Cell Biol.* **2013**, *14*, 153–165. [CrossRef] [PubMed]
12. Wang, E.T.; Sandberg, R.; Luo, S.; Khrebtukova, I.; Zhang, L.; Mayr, C.; Kingsmore, S.F.; Schroth, G.P.; Burge, C.B. Alternative isoform regulation in human tissue transcriptomes. *Nature* **2008**, *456*, 470–476. [CrossRef] [PubMed]
13. Takata, A.; Matsumoto, N.; Kato, T. Genome-wide identification of splicing QTLs in the human brain and their enrichment among schizophrenia-associated loci. *Nat. Commun.* **2017**, *8*, 14519. [CrossRef] [PubMed]
14. Reddy, A.; Marquez, Y.; Kalyna, M.; Barta, A. Complexity of the Alternative Splicing Landscape in Plants. *Plant Cell* **2013**, *25*, 3657–3683. [CrossRef] [PubMed]
15. Khokhar, W.; Hassan, M.A.; Reddy, A.; Chaudhary, S.; Jabre, I.; Byrne, L.J.; Syed, N.H. Genome-Wide Identification of Splicing Quantitative Trait Loci (sQTLs) in Diverse Ecotypes of Arabidopsis thaliana. *ISO4* **2019**, *10*, 1160. [CrossRef]
16. Iida, K.; Go, M. Survey of Conserved Alternative Splicing Events of mRNAs Encoding SR Proteins in Land Plants. *Mol. Biol. Evol.* **2006**, *23*, 1085–1094. [CrossRef]
17. Feng, J.; Li, J.; Gao, Z.; Lu, Y.; Yu, J.; Zheng, Q.; Yan, S.; Zhang, W.; He, H.; Ma, L. SKIP Confers Osmotic Tolerance during Salt Stress by Controlling Alternative Gene Splicing in Arabidopsis. *Mol. Plant* **2015**, *8*, 1038–1052. [CrossRef]
18. Jie, H.; Gu, L.; Ying, Z.; Yan, T.; Kong, G.; Liang, K.; Guo, B.; Min, Q.; Yang, W.; Jing, M. An oomycete plant pathogen reprograms host pre-mRNA splicing to subvert immunity. *Nat. Commun.* **2017**, *8*, 2051.
19. Delmont, T.O.; Robe, P.; Cecillon, S.; Clark, I.M.; Constancias, F.; Simonet, P.; Hirsch, P.R.; Vogel, T.M. Accessing the soil metagenome for studies of microbial diversity. *Appl. Environ. Microbiol.* **2011**, *77*, 1315–1324. [CrossRef]
20. Vivanco, L.; Rascovan, N.; Austin, A.T. Plant, fungal, bacterial, and nitrogen interactions in the litter layer of a native Patagonian forest. *PeerJ* **2018**, *6*, e4754. [CrossRef]
21. Teste, F.P.; Kardol, P.; Turner, B.L.; Wardle, D.A.; Zemunik, G.; Renton, M.; Laliberté, E. Plant-soil feedback and the maintenance of diversity in Mediterranean-climate shrublands. *Science* **2017**, *355*, 173–176. [CrossRef] [PubMed]
22. Hamonts, K.; Trivedi, P.; Garg, A.; Janitz, C.; Grinyer, J.; Holford, P.; Botha, F.C.; Anderson, I.C.; Singh, B.K. Field study reveals core plant microbiota and relative importance of their drivers. *Environ. Microbiol.* **2018**, *20*, 124–140. [CrossRef] [PubMed]
23. Xu, J.; Zhang, Y.; Zhang, P.; Trivedi, P.; Riera, N.; Wang, Y.; Liu, X.; Fan, G.; Tang, J.; Coletta-Filho, H.D.; et al. The structure and function of the global citrus rhizosphere microbiome. *Nat. Commun.* **2018**, *9*, 4894. [CrossRef] [PubMed]
24. Borrel, G.; Brugère, J.F.; Gribaldo, S.; Schmitz, R.A.; Moissl-Eichinger, C. The host-associated archaeome. *Nat. Rev. Microbiol.* **2020**, *18*, 622–636. [CrossRef]
25. Cordovez, V.; Dini-Andreote, F.; Carrión, V.; Raaijmakers, J.M. Ecology and Evolution of Plant Microbiomes. *Annu. Rev. Microbiol.* **2019**, *73*, 69. [CrossRef]
26. Van der Heijden, M.G.; Hartmann, M. Networking in the Plant Microbiome. *PLoS Biol.* **2016**, *14*, e1002378. [CrossRef]
27. Ashrafuzzaman; Hossen; Fa; Ismail; Mr; Hoque; Ma; Islam; Mz; Shahidullah, Efficiency of plant growth-promoting rhizobacteria (PGPR) for the enhancement of rice growth. *Afr. J. Biotechnol.* **2009**, *8*, 1247–1252.
28. Wei, G.; Li, M.; Shi, W.; Tian, R.; Gao, Z. Similar drivers but different effects lead to distinct ecological patterns of soil bacterial and archaeal communities. *Soil Biol. Biochem.* **2020**, *144*, 107759. [CrossRef]
29. Anna, K.; Pijl, A.S.; Van, V.J.A.; Kowalchuk, G.A. Differences in vegetation composition and plant species identity lead to only minor changes in soil-borne microbial communities in a former arable field. *FEMS Microbiol. Ecol.* **2010**, *63*, 372–382.

30. Hartmann, A.; Fischer, D.; Kinzel, L.; Chowdhury, S.P.; Hofmann, A.; Baldani, J.I.; Rothballer, M. Assessment of the structural and functional diversities of plant microbiota: Achievements and challenges—A review. *J. Adv. Res.* **2019**, *19*, 3–13. [CrossRef]

31. Pétriacq, P.; Williams, A.; Cotton, A.; McFarlane, A.E.; Rolfe, S.A.; Ton, J. Metabolite profiling of non-sterile rhizosphere soil. *Plant J.* **2017**, *92*, 147–162. [CrossRef] [PubMed]

32. Thomas, P.; Franco, C.M.M. Intracellular Bacteria in Plants: Elucidation of Abundant and Diverse Cytoplasmic Bacteria in Healthy Plant Cells Using In Vitro Cell and Callus Cultures. *Microorganisms* **2021**, *9*, 269. [CrossRef] [PubMed]

33. Armada, E.; Portela, G.; Roldán, A.; Azcón, R. Combined use of beneficial soil microorganism and agrowaste residue to cope with plant water limitation under semiarid conditions. *Geoderma* **2014**, *232–234*, 640–648. [CrossRef]

34. Calvo, P.; Nelson, L.; Kloepper, J.W. Agricultural uses of plant biostimulants. *Plant Soil* **2014**, *383*, 3–41. [CrossRef]

35. Bukhat, S.; Imran, A.; Javaid, S.; Shahid, M.; Majeed, A.; Naqqash, T. Communication of plants with microbial world: Exploring the regulatory networks for PGPR mediated defense signaling. *Microbiol. Res.* **2020**, *238*, 126486. [CrossRef]

36. Compant, S.; Duffy, B.; Nowak, J.; Clément, C.; Barka, E.A. Use of plant growth-promoting bacteria for biocontrol of plant diseases: Principles, mechanisms of action, and future prospects. *Appl. Environ. Microbiol.* **2005**, *71*, 4951–4959. [CrossRef]

37. Barea, J.M.; Pozo, M.J.; Azcón, R.; Azcón-Aguilar, C. Microbial co-operation in the rhizosphere. *J. Exp. Bot.* **2005**, *56*, 1761–1778. [CrossRef]

38. Yi; Zhou; Yong-Lark; Choi; Ming; SunZiniu; Yu, Novel roles of Bacillus thuringiensis to control plant diseases. *Appl. Microbiol. Biotechnol.* **2008**, *80*, 563–572. [CrossRef]

39. Lau, J.A.; Lennon, J.T. Rapid responses of soil microorganisms improve plant fitness in novel environments. *Proc. Natl. Acad. Sci. USA* **2012**, *109*, 14058–14062. [CrossRef]

40. Lee, B.; Lee, S.; Ryu, C.M. Foliar aphid feeding recruits rhizosphere bacteria and primes plant immunity against pathogenic and non-pathogenic bacteria in pepper. *Ann. Bot.* **2012**, *110*, 281–290. [CrossRef]

41. Chen, Y.; Wang, J.; Yang, N.; Wen, Z.; Sun, X.; Chai, Y.; Ma, Z. Wheat microbiome bacteria can reduce virulence of a plant pathogenic fungus by altering histone acetylation. *Nat. Commun.* **2018**, *9*, 3429. [CrossRef] [PubMed]

42. Berendsen, R.L.; Vismans, G.; Yu, K.; Song, Y.; de Jonge, R.; Burgman, W.P.; Burmølle, M.; Herschend, J.; Bakker, P.; Pieterse, C.M.J. Disease-induced assemblage of a plant-beneficial bacterial consortium. *ISME J.* **2018**, *12*, 1496–1507. [CrossRef] [PubMed]

43. Raaijmakers, J.M.; Bonsall, R.F.; Weller, D.M. Effect of Population Density of Pseudomonas fluorescens on Production of 2,4-Diacetylphloroglucinol in the Rhizosphere of Wheat. *Phytopathology* **1999**, *89*, 470–475. [CrossRef] [PubMed]

44. Yin, C.; Casa Vargas, J.M.; Schlatter, D.C.; Hagerty, C.H.; Hulbert, S.H.; Paulitz, T.C. Rhizosphere community selection reveals bacteria associated with reduced root disease. *Microbiome* **2021**, *9*, 86. [CrossRef]

45. Kwak, M.J.; Kong, H.G.; Choi, K.; Kwon, S.K.; Song, J.Y.; Lee, J.; Lee, P.A.; Choi, S.Y.; Seo, M.; Lee, H.J.; et al. Rhizosphere microbiome structure alters to enable wilt resistance in tomato. *Nat. Biotechnol.* **2018**, *36*, 1100. [CrossRef]

46. Dudenhffer, J.H.; Scheu, S.; Jousset, A. Systemic enrichment of antifungal traits in the rhizosphere microbiome after pathogen attack. *J. Ecol.* **2016**, *104*, 1566–1575. [CrossRef]

47. Timm, C.M.; Carter, K.R.; Carrell, A.A.; Jun, S.R.; Weston, D.J. Abiotic Stresses Shift Belowground Populus -Associated Bacteria Toward a Core Stress Microbiome. *mSystems* **2018**, *3*, e00070-17. [CrossRef]

48. Santos-Medellín, C.; Edwards, J.; Liechty, Z.; Bao, N.; Sundaresan, V. Drought Stress Results in a Compartment-Specific Restructuring of the Rice Root-Associated Microbiomes. *mBio* **2017**, *8*, e00764-17. [CrossRef]

49. Lucía, F.; Ana, F.S. Strong shift in the diazotrophic endophytic bacterial community inhabiting rice (Oryza sativa) plants after flooding. *FEMS Microbiol. Ecol.* **2015**, *91*, fiv104.

50. Gardener, M.S.; Weller, D.M. Changes in Populations of Rhizosphere Bacteria Associated with Take-All Disease of Wheat. *Appl. Environ. Microbiol.* **2001**, *67*, 4414–4425. [CrossRef]

51. Durán, P.; Thiergart, T.; Garrido-Oter, R.; Agler, M.; Kemen, E.; Schulze-Lefert, P.; Hacquard, S. Microbial Interkingdom Interactions in Roots Promote Arabidopsis Survival. *Cell* **2018**, *175*, 973–983. [CrossRef] [PubMed]

52. Finkel, O.M.; Castrillo, G.; Paredes, S.H.; González, I.; Dangl, J.L. Understanding and exploiting plant beneficial microbes— ScienceDirect. *Curr. Opin. Plant Biol.* **2017**, *38*, 155–163. [CrossRef] [PubMed]

53. Castrillo, G.; Teixeira, P.J.; Paredes, S.H.; Law, T.F.; de Lorenzo, L.; Feltcher, M.E.; Finkel, O.M.; Breakfield, N.W.; Mieczkowski, P.; Jones, C.D.; et al. Root microbiota drive direct integration of phosphate stress and immunity. *Nature* **2017**, *543*, 513–518. [CrossRef] [PubMed]

54. Maureen, B.; Britt, K. Nutrient- and Dose-Dependent Microbiome-Mediated Protection against a Plant Pathogen. *Curr. Biol.* **2018**, *28*, 2487.

55. Pieterse, C.M.; Zamioudis, C.; Berendsen, R.L.; Weller, D.M.; Van Wees, S.C.; Bakker, P.A. Induced systemic resistance by beneficial microbes. *Annu. Rev. Phytopathol.* **2014**, *52*, 347–375. [CrossRef]

56. Balmer, D.; Planchamp, C.; Mauch-Mani, B. On the move: Induced resistance in monocots. *J. Exp. Bot.* **2013**, *64*, 1249–1261. [CrossRef]

57. Gill, S.S.; Gill, R.; Trivedi, D.K.; Anjum, N.A.; Sharma, K.K.; Ansari, M.W.; Ansari, A.A.; Johri, A.K.; Prasad, R.; Pereira, E.; et al. Piriformospora indica: Potential and Significance in Plant Stress Tolerance. *Front. Microbiol.* **2016**, *7*, 332. [CrossRef]

58. Newitt, J.T.; Prudence, S.M.M.; Hutchings, M.I.; Worsley, S.F. Biocontrol of Cereal Crop Diseases Using Streptomycetes. *Pathogens* **2019**, *8*, 78. [CrossRef]

59. Aloo, B.N.; Makumba, B.A.; Mbega, E.R. The potential of Bacilli rhizobacteria for sustainable crop production and environmental sustainability. *Microbiol. Res.* **2019**, *219*, 26–39. [CrossRef]

60. Helfrich, E.J.N.; Vogel, C.M.; Ueoka, R.; Schäfer, M.; Ryffel, F.; Müller, D.B.; Probst, S.; Kreuzer, M.; Piel, J.; Vorholt, J.A. Bipartite interactions, antibiotic production and biosynthetic potential of the Arabidopsis leaf microbiome. *Nat. Microbiol.* **2018**, *3*, 909–919. [CrossRef]

61. Syed Ab Rahman, S.F.; Singh, E.; Pieterse, C.M.J.; Schenk, P.M. Emerging microbial biocontrol strategies for plant pathogens. *Plant Sci.* **2018**, *267*, 102–111. [CrossRef] [PubMed]

62. Chhabra, R.; Kaur, S.; Vij, L.; Gaur, K. Exploring Physiological and Biochemical Factors Governing Plant Pathogen Interaction: A Review. *Int. J. Curr. Microbiol. Appl. Sci.* **2020**, *9*, 1650–1666. [CrossRef]

63. Boyd, L.A.; Ridout, C.; O'Sullivan, D.M.; Leach, J.E.; Leung, H. Plant–pathogen interactions: Disease resistance in modern agriculture—ScienceDirect. *Trends Genet.* **2013**, *29*, 233–240. [CrossRef] [PubMed]

64. Jones, J.; Dangl, J.L. The plant immune system. *Nature* **2006**, *444*, 323–329. [CrossRef] [PubMed]

65. Spoel, S.H.; Dong, X. How do plants achieve immunity? Defence without specialized immune cells. *Nat. Rev. Immunol.* **2012**, *12*, 89–100. [CrossRef]

66. Vlot, A.C.; Dempsey, D.A.; Klessig, D.F. Salicylic Acid, a multifaceted hormone to combat disease. *Annu. Rev. Phytopathol.* **2009**, *47*, 177–206. [CrossRef] [PubMed]

67. Conrath, U.; Beckers, G.J.; Langenbach, C.J.; Jaskiewicz, M.R. Priming for enhanced defense. *Annu. Rev. Phytopathol.* **2015**, *53*, 97–119. [CrossRef] [PubMed]

68. Luna, E.; Bruce, T.J.; Roberts, M.R.; Flors, V.; Ton, J. Next-generation systemic acquired resistance. *Plant Physiol.* **2012**, *158*, 844–853. [CrossRef]

69. Carvalhais, L.C.; Dennis, P.G.; Badri, D.V.; Kidd, B.N.; Schenk, P.M. Linking Jasmonic Acid Signaling, Root Exudates, and Rhizosphere Microbiomes. *Mol. Plant-Microbe Interact.* **2015**, *28*, 1049. [CrossRef]

70. Lebeis, S.L.; Paredes, S.H.; Lundberg, D.S.; Breakfield, N.; Gehring, J.; McDonald, M.; Malfatti, S.; Glavina del Rio, T.; Jones, C.D.; Tringe, S.G.; et al. Salicylic acid modulates colonization of the root microbiome by specific bacterial taxa. *Science* **2015**, *349*, 860–864. [CrossRef]

71. Donn, S.; Kirkegaard, J.A.; Perera, G.; Richardson, A.E.; Watt, M. Evolution of bacterial communities in the wheat crop rhizosphere. *Environ. Microbiol.* **2014**, *17*, 610–621. [CrossRef] [PubMed]

72. Yin, C.; Schlatter, D.; Schroeder, K.; Mueth, N.; Paulitz, T.C. Bacterial Communities on Wheat Grown Under Long-Term Conventional Tillage and No-Till in the Pacific Northwest of the United States. *Phytobiomes J.* **2017**, *1*, 83–90. [CrossRef]

73. Hartman, K.; van der Heijden, M.G.; Wittwer, R.L.A.; Banerjee, S.; Walser, J.C.; Schlaeppi, K. Cropping practices manipulate abundance patterns of root and soil microbiome members paving the way to smart farming. *Microbiome* **2018**, *6*, 14. [CrossRef] [PubMed]

74. Mavrodi, D.V.; Mavrodi, O.V.; Elbourne, L.; Sasha, T.; Bonsall, R.F.; James, P.; Yang, M.; Paulsen, I.T.; Weller, D.M.; Thomashow, L.S. Long-Term Irrigation Affects the Dynamics and Activity of the Wheat Rhizosphere Microbiome. *Front. Plant Sci.* **2018**, *9*, 345. [CrossRef] [PubMed]

75. Lundberg, D.S.; Lebeis, S.L.; Paredes, S.H.; Yourstone, S.; Gehring, J.; Malfatti, S.; Tremblay, J.; Engelbrektson, A.; Kunin, V.; Del Rio, T.G.; et al. Defining the core Arabidopsis thaliana root microbiome. *Nature* **2012**, *488*, 86–90. [CrossRef]

76. Marie, S.; Cindy, D.; Valeria, T.; Ngonkeu, E.; Diégane, D.; Aboubacry, K.; Gilles, B.; Lionel, M. Influence of plant genotype and soil on the wheat rhizosphere microbiome: Evidences for a core microbiome across eight African and European soils. *FEMS Microbiol. Ecol.* **2020**, *96*, fiaa067.

77. Aira, M.; Gómez-Brandón, M.; Lazcano, C.; BaaTh, E.; Domínguez, J. Plant genotype strongly modifies the structure and growth of maize rhizosphere microbial communities. *Soil Biol. Biochem.* **2010**, *42*, 2276–2281. [CrossRef]

78. Lamit, L.J. COS 42-3: Different cottonwood genotypes support different aboveground fungal communities: A genetic basis to community structure. In Proceedings of the The ESA/SER Joint Meeting, San Jose, CA, USA, 5–10 August 2007.

79. Jiang, Y.; Li, S.; Li, R.; Zhang, J.; Liu, Y.; Lv, L.; Zhu, H.; Wu, W.; Li, W. Plant cultivars imprint the rhizosphere bacterial community composition and association networks. *Soil Biol. Biochem.* **2017**, *109*, 145–155. [CrossRef]

80. Macdonald, L.M.; Paterson, E.; Dawson, L.A.; Mcdonald, A. Defoliation and fertiliser influences on the soil microbial community associated with two contrasting Lolium perenne cultivars. *Soil Biol. Biochem.* **2006**, *38*, 674–682. [CrossRef]

81. Hou, D.; Wang, R.; Gao, X.; Wang, K.; Lin, Z.; Ge, J.; Liu, T.; Wei, S.; Chen, W.; Xie, R. Cultivar-specific response of bacterial community to cadmium contamination in the rhizosphere of rice (*Oryza sativa* L.). *Environ. Pollut.* **2018**, *241*, 63–73. [CrossRef]

82. Edwards, J.; Johnson, C.; Santos-Medellín, C.; Lurie, E.; Sundaresan, V. Structure, variation, and assembly of the root-associated microbiomes of rice. *Proc. Natl. Acad. Sci. USA* **2015**, *112*, E911–E920. [CrossRef]

83. Huang, X.P.; Mo, C.H.; Yu, J.; Zhao, H.M.; Meng, C.; Li, Y.W.; Li, H.; Cai, Q.Y.; Wong, M.H. Variations in microbial community and ciprofloxacin removal in rhizospheric soils between two cultivars of *Brassica parachinensis* L. *Sci. Total Environ.* **2017**, *603–604*, 66–76. [CrossRef] [PubMed]

84. Xue; Huang; Xd, Changes in soil microbial community structure with planting years and cultivars of tree peony (*Paeonia suffruticosa*). *World J. Microb. Biot.* **2014**, *30*, 389–397. [CrossRef] [PubMed]

85. Zhou, X.; Guan, S.; Wu, F. Composition of soil microbial communities in the rhizosphere of cucumber cultivars with differing nitrogen acquisition efficiency. *Appl. Soil Ecol.* **2015**, *95*, 90–98. [CrossRef]

86. Mahoney, A.K.; Yin, C.; Hulbert, S.H. Community Structure, Species Variation, and Potential Functions of Rhizosphere-Associated Bacteria of Different Winter Wheat (*Triticum aestivum*) Cultivars. *Front. Plant Sci.* **2017**, *8*, 132. [CrossRef] [PubMed]

87. Liu, H.; Brettell, L.E. Plant Defense by VOC-Induced Microbial Priming—ScienceDirect. *Trends Plant Sci.* **2019**, *24*, 187–189. [CrossRef]

88. Bakker, P.; Pieterse, C.; Jonge, R.D.; Berendsen, R.L. The Soil-Borne Legacy. *Cell* **2018**, *172*, 1178–1180. [CrossRef]

89. Carrión, V.; Perez-Jaramillo, J.; Cordovez, V.; Tracanna, V.; Raaijmakers, J.M. Pathogen-induced activation of disease-suppressive functions in the endophytic root microbiome. *Science* **2019**, *366*, 606–612. [CrossRef]

90. Mendes, L.W.; Raaijmakers, J.M.; de Hollander, M.; Mendes, R.; Tsai, S.M. Influence of resistance breeding in common bean on rhizosphere microbiome composition and function. *ISME J.* **2018**, *12*, 212–224. [CrossRef]

91. Chiang, Y.M.; Chang, S.L.; Oakley, B.R.; Wang, C.C. Recent advances in awakening silent biosynthetic gene clusters and linking orphan clusters to natural products in microorganisms. *Curr. Opin. Chem. Biol.* **2011**, *15*, 137–143. [CrossRef]

92. Fil, M.; Velasco-Rodríguez, Ó.; García-Calvo, L.; Sola-Landa, A.; Barreiro, C. Screening of Antibiotic Gene Clusters in Microorganisms Isolated from Wood. *Methods Mol. Biol.* **2021**, *2296*, 151–165. [PubMed]

93. Mathesius, H.U. The role of flavonoids in root-rhizosphere signalling: Opportunities and challenges for improving plant-microbe interactions. *J. Exp. Bot.* **2012**, *63*, 3429–3444.

94. Kudjordjie, E.N.; Sapkota, R.; Steffensen, S.K.; Fomsgaard, I.S.; Nicolaisen, M. Maize synthesized benzoxazinoids affect the host associated microbiome. *Microbiome* **2019**, *7*, 59. [CrossRef] [PubMed]

95. Lanoue, A.; Burlat, V.; Henkes, G.J.; Koch, I.; Schurr, U.; Röse, U.S. De novo biosynthesis of defense root exudates in response to Fusarium attack in barley. *New Phytol.* **2010**, *185*, 577–588. [CrossRef] [PubMed]

96. Huang, A.C.; Jiang, T.; Liu, Y.X.; Bai, Y.C.; Reed, J.; Qu, B.; Goossens, A.; Nützmann, H.W.; Bai, Y.; Osbourn, A. A specialized metabolic network selectively modulates Arabidopsis root microbiota. *Science* **2019**, *364*, 546. [CrossRef]

97. Van Deynze, A.; Zamora, P.; Delaux, P.M.; Heitmann, C.; Jayaraman, D.; Rajasekar, S.; Graham, D.; Maeda, J.; Gibson, D.; Schwartz, K.D.; et al. Nitrogen fixation in a landrace of maize is supported by a mucilage-associated diazotrophic microbiota. *PLoS Biol.* **2018**, *16*, e2006352. [CrossRef]

98. Schulz-Bohm, K.; Gerards, S.; Hundscheid, M.; Melenhorst, J.; de Boer, W.; Garbeva, P. Calling from distance: Attraction of soil bacteria by plant root volatiles. *ISME J.* **2018**, *12*, 1252–1262. [CrossRef]

99. Nchgesang, S.M.; Strehmel, N.; Schmidt, S.; Westphal, L.; Taruttis, F.; Müller, E.; Herklotz, S.; Neumann, S.; Scheel, D. Natural variation of root exudates in Arabidopsis thaliana-linking metabolomic and genomic data. *Sci. Rep.* **2016**, *6*, 29033. [CrossRef]

100. Carvalhais, L.C.; Dennis, P.G.; Fan, B.; Fedoseyenko, D.; Borriss, R. Linking Plant Nutritional Status to Plant-Microbe Interactions. *PLoS ONE* **2013**, *8*, e68555.

101. Jrg, Z.; Stephan, S.; Ranju, C.; Jens, M.; Christoph, B.; Nadine, S.; Dierk, S.; Steffen, A. Non-targeted profiling of semi-polar metabolites in Arabidopsis root exudates uncovers a role for coumarin secretion and lignification during the local response to phosphate limitation. *J. Exp. Bot.* **2016**, *67*, 1421–1432.

102. Ziegler, J.; Schmidt, S.; Strehmel, N.; Scheel, D.; Abel, S. Arabidopsis Transporter ABCG37/PDR9 contributes primarily highly oxygenated Coumarins to Root Exudation. *Sci. Rep.* **2017**, *7*, 3704. [CrossRef] [PubMed]

103. Brunner, H.I. Exudation of organic acid anions from poplar roots after exposure to Al, Cu and Zn. *Tree Physiol.* **2007**, *27*, 313.

104. Howard, B.E.; Hu, Q.; Babaoglu, A.C.; Chandra, M.; Borghi, M.; Tan, X.; He, L.; Winter-Sederoff, H.; Gassmann, W.; Veronese, P.; et al. High-Throughput RNA Sequencing of Pseudomonas-Infected Arabidopsis Reveals Hidden Transcriptome Complexity and Novel Splice Variants. *PLoS ONE* **2013**, *8*, e74183.

105. Mandadi, K.K.; Scholthof, K.B.G. Genome-Wide Analysis of Alternative Splicing Landscapes Modulated during Plant-Virus Interactions in Brachypodium distachyon. *Plant Cell* **2015**, *27*, 71–85. [CrossRef]

106. Juan, S.; Hui, L.; Huifu, Z.; Chunxia, Z.; Yuxing, X.; Shibo, W.; Jinfeng, Q.; Jing, L.; Christian, H.; Jianqiang, W. Transcriptomics and Alternative Splicing Analyses Reveal Large Differences between Maize Lines B73 and Mo17 in Response to Aphid Rhopalosiphum padi Infestation. *Front. Plant Sci.* **2017**, *8*, 1738.

107. Zheng, Y.; Wang, Y.; Ding, B.; Fei, Z. Comprehensive Transcriptome Analyses Reveal that Potato Spindle Tuber Viroid Triggers Genome-Wide Changes in Alternative Splicing, Inducible trans-Acting Activity of Phased Secondary Small Interfering RNAs, and Immune Responses. *J. Virol.* **2017**, *91*, e00247-17. [CrossRef]

108. Chen, M.X.; Zhang, K.L.; Zhang, M.; Das, D.; Zhu, F.Y. Alternative splicing and its regulatory role in woody plants. *Tree Physiol.* **2020**, *40*, 1475–1486. [CrossRef]

109. Chen, M.X.; Zhu, F.Y.; Gao, B.; Ma, K.L.; Zhang, Y.; Fernie, A.R.; Chen, X.; Dai, L.; Ye, N.H.; Zhang, X.; et al. Full-Length Transcript-Based Proteogenomics of Rice Improves Its Genome and Proteome Annotation. *Plant Physiol.* **2020**, *182*, 1510–1526. [CrossRef]

110. Kuang, Z.; Boeke, J.D.; Canzar, S. The dynamic landscape of fission yeast meiosis alternative-splice isoforms. *Genome Res.* **2017**, *27*, 145–156. [CrossRef]

111. Chen, M.X.; Mei, L.C.; Wang, F.; Boyagane Dewayalage, I.K.W.; Yang, J.F.; Dai, L.; Yang, G.F.; Gao, B.; Cheng, C.L.; Liu, Y.G.; et al. PlantSPEAD: A web resource towards comparatively analysing stress-responsive expression of splicing-related proteins in plant. *Plant Biotechnol. J.* **2021**, *19*, 227–229. [CrossRef]

112. Gillet, L.C.; Navarro, P.; Tate, S.; Röst, H.; Selevsek, N.; Reiter, L.; Bonner, R.; Aebersold, R. Targeted data extraction of the MS/MS spectra generated by data-independent acquisition: A new concept for consistent and accurate proteome analysis. *Mol. Cell. Proteom.* **2012**, *11*, O111.016717. [CrossRef] [PubMed]

113. Chen, M.X.; Zhang, Y.; Fernie, A.R.; Liu, Y.G.; Zhu, F.Y. SWATH-MS-Based Proteomics: Strategies and Applications in Plants. *Trends Biotechnol.* **2021**, *39*, 433–437. [CrossRef] [PubMed]

114. Zhu, F.Y.; Chen, M.X.; Ye, N.H.; Shi, L.; Ma, K.L.; Yang, J.F.; Cao, Y.Y.; Zhang, Y.; Yoshida, T.; Fernie, A.R.; et al. Proteogenomic analysis reveals alternative splicing and translation as part of the abscisic acid response in Arabidopsis seedlings. *Plant J.* **2017**, *91*, 518–533. [CrossRef] [PubMed]

115. Park, C.J.; Han, S.W.; Chen, X.; Ronald, P.C. Elucidation of XA21-mediated innate immunity. *Cell Microbiol.* **2010**, *12*, 1017–1025. [CrossRef] [PubMed]

116. Thomas, N.C.; Schwessinger, B.; Liu, F.; Chen, H.; Wei, T.; Nguyen, Y.P.; Shaker, I.W.F.; Ronald, P.C. XA21-specific induction of stress-related genes following Xanthomonas infection of detached rice leaves. *PeerJ* **2016**, *4*, e2446. [CrossRef] [PubMed]

117. Wang, X.; Zafian, P.; Choudhary, M.; Lawton, M. The PR5K receptor protein kinase from Arabidopsis thaliana is structurally related to a family of plant defense proteins. *Proc. Natl. Acad. Sci. USA* **1996**, *93*, 2598–2602. [CrossRef] [PubMed]

118. Di Gaspero, G.; Cipriani, G. Nucleotide binding site/leucine-rich repeats, Pto-like and receptor-like kinases related to disease resistance in grapevine. *Mol. Genet. Genom.* **2003**, *269*, 612–623. [CrossRef]

119. Pilotti, M.; Brunetti, A.; Uva, P.; Lumia, V.; Tizzani, L.; Gervasi, F.; Iacono, M.; Pindo, M. Kinase domain-targeted isolation of defense-related receptor-like kinases (RLK/Pelle) in Platanus × acerifolia: Phylogenetic and structural analysis. *BMC Res. Notes* **2014**, *7*, 884. [CrossRef]

120. Cheng, Q.; Xiao, H.; Xiong, Q. Conserved exitrons of FLAGELLIN-SENSING 2 (FLS2) across dicot plants and their functions. *Plant Sci.* **2020**, *296*, 110507. [CrossRef]

121. Li, L.; Li, M.; Yu, L.; Zhou, Z.; Liang, X.; Liu, Z.; Cai, G.; Gao, L.; Zhang, X.; Wang, Y.; et al. The FLS2-associated kinase BIK1 directly phosphorylates the NADPH oxidase RbohD to control plant immunity. *Cell Host Microbe* **2014**, *15*, 329–338. [CrossRef]

122. Zhanga, Y.; Li, X. A Putative Nucleoporin 96 Is Required for Both Basal Defense and Constitutive Resistance Responses Mediated by suppressor of npr1-1, constitutive 1. *Plant Cell* **2005**, *17*, 1306–1316. [CrossRef] [PubMed]

123. Zhang, Z.; Liu, Y.; Ding, P.; Yan, L.; Kong, Q.; Zhang, Y. Splicing of Receptor-Like Kinase-Encoding SNC4 and CERK1 is Regulated by Two Conserved Splicing Factors that Are Required for Plant Immunity. *Mol. Plant* **2014**, *7*, 1766–1775. [CrossRef] [PubMed]

124. Zhao, Z.; Li, Y.; Liu, H.; Zhai, X.; Deng, M.; Dong, Y.; Fan, G. Genome-wide expression analysis of salt-stressed diploid and autotetraploid Paulownia tomentosa. *PLoS ONE* **2017**, *12*, e0185455. [CrossRef] [PubMed]

125. Ding, Y.; Wang, Y.; Qiu, C.; Qian, W.; Xie, H.; Ding, Z. Alternative splicing in tea plants was extensively triggered by drought, heat and their combined stresses. *PeerJ* **2020**, *8*, e8258. [CrossRef]

126. Zhang, X.C.; Gassmann, W. RPS4-mediated disease resistance requires the combined presence of RPS4 transcripts with full-length and truncated open reading frames. *Plant Cell* **2003**, *15*, 2333–2342. [CrossRef]

127. Whitham, S.; Dinesh-Kumar, S.P.; Choi, D.; Hehl, R.; Corr, C.; Baker, B. The product of the tobacco mosaic virus resistance gene N: Similarity to toll and the interleukin-1 receptor. *Cell* **1994**, *1*, S163–S164. [CrossRef]

128. Erickson, F.L.; Holzberg, S.; Calderon-Urrea, A.; Handley, V.; Baker, B. The helicase domain of the TMV replicase proteins induces the N-mediated defence response in tobacco. *Plant J.* **2010**, *18*, 67–75. [CrossRef]

129. Chen, M.X.; Zhang, K.L.; Gao, B.; Yang, J.F.; Tian, Y.; Das, D.; Fan, T.; Dai, L.; Hao, G.F.; Yang, G.F.; et al. Phylogenetic comparison of 5′ splice site determination in central spliceosomal proteins of the U1-70K gene family, in response to developmental cues and stress conditions. *Plant J.* **2020**, *103*, 357–378. [CrossRef]

130. Shen, C.C.; Chen, M.X.; Xiao, T.; Zhang, C.; Shang, J.; Zhang, K.L.; Zhu, F.Y. Global proteome response to Pb(II) toxicity in poplar using SWATH-MS-based quantitative proteomics investigation. *Ecotoxicol. Environ. Saf.* **2021**, *220*, 112410. [CrossRef]

131. Kissen, R.; Hyldbakk, E.; Wang, C.W.V.; Sørmo, C.G.; Rossiter, J.T.; Bones, A.M. Ecotype dependent expression and alternative splicing of epithiospecifier protein (ESP) in Arabidopsis thaliana. *Plant Mol. Biol.* **2012**, *78*, 361–375. [CrossRef]

132. Chen, Q.; Han, Y.; Liu, H.; Wang, X.; Sun, J.; Zhao, B.; Li, W.; Tian, J.; Liang, Y.; Yan, J.; et al. Genome-Wide Association Analyses Reveal the Importance of Alternative Splicing in Diversifying Gene Function and Regulating Phenotypic Variation in Maize. *Plant Cell* **2018**, *30*, 1404–1423. [CrossRef] [PubMed]

International Journal of
Molecular Sciences

MDPI

Review

The Importance of a Genome-Wide Association Analysis in the Study of Alternative Splicing Mutations in Plants with a Special Focus on Maize

Zi-Chang Jia [1,2,†], Xue Yang [2,†], Xuan-Xuan Hou [2], Yong-Xin Nie [2,*] and Jian Wu [1,*]

1   State Key Laboratory Breeding Base of Green Pesticide and Agricultural Bioengineering, Key Laboratory of Green Pesticide and Agricultural Bioengineering, Ministry of Education, Research and Development Center for Fine Chemicals, Guizhou University, Guiyang 550025, China; jiazc973@163.com
2   State Key Laboratory of Crop Biology, College of Life Science, Shandong Agricultural University, Taian 271018, China; xueyang202001@163.com (X.Y.); 18854806295@163.com (X.-X.H.)
*   Correspondence: nyx03@163.com (Y.-X.N.); jwu6@gzu.edu.cn (J.W.)
†   These authors contributed equally to this work.

Abstract: Alternative splicing is an important mechanism for regulating gene expressions at the post-transcriptional level. In eukaryotes, the genes are transcribed in the nucleus to produce pre-mRNAs and alternative splicing can splice a pre-mRNA to eventually form multiple different mature mRNAs, greatly increasing the number of genes and protein diversity. Alternative splicing is involved in the regulation of various plant life activities, especially the response of plants to abiotic stresses and is also an important process of plant growth and development. This review aims to clarify the usefulness of a genome-wide association analysis in the study of alternatively spliced variants by summarizing the application of alternative splicing, genome-wide association analyses and genome-wide association analyses in alternative splicing, as well as summarizing the related research progress.

Keywords: alternative splicing; GWAS; QTL; proteogenomics; *Zea mays* L.

Citation: Jia, Z.-C.; Yang, X.; Hou, X.-X.; Nie, Y.-X.; Wu, J. The Importance of a Genome-Wide Association Analysis in the Study of Alternative Splicing Mutations in Plants with a Special Focus on Maize. *Int. J. Mol. Sci.* **2022**, 23, 4201. https://doi.org/10.3390/ijms23084201

Academic Editors: Melvin J. Oliver, Bei Gao and Moxian Chen

Received: 9 March 2022
Accepted: 8 April 2022
Published: 11 April 2022

1. Introduction

During gene expressions, a process may occur that results in the acquisition of different proteins starting from a single gene. This process, commonly known in biological sciences as alternative splicing (AS), occurs when certain exons are included or excluded from the final form of the mRNA produced by the gene. This process is an important way to achieve functional diversity of genes and explains that the existing diversity of the transcriptome and proteome exceeds the actual number of genes present in the genome of a given species [1]. Although its occurrence varies across the plant and animal kingdoms, it remains key to understanding growth, health and disease, as well as evolution and adaptations across species. Alternative splicing is involved in all stages of plant growth and development and plays an important role in seed development and flowering transition. Alternative splicing is also involved in the process of plant stress resistance, mainly playing a key role in abiotic stresses such as drought, extreme temperature and salinity.

At present, there are many methods to study alternative splicing variants: the RT-PCR method [2]; the expression sequence tag and cDNA sequence analysis method [3]; and single-molecule sequencing technology [4]. Genome-wide association analyses have been widely used in animal and plant research since they were first proposed. A population-level genome-wide association analysis has many advantages such as a wide detection range and high accuracy. Currently, in plants, a genome-wide association analysis is mainly used to locate genes or QTLs corresponding with a few complex traits. However, there are relatively few related studies on its alternative splicing in plants. The following is an overview of alternative splicing, genome-wide association analyses and the application of genome-wide association analyses in alternative splicing, aiming to highlight the application value of genome-wide association analyses in alternative splicing research.

2. Alternative Splicing: A Ubiquitous Regulatory Mechanism in Eukaryotes

2.1. Definition and Classification of Alternative Splicing

Alternative splicing is a regulatory mechanism that has been developed by organisms during evolution. It is a process by which pre-mRNA selects different splicing sites, excises introns and joins different exons together to generate multiple mature mRNA splicing isoforms [5,6]. A widespread natural variation in alternative splicing is observed in organisms, a process that increases the diversity of their transcriptomes and proteomes to improve survival. Alternative splicing events mainly include intron retention (IR), an alternative 5′ splice site (AE5′), an alternative first exon (AFE), exon skipping (ES), an alternative 3′ splice site (AE3′) and an alternative last exon (ALE). In addition, there is a more complex type called mutually exclusive alternative splicing of exons (MEE) (Figure 1) [7,8]. Alternative splicing is ubiquitous in eukaryotes. In plants, the predominant form of alternative splicing is intron retention [9–11]. In dicots, nearly 60% of *Arabidopsis* genes contain introns [12]. Among monocots, more than 70% of the intronic genes in rice give rise to different isoforms [11]. In maize, more than 50% of the genes are alternatively spliced; the genetic structure leading to a natural variation in alternative splicing is relatively simple and the cis-regulation effect is much higher than the trans-regulation effect [13]. This demonstrates that alternative splicing is essential for regulating the gene expression in plants.

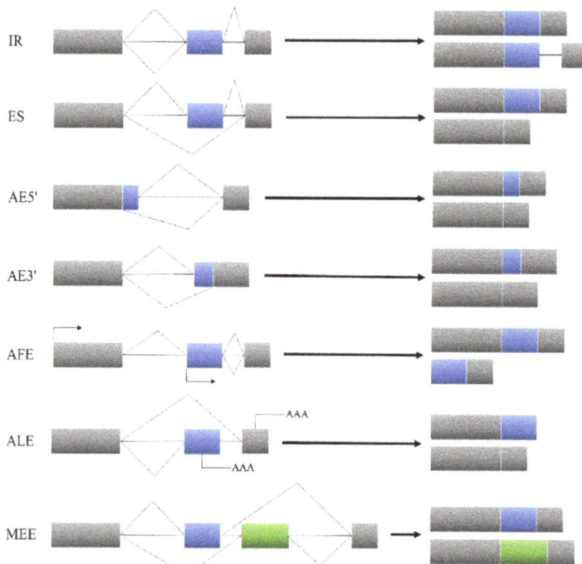

Figure 1. Alternative splicing species.

2.2. Generation Mechanism of Alternative Splicing

The splicing process in organisms requires the participation of the spliceosome and its formation requires the complex synergy of various trans-acting factors. This includes the participation of more than 150 proteins such as snRNPs, U1, U2, U4/U6 and U5 [14]. The splicing process in the nucleus is mainly accomplished by two consecutive steps of transesterification. In the first transesterification reaction, the 5′ end of the intron is connected to the A base on the branch site to form a lariat structure. In the second transesterification reaction, the 3′ splice site is cut open, the left and right exons are connected by phosphodiester bonds and the intron is released in the form of a lariat and quickly degrades (Figure 2) [15–17]. Alternative splicing is primarily regulated by cis-acting elements and trans-acting factors.

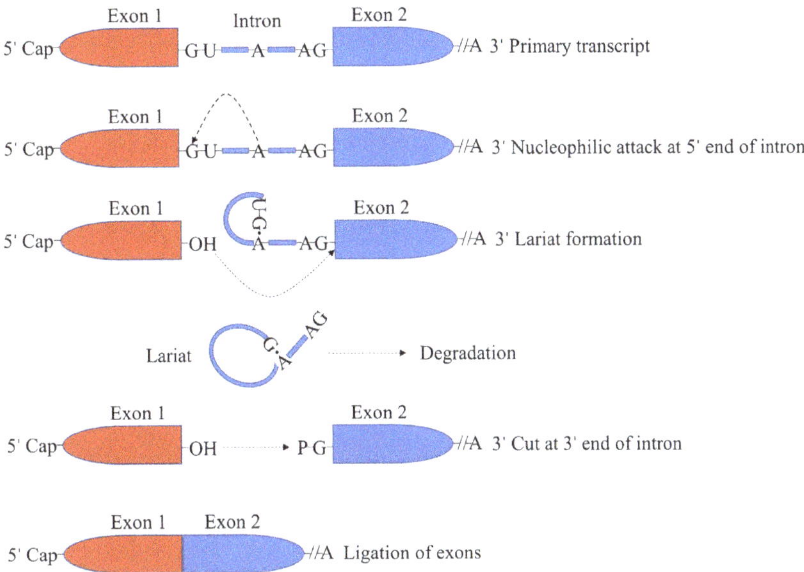

Figure 2. mRNA splicing mechanism.

RNA-binding proteins can regulate alternative splicing. For example, most S-serine, R-arginine (SR) proteins exist only in the nucleus, but a few can shuttle back and forth between the nucleus and cytoplasm. Typically, SR proteins primarily bind to exon splicing enhancers, promoting splicing by recruiting spliceosome proteins [18]. In contrast, heterogeneous nuclear ribonucleoprotein(hnRNP) family proteins antagonize the action of SR proteins by binding to exon splice silencers or intron splice silencers, thereby inhibiting splicing [9,19].

In addition, epigenetic factors regulate alternative splicing. Transcription and alternative splicing are not independent processes, as co-transcription (where RNA polymerase II acts as a link between transcription and splicing) can occur. The chromatin structure and histone modifications mainly affect RNA polymerase and splicing factors to regulate mRNA splicing. Furthermore, DNA methylation, the RNA secondary structure and non-coding RNAs also affect alternative splicing [9,16,20–23].

2.3. Output of Alternative Splicing

Importantly, alternative splicing increases the diversity of life [24,25]. Alternative splicing of genes can lead to the production of many isoforms, which indirectly leads to an increase in the diversity of proteins produced by translation. Furthermore, nonsense-mediated mRNA decay can affect alternative splicing by regulating mRNA stability [26–28]. In addition, alternative splicing competitively inhibits the function of transcription factors through a polypeptide interference mechanism and negatively regulates the expression of genes [29]. Finally, alternative splicing can also regulate the gene expression through miRNA, cutting and degrading the mRNA or blocking mRNA translation [30–32]. The detailed alternative splicing process in plants is shown in Figure 3.

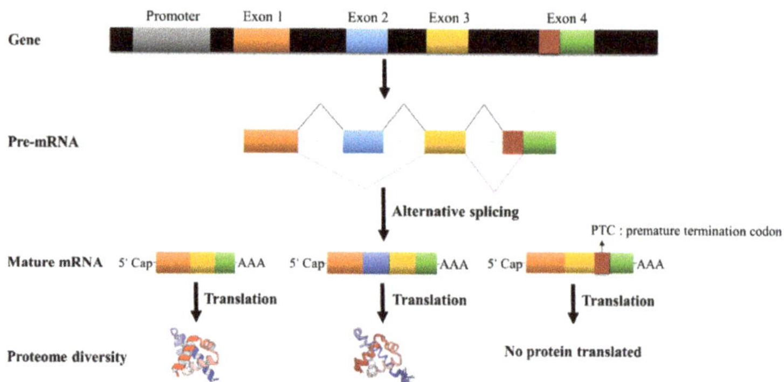

Figure 3. Alternative splicing process.

2.4. Biological Function of Alternative Splicing in Plants

Alternative splicing is involved in the regulation of various plant life activities, especially the response of plants to abiotic stresses, and is also an important process of plant growth and development.

2.4.1. Alternative Splicing Is Involved in the Regulation of Plant Growth and Development

Alternative splicing occurs in all stages of plant growth and development, mainly in the seed development stage and the flowering transition stage. *GRMZM2G104658* encodes a kinase of an unknown function that is alternatively spliced in three tissues of the seed, embryo and endosperm during the seed development, leading to its transition called a regulatory protein with a binding site that plays a role in seed development. NAC transcription factor 109 (*NACTF109*; *GRMZM2G014653*) is a regulator of abscisic acid, which is alternatively spliced during embryo development to regulate the ABA content in seeds and thus regulate seed dormancy. *LEAF RUST 10 DISEASE-RESISTANCE LOCUS RECEPTOR-LIKE PROTEIN KINASE-LIKE* (*LRK10L, GRMZM2G028568*) is alternatively spliced during the endosperm development, thereby assisting the ability of maize to regulate environmental stress during the developmental stages (Table 1) [33]. Pan et al. [34] showed that a reduced expression of *ZmSmk3* resulted in the defective splicing of mitochondrial nad4, resulting in mitochondrial damage, impaired embryonic and endosperm development and a smaller grain size. Xie et al. [35] analyzed the results of RNA sequencing of maize endosperm and identified 30 genes with intron retention that may be involved in maize endosperm development. Xiu et al. [36] found that EMP16 affected mitochondrial nad2 intron 4 splicing, thereby affecting embryogenesis and endosperm development in maize. Chen et al. [37] concluded that *ZmnMAT3* caused empty husks by regulating the splicing of mitochondrial group II introns during maize kernel development.

An important repressor of *Arabidopsis* flowering, *FLOWERING LOCUS C* (*FLC*), undergoes alternative splicing after vernalization [38]. The splicing factor AtU2AF65b is involved in the ABA-mediated regulation of the flowering time in *Arabidopsis* by splicing the *FLC* pre-mRNA [39]. As a splicing factor, SKI-interacting protein can affect the flowering time by regulating alternative splicing of splicing of early flowering (SEF) pre-mRNA in *Arabidopsis* [40].

2.4.2. Alternative Splicing and Abiotic Stress in Plants

When the living environment changes, the property of plant sessile growth determines that plants must respond to environmental stress by adjusting their own physiological state. Alternative splicing plays an important role in this process. The abiotic stresses faced by plants include drought, extreme temperature and salinity. Abiotic stress can cause alternative splicing of related functional genes to produce gene products with different functions in plants, which are then used to combat abiotic stress.

Thatcher et al. [33] found that drought-induced AS in maize (*Zea mays* L.) mainly occurred in the leaves and ears; 1060 and 932 AS events were identified in those tissues, respectively. Under normal developmental conditions of the maize leaves, there were relatively few morphological changes and few differences in developmental splicing. However, under drought-treated conditions, the splicing events in the maize leaves increased. As the drought progressed, the splicing events gradually increased. This suggests that alternative splicing responds to abiotic stress. A transcriptome analysis of maize seeds, embryos and endosperm at different developmental stages revealed that *GRMZM2G104658* is an unknown kinase that is highly alternatively spliced and produces nine different known transcripts in three tissues. *GRMZM2G104658* reduces isoform 1 and increases isoform 7 in three tissue types the maize seed, embryo and endosperm resulting in a protein with regulatory functions. However, its role in seed development requires further investigation. *CORONATINE INSENSITIVE1* (*COI1*; *GRMZM2G353209*) is a jasmonic acid component involved in the stress response. Under drought conditions, COI1 produces isoforms containing Leu-rich repeat regions, resulting in an increased leaf sensitivity to jasmonic acid. *HSP93-V* (*GRMZM2G009443*) has increased Clp domain-containing transcripts under drought stress, which can increase protein degradation. Monodehydroascorbate reductase4 (*MDAR4*; *GRMZM2G320307*) encodes a monodehydroascorbate reductase that removes hydrogen peroxide. Under drought stress, MDAR4 increases hydrogen peroxide in response to stress through selective splicing. RNA-binding protein-defense related 1 (*BRN1*; *GRMZM2G005459*) regulates drought stress as a splicing factor. BRN1 undergoes alternative splicing under drought stress and participates in the splicing of other genes in response to stress. *U2AF65A* (*GRMZM5G813627*) is a U2 small nuclear ribonucleoprotein cofactor. *U2AF65A* undergoes drought stress with altered splicing patterns, encompassing isoforms of all three RNA recognition motifs (RRMs) in response to stress. The expression level of the splicing factor PRP18 (*GRMZM2G102711*) is closely related to alternative splicing in response to drought. Under well-watered conditions, the expression level of this gene was reduced. However, under drought conditions, the expression of this gene was greatly increased and the splicing of the ears and leaves significantly increased (Table 1). Tian et al. [41] demonstrated that the variable splicing of *ZmCCA1* was affected by the maize tissue type, as well as the photoperiod and drought stress. Alternative splicing of this gene mediated the response of the maize to heat and drought stress (Table 1). Li et al. [42] found that a high temperature enhanced alternative RNA splicing in maize and amplified the plant response to heat stress. An elevated temperature increased the frequency of major patterns of alternative splicing (AS) and particularly retained introns and skipped exons.

The *Arabidopsis* clock gene *CCA1* is alternatively spliced to generate CCA1β. CCA1β interferes with the formation of the dimers of *CCA1* and *LHY*, thereby inhibiting their transcription [43]. Yang et al. [44] demonstrated by methods such as RNA-seq that the *Arabidopsis* circadian clock not only controls the transcription of genes, but can also affect their post-transcriptional regulation by affecting alternative splicing. DREB2s in rice are dehydration response element (DRE)-binding proteins that regulate the expression of downstream genes involved in the drought response. Under normal growth conditions, *OsDREB2B* is mainly transcribed to produce *OsDREB2B1*. When rice is subjected to drought stress or exposed to a high temperature, OsDREB2B is mainly transcribed to produce OsDREB2B2, indicating that the resistance of rice to drought stress is affected by alternative splicing [45].

Table 1. Summary of splicing-associated genes [33,41].

Gene ID	Gene Name	Gene Function
GRMZM2G353209	COI1	Response to drought stress
GRMZM2G009443	HSP93-V	Heat shock protein
GRMZM2G320307	MDAR4	Removes hydrogen peroxide
GRMZM2G005459	BRN1	Salicylic acid responsive
GRMZM5G813627	U2AF65A	Splicing factor
GRMZM2G104658		Seed development
GRMZM2G014653	NACTF109	Abscisic acid (ABA) synthesis regulator
GRMZM2G028568	LRK10L	Endosperm development
	CCA1 [41]	Drought response

Alternative splicing has been shown to play a role in a variety of plant processes including seed germination and the transition to flowering. Alternative splicing is also thought to play an important role in abiotic stress processes including drought, extreme temperature and salinity.

3. Overview of a Genome-Wide Association Analysis

3.1. Definition and Principles of a Genome-Wide Association Analysis

Due to the presence of a large number of SNPs in the genome, these variants are often used as molecular markers for a genome-wide association analysis (GWAS). A control analysis or correlation analysis at a genome-wide level can be achieved, which assists in the identification of genetic variations affecting complex traits. This method was first proposed in 1996 by Neil Risch and Kathleen Merikangas [46]. Klein et al. were the first to use this method to identify polymorphisms associated with age-related macular degeneration [47]. Since then, it has been applied to research on human diseases such as coronary heart disease [48], obesity [49] and type 2 diabetes [50]. In addition, an association analysis is widely used in plants as a popular method for understanding complex quantitative traits in natural populations.

There are two methods for the identification of plant quantitative trait loci (QTLs): a linkage analysis and an association analysis. Compared with a linkage analysis, which is based on a population construction, GWAS is based on a linkage disequilibrium (LD), which makes full use of the historical recombination accumulated during plant evolution to correlate genotypes and phenotypes. This increased recombination increases the genetic diversity of the population, which greatly improves the accuracy of the localization and reduces the tedious work of group building [51].

LD is a measure of the degree of correlation between two markers. Similar to chromosomal linkage, when two linked markers recombine, the linkage state between the two is broken and a linked or independent state appears in the population. This recombination contributes to the LD between the two markers [52,53]. The strength of the LD is closely associated with the accuracy of a GWAS. Generally, the breakdown of the LD across different species is used to accurately determine the gene positions within the species during the association analysis. For example, the attenuation distance of *Arabidopsis* is approximately 250 kb, that of rice is approximately 100 kb and that of maize is only approximately 1 kb, thus association analyses in maize can reach the gene level [52,54].

3.2. General Pipeline of a GWAS in Plants

The process of GWAS research can be roughly divided into five steps (Figure 4). First, an appropriate association group is selected. A GWAS in crops can directly use natural variations within populations. When constructing a population, it is necessary to select materials with rich genetic variations, large morphological differences, broad genetic bases and regional representativeness. The sample size is generally approximately 300 to 500 individuals. Second, the phenotype of the target trait needs to be investigated. To reduce the influence of environmental and natural factors, the examination of phenotypes requires repeated experiments in multiple locations over many years. Third, the genotyping of the whole genomes of associated populations is performed using methods such as SNP arrays or resequencing. Fourth, the use of software such as STRUCTURE to analyze the population structure of the associated groups reduces the problem of high false-positive rates in the GWAS. Finally, analysis software such as TASSEL can be used to perform the GWAS on the genotypes and phenotypic traits of associated populations and to obtain loci that are significantly related to the phenotypic traits, which is conducive to the determination of subsequent candidate genes and the development of linked markers.

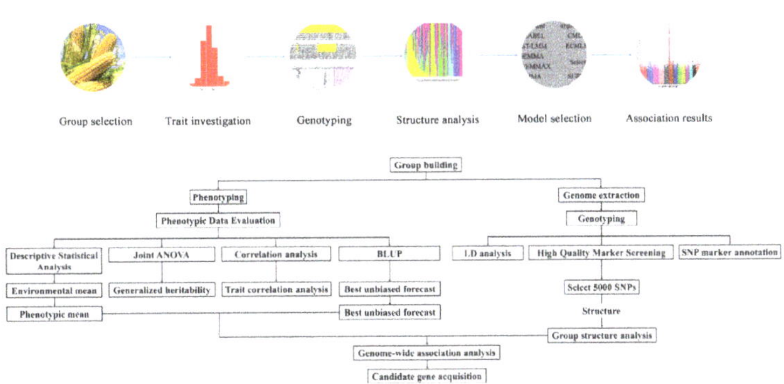

Figure 4. Schematic diagram of the GWAS process.

3.3. Progress of the GWAS in Maize Research

With the completion of the maize genome sequencing and the construction of multiple association groups, maize has become a model plant for association analyses. Maize has its own unique genetic characteristics and rich genetic diversity. In recent years, a GWAS has been widely used in the genetic analysis of various functional genes in maize. The following summarizes the application and research progress of the GWAS in maize in recent years.

There has been progress in the use of a GWAS in maize growth and development research. Liu et al. [55] genotyped 263 maize inbred lines using the SNP50 BeadChip maize array and confirmed that 4 SNPs were significantly associated with the starch content. Among the four candidate genes, *Glucose-1-phosphate adenylyltransferase* (*APS1*) was thought to be an important regulator of the starch content in the grain. Zheng et al. [56] performed a GWAS of 248 different maize inbred lines. It was found that the maize inbred lines showed an obvious natural variation regarding the grain quality under different environments. Further research identified a total of 29 genes related to the grain quality. Lu et al. [57] performed a GWAS of the total above-ground biomass and dry matter of different organs in 412 maize inbred lines. As a result, 1103 candidate genes were identified, after which a total of 224 genes detected by various GWAS models were considered to be of high confidence. Zhang et al. [58] studied the flowering traits in 310 maize inbred lines. Three flowering traits were mapped to SNP molecular markers through a GWAS. The results showed that there were 22 SNP markers associated with DTT (days to tasseling) and 234 candidate genes were identified near these SNP markers. Dong et al. [59] used an Illumina PE150 sequencer to perform the whole genome resequencing of 80 maize core inbred lines and obtained 1,490,007 SNPs for the subsequent association analysis. After two years of the GWAS, 10 SNP loci were significantly associated with the grain fat content. Six candidate genes were found in the linkage disequilibrium region of the most significant SNP site and may be closely related to the anabolism of fat.

In addition, a GWAS has been applied to the study of the maize stress response. Zhang et al. [60] studied the phenotypes associated with seed germination at a low temperature of 10 °C in 5 out of 300 maize inbred lines. Based on the FarmCPU model in the GWAS, 15 significant SNPs were identified that were associated with seed germination under cold stress and three of them were associated with multiple characteristics. Ma et al. [61] performed a GWAS on 305 maize inbred lines under salt stress and identified 7 significant SNPs; 120 genes were obtained by searching the LD regions of these loci. These 120 genes were analyzed by transcriptome data and *GRMZM2G333183* and *GRMZM2G075104* were identified as the key genes in response to salt stress. Xiang et al. [62] performed a GWAS on maize seedlings in arid environments and found 83 genetic variants that might be related to drought tolerance in seedlings. They then analyzed these variant loci and identified 42 candidate genes. The *ZmVPP1* gene, encoding vacuolar-type H^+ pyrophosphatase, had the largest phenotypic contribution as indicated by the peak GWAS signal.

In summary, we found that the application of a GWAS in the genetic analysis of functional genes in maize could help us to obtain information regarding a large number of genes related to maize growth and development such as grain quality, grain fat synthesis, grain starch content, plant flowering traits and biomass. Several genes related to maize resistance were also found; for example, genes related to cold stress, salt stress and drought stress. The acquisition of these genes may help to improve the yield of maize from two aspects of plant growth and development, as well as providing an improvement in stress resistance.

4. Application of a GWAS in Alternative Splicing Studies

The expression of intracellular genes is inseparable from alternative splicing and is a key link in the central dogma of the transmission of genetic information from DNA to proteins. A differential splicing analysis describes the difference in the use of alternatively spliced transcripts between two or more samples. When the differential splicing analysis is performed at the population level, it can be used to assess the genetic factors involved in differential splicing changes. We refer to the sites of genetic variation that regulate the alternative splicing variants as splicing quantitative trait loci (sQTLs) [13]. A GWAS based on a linkage disequilibrium has been widely used to examine complex quantitative trait loci from the transcriptional level to the metabolic level and even to the level of complex agronomic traits and epigenetics. Researchers have analyzed splicing-level RNA-seq data using genome sequence and gene annotation information to identify the quantitative trait loci that control genetic variations at the genome-wide level.

With the development of sequencing technology and bioinformatics, researchers have used quantitative genetics to study the genetic variation of gene expressions. However, there are few reports on the genetic regulatory mechanism of plant alternative splicing variations. The following summarizes the progress of the GWAS in plant alternative splicing variation research and compares it with animal-related research to provide a reference for the application of a GWAS in plant alternative splicing variation research. It may contribute to the in-depth study of a GWAS in the study of plant alternative splicing.

In plant research, there are relatively few reports on AS using GWAS methods. In *Arabidopsis*, Yoo et al. [63] investigated sQTLs in 141 *Arabidopsis* individuals by a GWAS. The study identified 1694 candidate sQTLs. These candidate sQTLs were then interpreted using 107 trait-related published SNPs. As a result, 96 candidate sQTLs were identified that could explain the mechanism of trait-related SNPs. Among them, 25 sQTLs that affect the high differential expression of alternative splicing target exons between genotypes were further screened. They are distributed on different chromosomes and are associated with the growth, flowering and stress of *Arabidopsis* thaliana. Khokhar et al. [64], through a GWAS of 666 geographically distinct *Arabidopsis* ecotypes, found that trans-sQTLs were abundantly accumulated on chromosome 1. In addition, sQTLs co-localized with trait-associated SNPs were also identified. Among them, a large number of sQTLs were enriched in the genes that regulate the stress response, as well as flowering and the circadian clock, mediating the expression of multiple genes including circadian clock-associated 1 (*CCA1*), late elongated hypocotyl (*LHY*), phytochrome-interacting factor 5 (*PIF5*) and b-box domain protein 19 (*BBX19*). It indicated that the existence of differential isoforms is important in the regulation of different ecological types in *Arabidopsis*. An analysis of alternative splicing types also showed that although all types of AS contributed to sQTLs, intron retention (IR) was more prevalent than any other AS type. Yu et al. [65] identified a total of 764 genotype-specific splicing (GSS) events in rice–salt tolerance studies. Among them, five genes were significantly associated with the Na$^+$ content. Ultimately, *OsNUC1* and *OsRAD23* emerged as the most likely candidate genes with splice variants. In studies of alternative splicing in maize, Mei et al. [66] analyzed AS differences between B73 and Mo17 via a linkage population. Chen et al. [13] used mixed linear models in a GWAS and combined 1.25 million SNPs to identify the sQTLs in 368 maize inbred lines. By performing an LD analysis of the associated SNPs in each gene detected by the sQTLs and examining and screening for changes in the splicing rate for each specific sQTL (rate of splicing difference exceeding 5%), a total of 19,554 unique sQTLs corresponding with 6570 genes were detected. The AS types involved in each sQTL were then analyzed. In total, 20,317 sQTL-related AS events were identified with IR being the most common (36%). In addition, the results of a functional enrichment analysis of the genes mapped to the sQTLs showed that the genes with alternative splicing variants mainly functioned in the cells and were involved in protein metabolism and RNA processing and binding in response to external environmental stimuli.

The use of a GWAS is more common in the study of alternative splicing in animals than in plants, especially in the study of human diseases. For example, existing GWAS studies have identified the existence of a large number of SNPs that are closely related to breast cancer (BC); these SNPs are mostly located in non-coding regions, suggesting a regulatory function. Machado et al. [67] identified BC-related SNPs through screening and the analysis of the genotype and expression data of breast tissue samples from the GTEx project and finally identified four sQTLs. Saharali et al. [68] performed RNA-seq on 376 whole blood samples from different individuals in the COPD (chronic obstructive pulmonary disease) Genetic Study. Through the GWAS linear model, 561,060 specific SNPs were identified and were closely related to 30,333 splice sites, which corresponded with 6419 genes. A GWAS has also been used to study various diseases including human multiple sclerosis [69], gliomas [70] and placental disorders [71]. The GWAS-related sQTLs are summarized in Table 2.

Table 2. Summary of GWAS-associated sQTLs.

SNP rsID	SNP Location	Gene	References
rs11249433	chr1	*SRGAP2D*	[67]
rs11552449	chr1	*DCLRE1B*	[67]
rs11552449	chr1	*PHTF1*	[67]
rs6504950	chr17	*STXBP4*	[67]
rs931794	chr15	*CHRNA5*	[68]
rs2202507	chr4	*GYPE*	[68]
rs10470936	chr4	*FAM13A*	[68]
rs2843128	chr1	*CDK11A*	[68]
rs4788084	chr16	*SULT1A2*	[68]
rs7740107	chr6	*L3MBTL3*	[69]
rs7572263	chr2	*C2orf80*	[70]
rs2699247	chr7	*SEC61G*	[70]
rs11216924	chr11	*TMEM25*	[70]
GRMZM2G041223	chr2	*ZmGRF8*	[13]
GRMZM2G171365	chr9	*ZmMADS1*	[13]
GRMZM2G348551	chr6	*sugary2*	[13]
GRMZM2G005887	chr1	*Cys2*	[13]
LOC_Os04g52960	chr4	*OsNUC1*	[65]
LOC_Os09g24200	chr9	*OsRAD23*	[65]

Combined with the research progress of the GWAS in maize in Section 2 and the study of the GWAS in plant alternative splicing variations in this section, it was found that the application of a GWAS in plants can be divided into two categories. First, a GWAS can be used to map several genes or QTLs corresponding with complex traits such as the mapping of plant growth, as well as development-related genes and stress resistance-related genes. Second, a GWAS can also be used to study alternatively spliced variants in plants. Through the detection and screening of the splicing ratio at the whole genome level, a large number of SNP variant sites related to alternative splicing have been found and these sites were associated with the corresponding traits. This contributes to our analysis of the possible regulatory mechanisms of alternative splicing to assess the contribution of the relevant phenotypic variation.

A comparison between the animal- and the plant-related studies revealed that the genetic structure of alternatively spliced variants in plants is simpler than in animal studies. In the regulation of splicing events, the proportion or priority of cis-acting and trans-acting factors detected by a GWAS is different in plants. For example, in *Arabidopsis*, the trans-acting elements are more abundant than the cis-acting elements [64]. However, in maize, alternative splicing is preferentially regulated by various cis-acting elements and cis-sQTLs explain more splicing variations than trans-sQTLs [13]. A GWAS is widely used in the study of alternative splicing in animals and has a relatively complete system. However, few studies in plants suggest that the application of a GWAS in the study of plant alternative splicing variants needs to be further explored. It may be possible to guide experimental research on plants with the help of existing research models in animals.

5. Conclusions and Prospects

Alternative splicing is common in eukaryotes. Relevant research advances in plants have shown that it plays a role in the growth and development of a variety of plants including seed germination and flowering. It also plays an important role in abiotic stresses such as drought, extreme temperature and salinity.

The application of a GWAS in the genetic analysis of the functional genes in maize showed that a GWAS can be used as a means of mapping genes or QTLs corresponding with complex traits; a large number of genes related to growth and development can be mapped through a GWAS such as grain quality, flowering and biomass, as well as the location of the genes related to stress responses (for example, salt stress, drought stress and low temperature stress). The acquisition of these genes helps to improve plants in terms of the plant growth process and stress resistance.

The application of a GWAS in plant alternative splicing can assist with quickly obtaining a large number of sQTLs in the whole genome of plants. These sQTLs with different traits can then be associated through identification and screening. The sQTLs associated with plant flowering, the circadian clock and other growth and development aspects distributed on different chromosomes, as well as the sQTLs associated with stress response, were obtained.

Due to advances in chip technology and sequencing technology, the cost of whole genome sequencing and gene chip technology should decrease and a GWAS may be more widely used in the study of alternative splicing variants. In addition, based on the application and development of proteomics such as SWATH-MS in plants [72,73], a GWAS may be combined with proteomics to further study the relationship between gene alternative splicing and trait regulations to discover the relevant regulatory networks to provide a theoretical basis for the treatment of animal diseases and increase crop yields.

Author Contributions: Conceptualization, J.W.; writing—original draft preparation, Z.-C.J., X.Y. and X.-X.H.; writing—review and editing, J.W. and Y.-X.N.; funding, J.W. All authors have read and agreed to the published version of the manuscript.

Funding: This work was supported by National Natural Science Foundation of China (No. 32072445), the Program of Introducing Talent to Chinese Universities (111 Program, D20023), the Natural Science

Research Project of the Guizhou Education Department (KY(2018)009) and the Scientific Research Innovation Team of Young Scholars in Colleges and Universities of Shandong Province (2019KJE011).

Institutional Review Board Statement: Not applicable.

Informed Consent Statement: Not applicable.

Data Availability Statement: Not applicable.

Conflicts of Interest: The authors declare no conflict of interest.

References

1. Verta, J.P.; Jacobs, A. The role of alternative splicing in adaptation and evolution. *Trends Ecol. Evol.* **2022**, *37*, 299–308. [CrossRef]
2. Harvey, S.E.; Lyu, J.; Cheng, C. Methods for Characterization of Alternative RNA Splicing. *Methods Mol. Biol.* **2021**, *2372*, 209–222.
3. Hirsch, J.; Lefort, V.; Vankersschaver, M.; Boualem, A.; Lucas, A.; Thermes, C.; d'Aubenton-Carafa, Y.; Crespi, M. Characterization of 43 non-protein-coding mRNA genes in *Arabidopsis*, including the MIR162a-derived transcripts. *Plant Physiol.* **2006**, *140*, 1192–1204. [CrossRef]
4. Ding, N.; Cui, H.; Miao, Y.; Tang, J.; Cao, Q.; Luo, Y. Single-molecule real-time sequencing identifies massive full-length cDNAs and alternative-splicing events that facilitate comparative and functional genomics study in the hexaploid crop sweet potato. *PeerJ* **2019**, *7*, e7933. [CrossRef]
5. Ellis, J.D.; Llères, D.; Denegri, M.; Lamond, A.I.; Cáceres, J.F. Spatial mapping of splicing factor complexes involved in exon and intron definition. *J. Cell Biol.* **2008**, *181*, 921–934. [CrossRef]
6. Chen, M.X.; Zhang, K.L.; Gao, B.; Yang, J.F.; Tian, Y.; Das, D.; Fan, T.; Dai, L.; Hao, G.F.; Yang, G.F. Phylogenetic comparison of 5' splice site determination in central spliceosomal proteins of the U1–70K gene family, in response to developmental cues and stress conditions. *Plant J.* **2020**, *103*, 357–378. [CrossRef]
7. Chen, M.X.; Zhang, K.L.; Zhang, M.; Das, D.; Zhu, F.Y. Alternative splicing and its regulatory role in woody plants. *Tree Physiol.* **2020**, *40*, 1475–1486. [CrossRef]
8. Fan, T.; Zhao, Y.Z.; Yang, J.F.; Liu, Q.L.; Tian, Y.; Debatosh, D.; Liu, Y.G.; Zhang, J.; Chen, C.; Chen, M.X.; et al. Phylogenetic comparison and splice site conservation of eukaryotic U1 snRNP-specific U1–70K gene family. *Sci Rep.* **2021**, *11*, 12760. [CrossRef]
9. Brown, J.; Kalyna, M.; Syed, N.H.; Marquez, Y.; Barta, A. Alternative splicing in plants—Coming of age. *Trends Plant Sci.* **2013**, *17*, 616–623.
10. Zhu, F.Y.; Chen, M.X.; Ye, N.H.; Shi, L.; Ma, K.L.; Yang, J.F.; Cao, Y.Y.; Zhang, Y.; Yoshida, T.; Fernie, A.R. Proteogenomic analysis reveals alternative splicing and translation as part of the abscisic acid response in *Arabidopsis* seedlings. *Plant J.* **2017**, *91*, 518–533. [CrossRef]
11. Chen, M.X.; Mei, L.C.; Wang, F.; Boyagane Dewayalage, I.K.W.; Yang, J.F.; Dai, L.; Yang, G.F.; Gao, B.; Cheng, C.L.; Liu, Y.G.; et al. PlantSPEAD: A web resource towards comparatively analysing stress-responsive expression of splicing-related proteins in plant. *Plant Biotechnol. J.* **2021**, *19*, 227–229. [CrossRef]
12. Zhou, K. The alternative splicing of SKU5-Similar3 in *Arabidopsis*. *Plant Signal Behav.* **2019**, *14*, e1651182. [CrossRef]
13. Chen, Q.; Han, Y.; Liu, H.; Wang, X.; Sun, J.; Zhao, B.; Li, W.; Tian, J.; Liang, Y.; Yan, J.; et al. Genome-Wide Association Analyses Reveal the Importance of Alternative Splicing in Diversifying Gene Function and Regulating Phenotypic Variation in Maize. *Plant Cell* **2018**, *30*, 1404–1423. [CrossRef]
14. Ule, J.; Blencowe, B.J. Alternative Splicing Regulatory Networks: Functions, Mechanisms, and Evolution. *Mol. Cell* **2019**, *76*, 329–345. [CrossRef]
15. Black, D.L. Mechanisms of alternative pre-messenger RNA splicing. *Annu. Rev Biochem.* **2003**, *72*, 291–336. [CrossRef]
16. Hertel; Klemens, J. Spliceosomal Pre-mRNA Splicing. *Biol. Methods* **2014**, 1126.
17. Lee, Y.; Rio, D.C. Mechanisms and Regulation of Alternative Pre-mRNA Splicing. *Annu. Rev. Biochem.* **2015**, *84*, 291–323. [CrossRef]
18. Jeong, S. SR Proteins: Binders, Regulators, and Connectors of RNA. *Mol. Cells* **2017**, *40*, 1–9. [CrossRef]
19. Geuens, T.; Bouhy, D.; Timmerman, V. The hnRNP family: Insights into their role in health and disease. *Hum. Genet.* **2016**, *135*, 851–867. [CrossRef]
20. Brown, S.J.; Stoilov, P.; Xing, Y. Chromatin and epigenetic regulation of pre-mRNA processing. *Hum. Mol. Genet.* **2012**, *21*, R90–R96. [CrossRef]
21. Luco, R.F.; Allo, M.; Schor, I.E.; Kornblihtt, A.R.; Misteli, T. Epigenetics in alternative pre-mRNA splicing. *Cell* **2011**, *144*, 16–26. [CrossRef]
22. Naftelberg, S.; Schor, I.E.; Ast, G.; Kornblihtt, A.R. Regulation of alternative splicing through coupling with transcription and chromatin structure. *Annu. Rev Biochem.* **2015**, *84*, 165–198. [CrossRef]
23. Masashi, Y. Functions of long intergenic non-coding (linc) RNAs in plants. *J. Plant Res.* **2017**, *130*, 67–73.
24. Severing, E.I.; Dijk, A.D.J.V.; Ham, R.C.H.J.V. Assessing the contribution of alternative splicing to proteome diversity in *Arabidopsis* thaliana using proteomics data. *BMC Plant Biol.* **2011**, *11*, 82. [CrossRef]
25. Yang, X.; Coulombe-Huntington, J.; Kang, S.; Sheynkman, G.M.; Vidal, M. Widespread Expansion of Protein Interaction Capabilities by Alternative Splicing. *Cell* **2016**, *164*, 805–817. [CrossRef]

26. Chang, Y.F.; Imam, J.S.; Wilkinson, M.F. The nonsense-mediated decay RNA surveillance pathway. *Annu. Rev Biochem.* **2007**, *76*, 51–74. [CrossRef]

27. Lykke-Andersen, S.; Jensen, T.H. Nonsense-mediated mRNA decay: An intricate machinery that shapes transcriptomes. *Nat. Rev. Mol. Cell Biol.* **2015**, *16*, 665–677. [CrossRef]

28. Nasim, Z.; Fahim, M.; Hwang, H.; Susila, H.; Jin, S.; Youn, G.; Ahn, J.H. Nonsense-mediated mRNA decay modulates *Arabidopsis* flowering time via the SET DOMAIN GROUP 40 – FLOWERING LOCUS C module. *J. Exp. Bot.* **2021**, *72*, 7049–7066. [CrossRef]

29. Seo, P.J.; Park, M.J.; Park, C.M. Alternative splicing of transcription factors in plant responses to low temperature stress: Mechanisms and functions. *Planta* **2013**, *237*, 1415–1424. [CrossRef]

30. Stepien, A.; Knop, K.; Dolata, J.; Taube, M.; Bajczyk, M.; Barciszewska-Pacak, M.; Pacak, A.; Jarmolowski, A.; Szweykowska-Kulinska, Z. Posttranscriptional coordination of splicing and miRNA biogenesis in plants. *Wiley Interdiscip. Rev. RNA* **2017**, *8*, e1403. [CrossRef]

31. Pradillo, M.; Santos, J.L. Genes involved in miRNA biogenesis affect meiosis and fertility. *Chromosome Res.* **2018**, *26*, 233–241. [CrossRef]

32. Li, M.; Yu, B. Recent advances in the regulation of plant miRNA biogenesis. *RNA Biol.* **2021**, *18*, 2087–2096. [CrossRef]

33. Thatcher, S.R.; Danilevskaya, O.N.; Meng, X.; Beatty, M.; Zastrow-Hayes, G.; Harris, C.; Van Allen, B.; Habben, J.; Li, B. Genome-Wide Analysis of Alternative Splicing during Development and Drought Stress in Maize. *Plant Physiol* **2016**, *170*, 586–599. [CrossRef]

34. Pan, Z.; Ren, X.; Zhao, H.; Liu, L.; Tan, Z.; Qiu, F. A Mitochondrial Transcription Termination Factor, ZmSmk3, Is Required for nad1 Intron4 and nad4 Intron1 Splicing and Kernel Development in Maize. *G3* **2019**, *9*, 2677–2686. [CrossRef]

35. Xie, S.; Zhang, X.; Zhou, Z.; Li, X.; Huang, Y.; Zhang, J.; Weng, J. Identification of genes alternatively spliced in developing maize endosperm. *Plant Biol.* **2018**, *20*, 59–66. [CrossRef]

36. Xiu, Z.; Sun, F.; Shen, Y.; Zhang, X.; Jiang, R.; Bonnard, G.; Zhang, J.; Tan, B.C. EMPTY PERICARP16 is required for mitochondrial nad2 intron 4 cis-splicing, complex I assembly and seed development in maize. *Plant J.* **2016**, *85*, 507–519. [CrossRef]

37. Chen, W.; Cui, Y.; Wang, Z.; Chen, R.; He, C.; Liu, Y.; Du, X.; Liu, Y.; Fu, J.; Wang, G.; et al. Nuclear-Encoded Maturase Protein 3 Is Required for the Splicing of Various Group II Introns in Mitochondria during Maize (*Zea mays* L.) Seed Development. *Plant Cell Physiol.* **2021**, *62*, 293–305. [CrossRef]

38. Sharma, N.; Geuten, K.; Giri, B.S.; Varma, A. The molecular mechanism of vernalization in *Arabidopsis* and Cereals: Role of Flowering Locus C and its homologs. *Physiol. Plant.* **2020**, *170*, 373–383. [CrossRef]

39. Xiong, F.; Ren, J.J.; Yu, Q.; Wang, Y.Y.; Lu, C.C.; Kong, L.J.; Otegui, M.S.; Wang, X.L. AtU2AF65b functions in abscisic acid mediated flowering via regulating the precursor messenger RNA splicing of ABI5 and FLC in *Arabidopsis*. *New Phytol.* **2019**, *223*, 277–292. [CrossRef]

40. Cui, Z.; Tong, A.; Huo, Y.; Yan, Z.; Yang, W.; Yang, X.; Wang, X.X. SKIP controls flowering time via the alternative splicing of SEF pre-mRNA in *Arabidopsis*. *BMC Biol.* **2017**, *15*, 80. [CrossRef]

41. Tian, L.; Zhao, X.; Liu, H.; Ku, L.; Wang, S.; Han, Z.; Wu, L.; Shi, Y.; Song, X.; Chen, Y. Alternative splicing of ZmCCA1 mediates drought response in tropical maize. *PLoS ONE* **2019**, *14*, e0211623. [CrossRef] [PubMed]

42. Li, Z.; Tang, J.; Bassham, D.C.; Howell, S.H. Daily temperature cycles promote alternative splicing of RNAs encoding SR45a, a splicing regulator in maize. *Plant Physiol.* **2021**, *186*, 1318–1335. [CrossRef] [PubMed]

43. Park, M.J.; Seo, P.J.; Park, C.M. CCA1 alternative splicing as a way of linking the circadian clock to temperature response in *Arabidopsis*. *Plant Signal. Behav.* **2012**, *7*, 1194–1196. [CrossRef]

44. Yang, Y.; Li, Y.; Sancar, A.; Oztas, O. The circadian clock shapes the *Arabidopsis* transcriptomeby regulating alternative splicing and alternative polyadenylation. *J. Biol. Chem.* **2020**, *295*, 7608–7619. [CrossRef]

45. Matsukura, S.; Mizoi, J.; Yoshida, T.; Todaka, D.; Ito, Y.; Maruyama, K.; Shinozaki, K.; Yamaguchi-Shinozaki, K. Comprehensive analysis of rice DREB2-type genes that encode transcription factors involved in the expression of abiotic stress-responsive genes. *Mol Genet. Genom.* **2010**, *283*, 185–196. [CrossRef] [PubMed]

46. Risch, N.; Merikaringas, K. The Future of Genetic Studies of Complex Human Diseases. *Science* **1996**, *273*, 1516–1517. [CrossRef] [PubMed]

47. Klein, R.J.; Zeiss, C.; Chew, E.Y.; Tsai, J.Y.; Sackler, R.S.; Haynes, C.; Henning, A.K.; Sangiovanni, J.P.; Manne, S.M.; Mayne, S.T. Complement factor H polymorphism in age-related macular degeneration. *Science* **2005**, *308*, 385–389. [CrossRef]

48. Assimes, T.; Tcheandjieu, C.; Zhu, X.; Hilliard, A.; Huang, J. A large-scale multi-ethnic genome-wide association study of coronary artery disease. *Res. Sq.* **2021**.

49. Herbert, A.; Gerry, N.P.; McQueen, M.B.; Heid, I.M.; Pfeufer, A.; Illig, T.; Wichmann, H.E.; Meitinger, T.; Hunter, D.; Hu, F.B.; et al. A common genetic variant is associated with adult and childhood obesity. *Science* **2006**, *312*, 279–283. [CrossRef]

50. Saxena, R.; Voight, B.F.; Lyssenko, V.; Burtt, N.P.; de Bakker, P.I.; Chen, H.; Roix, J.J.; Kathiresan, S.; Hirschhorn, J.N.; Daly, M.J.; et al. Genome-wide association analysis identifies loci for type 2 diabetes and triglyceride levels. *Science* **2007**, *316*, 1331–1336. [CrossRef]

51. Flint-Garcia, S.A.; Thuillet, A.C.; Yu, J.; Pressoir, G.; Romero, S.M.; Mitchell, S.E.; Doebley, J.; Kresovich, S.; Goodman, M.M.; Buckler, E.S. Maize association population: A high-resolution platform for quantitative trait locus dissection. *Plant J.* **2005**, *44*, 1054–1064. [CrossRef] [PubMed]

52. Flint-Garcia, S.A.; Thornsberry, J.M.; Buckler, E.S.; Thornsberry, J.M.; Buckler, E.S. Structure of Linkage Disequilibrium in Plants. *Annu. Rev. Plant Biol.* **2003**, *54*, 357–374. [CrossRef] [PubMed]
53. Remington, D.L.; Thornsberry, J.M.; Matsuoka, Y.; Wilson, L.M.; Whitt, S.R.; Doebley, J.; Kresovich, S.; Goodman, M.M.; Buckler, E.S.t. Structure of linkage disequilibrium and phenotypic associations in the maize genome. *Proc. Natl. Acad. Sci. USA* **2001**, *98*, 11479–11484. [CrossRef] [PubMed]
54. Kim, S.; Plagnol, V.; Hu, T.T.; Toomajian, C.; Clark, R.M.; Ossowski, S.; Ecker, J.R.; Weigel, D.; Nordborg, M. Recombination and linkage disequilibrium in *Arabidopsis* thaliana. *Nat. Genet.* **2007**, *39*, 1151. [CrossRef] [PubMed]
55. Liu, N.; Xue, Y.; Guo, Z.; Li, W.; Tang, J. Genome-Wide Association Study Identifies Candidate Genes for Starch Content Regulation in Maize Kernels. *Front. Plant Sci.* **2016**, *7*, 1046. [CrossRef] [PubMed]
56. Zheng, Y.; Yuan, F.; Huang, Y.; Zhao, Y.; Guo, J. Genome-wide association studies of grain quality traits in maize. *Sci. Rep.* **2021**, *11*, 9797. [CrossRef]
57. Lu, X.; Wang, J.; Wang, Y.; Wen, W.; Guo, X. Genome-Wide Association Study of Maize Aboveground Dry Matter Accumulation at Seedling Stage. *Front. Genet.* **2021**, *11*, 571236. [CrossRef]
58. Zhang, H.; Gao, S.; Binyang, L.I.; Zhong, H.; Zhang, Z.; Luo, B. Genome-wide Association Analysis of Maize Flowering Traits. *Asian Agric. Res.* **2020**, *12*, 5.
59. Dong, Q.S.; Dai, L.Q.; Lu, W.U.; Shi, T.T.; Wang, Y.W.; University, J.A. Genome-wide Association Analysis of Fat Content in Maize. *J. Maize Sci.* **2018**, *26*, 58–63.
60. Zhang, Y.; Liu, P.; Wang, C.; Zhang, N.; Shen, Y. Genome-wide association study uncovers new genetic loci and candidate genes underlying seed chilling-germination in maize. *PeerJ* **2021**, *9*, e11707. [CrossRef]
61. Ma, L.; Zhang, M.; Chen, J.; Qing, C.; Shen, Y. GWAS and WGCNA uncover hub genes controlling salt tolerance in maize (*Zea mays* L.) seedlings. *Theor. Appl. Genet.* **2021**, *134*, 3305–3318. [CrossRef] [PubMed]
62. Wang, X.; Wang, H.; Liu, S.; Ferjani, A.; Li, J.; Yan, J.; Yang, X.; Qin, F. Genetic variation in ZmVPP1 contributes to drought tolerance in maize seedlings. *Nat. Genet.* **2016**, *48*, 1233–1241. [CrossRef] [PubMed]
63. Yoo, W.; Kyung, S.; Han, S.; Kim, S. Investigation of Splicing Quantitative Trait Loci in *Arabidopsis* thaliana. *Genom. Inf.* **2016**, *14*, 211–215. [CrossRef] [PubMed]
64. Khokhar, W.; Hassan, M.A.; Reddy, A.S.N.; Chaudhary, S.; Jabre, I.; Byrne, L.J.; Syed, N.H. Genome-Wide Identification of Splicing Quantitative Trait Loci (sQTLs) in Diverse Ecotypes of *Arabidopsis* thaliana. *Front. Plant Sci.* **2019**, *10*, 1160. [CrossRef]
65. Yu, H.; Du, Q.; Campbell, M.; Yu, B.; Walia, H.; Zhang, C. Genome-wide discovery of natural variation in pre-mRNA splicing and prioritising causal alternative splicing to salt stress response in rice. *New Phytol.* **2021**, *230*, 1273–1287. [CrossRef]
66. Mei, W.; Liu, S.; Schnable, J.C.; Yeh, C.T.; Springer, N.M.; Schnable, P.S.; Barbazuk, W.B. A Comprehensive Analysis of Alternative Splicing in Paleopolyploid Maize. *Front. Plant Sci.* **2017**, *8*, 694. [CrossRef]
67. Machado, J.; Magno, R.; Xavier, J.M.; Maia, A.T. Alternative splicing regulation by GWAS risk loci for breast cancer. *BioRxiv* **2019**, 766394.
68. Saferali, A.; Yun, J.H.; Parker, M.M.; Sakornsakolpat, P.; Chase, R.P.; Lamb, A.; Hobbs, B.D.; Boezen, M.H.; Dai, X.; de Jong, K.; et al. Analysis of genetically driven alternative splicing identifies FBXO38 as a novel COPD susceptibility gene. *PLoS Genet.* **2019**, *15*, e1008229. [CrossRef]
69. Alcina, A.; Fedetz, M.; Vidal-Cobo, I.; Andrés-León, E.; García-Sánchez, M.-I.; Barroso-del-Jesus, A.; Eichau, S.; Gil-Varea, E.; Luisa-Maria, V.; Saiz, A.; et al. Identification of the genetic mechanism that associates L3MBTL3 to multiple sclerosis. *Hum. Mol. Genet.* **2022**, ddac009. [CrossRef]
70. Patro, C.P.K.; Nousome, D.; Lai, R.K. Meta-Analyses of Splicing and Expression Quantitative Trait Loci Identified Susceptibility Genes of Glioma. *Front. Genet.* **2021**, *12*, 609657. [CrossRef]
71. Ruano, C.S.M.; Apicella, C.; Jacques, S.; Gascoin, G.; Gaspar, C.; Miralles, F.; Méhats, C.; Vaiman, D. Alternative splicing in normal and pathological human placentas is correlated to genetic variants. *Hum Genet.* **2021**, *140*, 827–848. [CrossRef] [PubMed]
72. Chen, M.X.; Zhang, Y.; Fernie, A.R.; Liu, Y.G.; Zhu, F.Y. SWATH-MS-Based Proteomics: Strategies and Applications in Plants. *Trends Biotechnol.* **2021**, *39*, 433–437. [CrossRef] [PubMed]
73. Shen, C.C.; Chen, M.X.; Xiao, T.; Zhang, C.; Zhu, F.Y. Global proteome response to Pb(II) toxicity in poplar using SWATH-MS-based quantitative proteomics investigation. *Ecotoxicol. Environ. Saf.* **2021**, *220*, 112410. [CrossRef] [PubMed]

MDPI

St. Alban-Anlage 66

4052 Basel

Switzerland

Tel. +41 61 683 77 34

Fax +41 61 302 89 18

www.mdpi.com

International Journal of Molecular Sciences Editorial Office

E-mail: ijms@mdpi.com

www.mdpi.com/journal/ijms

www.ingramcontent.com/pod-product-compliance
Lightning Source LLC
LaVergne TN
LVHW070922170726
843515LV00005B/847